Theory of Dielectric Optical Waveguides

QUANTUM ELECTRONICS — PRINCIPLES AND APPLICATIONS

A Series of Monographs

EDITED BY
YOH-HAN PAO
Case Western Reserve University
Cleveland, Ohio

N. S. Kapany and J. J. Burke. OPTICAL WAVEGUIDES, 1972

Dietrich Marcuse. THEORY OF DIELECTRIC OPTICAL WAVEGUIDES, 1974

Benjamin Chu. LASER LIGHT SCATTERING, 1974

Bruno Crosignani, Paolo Di Porto, and Mario Bertolotti. STATISTICAL PROPERTIES OF SCATTERED LIGHT, 1975

John D. Anderson, Jr. GASDYNAMIC LASERS: AN INTRODUCTION, 1976

In preparation

W. W. Duley. CO_2 LASERS: EFFECTS AND APPLICATIONS

Theory of Dielectric Optical Waveguides

DIETRICH MARCUSE

Bell Laboratories
Crawford Hill Laboratory
Holmdel, New Jersey

ACADEMIC PRESS New York San Francisco London 1974

A Subsidiary of Harcourt Brace Jovanovich, Publishers

ACADEMIC PRESS, INC.
111 Fifth Avenue, New York, New York 10003

United Kingdom Edition published by
ACADEMIC PRESS, INC. (LONDON) LTD.
24/28 Oval Road, London NW1

Marcuse, Dietrich, Date
Theory of dielectric optical waveguides.

(Quantum electronics–principles and applications)
Bibliography: p.
1. Fiber optics. 2. Optical waveguides.
3. Dielectrics. I. Title.
QC448.M37 535'.89 73-21718
ISBN 0-12-470950-8

PRINTED IN THE UNITED STATES OF AMERICA

To Christina and Mikel

Contents

Chapter 3. **Coupled Mode Theory**

Chapter 4. **Applications of the Coupled Mode Theory**

Chapter 5. **Coupled Power Theory**

Preface

The history of communications technology has seen a steady increase of the carrier frequency used for the transmission of information. With the invention of the laser this steady climb made a huge jump from millimeter wavelength to the optical frequency range—an increase by three to four orders of magnitude.

However, the source of coherent light energy is only one of several steps that lead to a workable communications system. The next step is the search for a suitable transmission medium. The early stages of this search were dominated by attempts to utilize lens systems and mirrors to build light waveguides. Of the several problems that arise with such waveguides the cost factor appears at present the most serious detriment for the actual use of lens waveguides in optical communications systems.

When I wrote my book "Light Transmission Optics," lens waveguides were still competing successfully with other types of optical waveguides. The glass fiber guide was just beginning to appear as a serious competitor, but its high losses made it appear as though its prospects of winning out over the lens waveguide were not too bright.

Four years have passed since the "Light Transmission Optics" manuscript was written. During this time a revolution has taken place. The losses of optical glass fibers have been reduced so much (they are presently well below 10 dB/km) that the advantages of the optical fiber for communications purposes far outweigh all its competitors. The interest of communications engineers has consequently turned almost exclusively to optical fibers. Much new knowledge has been accumulated so that a book exclusively devoted to the theory of dielectric optical waveguides has become necessary.

A second reason for concentrating exclusively on dielectric optical waveguides is the rapid growth of integrated optics. This new field is devoted to the development of microscopic optical circuits based on thin film technology. The dielectric waveguides used in integrated optics are asymmetric slabs which have not previously been covered in textbooks. I am presenting the theory of light transmission through asymmetric slab waveguides from the point of view of geometrical as well as wave optics. The guided and radiation modes are derived and leaky waves and an inverted slab waveguide—the hollow dielectric waveguide which supports only leaky waves—are discussed.

Since "Light Transmission Optics" was written, a simplified analytical treatment of round optical fibers has been developed by Snyder and Gloge. The complexity of the theory of round fibers forced me to limit the discussion in "Light Transmission Optics" to the description of its guided modes. The radiation modes and mode conversion and radiation phenomena of this important structure could only be discussed with the help of the analogy to the slab waveguide. The new technique of approximating the description of the round optical fiber is based on the fact that all practical fibers are made with core and cladding materials whose refractive indices are very nearly identical. The approximate theory enables us to treat radiation and mode conversion phenomena of slightly imperfect optical fibers.

Optical communications systems using lasers will probably use single-mode fibers as light transmission media. However, cheaper and simpler systems are likely to use luminescent diodes instead of lasers making it necessary to employ multimode fibers for efficient transmission of light power. Multimode dielectric optical waveguides have many interesting properties that can be understood with the help of coupled mode theory. Chapter 3 develops two approaches to the coupled mode theory. An expansion of the electromagnetic field in terms of modes of the ideal waveguide as well as the use of local normal modes are discussed. Both approaches have advantages for certain applications. The theory of Chapters 1 and 2 is used to derive coupling coefficients for core–cladding boundary irregularities of the asymmetric slab waveguide and the round optical fiber.

Chapter 4 presents applications of the coupled mode theory. The radiation losses of asymmetric slab waveguides and optical fibers caused by core–cladding boundary imperfections are treated. This chapter also contains a description of a useful approximate method for deriving coupling coefficients for the usual and the hollow dielectric slab waveguides. The coupled mode theory is finally applied to the problem of Rayleigh scattering by random fluctuations of the refractive index.

The problem of the distribution of power over the many possible guided modes of the multimode waveguide and the effect of mode coupling on pulse transmission are the subject of the last chapter. We begin with a derivation of coupled equations for the average power carried by each mode. This coupled power theory is essential for an understanding of multimode waveguides with mode coupling since the complexity of this problem precludes its treatment by means of the theory of coupled mode amplitudes. However, the results of the coupled mode theory are necessary for the derivation of coupled power equations. We treat coupled power equations for forward and backward modes, discuss the problem of power distribution versus mode number, and study pulse propagation. A very useful approximate diffusion theory of multimode power coupling is presented as an example of analytical methods for solving the coupled power equations.

The present book contains up to date information necessary for an understanding of single as well as multimode optical fibers and the waveguides of integrated optics. This material is essential for the engineer engaged in the development of optical communications systems. This book should also be useful for classroom instruction in universities since the ever increasing importance of light communications will certainly require its coverage in engineering colleges.

I would like to thank Mrs. Ann Flemer who supervised the various stages of typing and reproduction of the manuscript. Many thanks are also due Mrs. Elizabeth Goldsmith who typed most of the manuscript.

1

The Asymmetric Slab Waveguide

1.1 Introduction

Dielectric slabs are the simplest optical waveguides. Because of their simple geometry, guided and radiation modes of slab waveguides can be described by simple mathematical expressions. The study of slab waveguides and their properties is thus often useful in gaining an understanding of the waveguiding properties of more complicated dielectric waveguides. However, slab waveguides are not only useful as models for more general types of optical waveguides, but they are actually employed for light guidance in integrated optics circuits [Mr1, Mr3].

A dielectric slab waveguide is shown schematically in Fig. 1.1.1. The figure shows a slab waveguide as it would be used in a typical integrated optics application. The core region of the waveguide is assumed to have refractive index n_1 and is deposited on a substrate with refractive index n_2. The refractive index of the medium above the core is indicated as n_3. The refractive index n_3 may be unity if the region above the core is air, or it may have some other

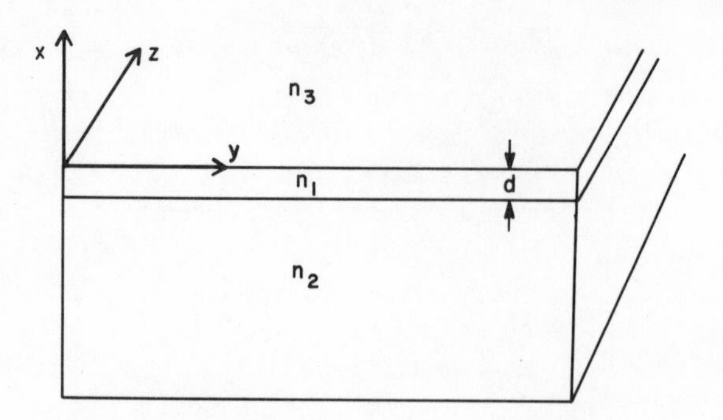

Fig. 1.1.1 *Schematic of a dielectric slab waveguide.*

value if the guiding region of index n_1 is surrounded by dielectric materials on both sides. In order to achieve true mode guidance it is necessary that n_1 be larger than n_2 and n_3. In order to have a specific example in mind we shall assume that

$$n_1 > n_2 \geqslant n_3 \tag{1.1-1}$$

If $n_2 = n_3$, we speak of a symmetric slab waveguide. In case that $n_2 \neq n_3$, the slab waveguide is asymmetric. The modes of symmetric slab waveguides are simpler than those of asymmetric slabs because they can be expressed either as even or odd field distributions [Me1]. The lowest-order mode of a symmetric slab waveguide does not have a cutoff frequency, which means that, in principle, this mode can propagate at arbitrarily low frequencies. By contrast, all modes of asymmetric slabs become cutoff if the frequency of operation is sufficiently low.

This chapter is devoted to the description of the optical properties of asymmetric slab waveguides. Since symmetric slabs are only limiting cases of asymmetric slabs, the description of symmetric slab waveguides is necessarily included in our present treatment.

Like all dielectric waveguides the asymmetric slab supports a finite number of guided modes which is supplemented by an infinite continuum of unguided radiation modes. Both types of modes are obtained as solutions of a boundary value problem. However, the guided modes can also be considered from the point of view of ray optics. Since ray optics is intuitively more appealing than wave optics, we start the discussion by deriving the eigenvalue equation of the guided modes from geometrical optics, which is supplemented by some simple results of plane wave reflection and refraction at plane dielectric interfaces.

The boundary value problem is discussed and solved in Sects. 1.3 and 1.4 on wave optics of the guided modes of the slab waveguide. The chapter then continues with a discussion of leaky waves and of an inverted slab waveguide with $n_1 < n_2, n_3$ which supports only leaky waves.

In integrated optics applications, slab waveguides are formed by various means, the simplest of which use the deposition of glass or plastic films on glass or plastic substrates. These films can be deposited by evaporation, sputtering, or by epitaxial growth techniques. The last method is restricted to the deposition of thin single crystalline films on crystal substrates. Another method of forming dielectric optical waveguides for integrated optics applications employs ion implantation techniques. By bombarding the substrate material with suitable ions it is possible to alter the refractive index of the substrate so that a dielectric slab waveguide results. The depth at which the guiding region appears below the substrate surface can be controlled by the choice of the energy that is used to accelerate the ion beam.

Many integrated optics applications use narrow dielectric strip waveguides instead of a continuous two-dimensional film. The modes of such structures are discussed in Sect. 1.7. Such waveguides are formed by ion implantation techniques or by the deposition of a thin film on top of a substrate which is subsequently etched away, so that only the narrow strip waveguides are left.

The study of asymmetric slab waveguides serves as a valuable introduction to the entire field of dielectric optical waveguides. Because of their simplicity, slab waveguides provide insight into the mechanism of waveguidance by dielectric optical waveguides.

1.2 Geometrical Optics Treatment of Slab Waveguides

Geometrical (or ray) optics describes the propagation of light fields by defining rays as the lines that cross the surfaces of constant phase of the light field at right angles. Light rays have intuitive appeal since a narrow beam of light is a good approximation to the more abstract notion of light rays.

The laws of ray optics, needed for our present purpose, are simple. We need only assume that a light ray in a homogeneous optical medium follows a straight path. In addition, we need to know Snell's law, which relates the angles with respect to the normal to a dielectric interface that a beam forms that passes through this interface. With the definition of the angles shown in Fig. 1.2.1 Snell's law can be expressed in the form [Me1]

$$n_1 \sin\alpha_1 = n_2 \sin\alpha_2 \tag{1.2-1}$$

For our purposes it is more convenient to use the angle between the ray and the dielectric interface, so that Snell's law assumes the form

$$n_1 \cos\theta_1 = n_2 \cos\theta_2 \tag{1.2-2}$$

If $n_1 > n_2$ it is apparent from Eq. (1.2-2) that there is no real angle θ_2 if $n_1 \cos\theta_1 > n_2$. The absence of a real angle satisfying Eq. (1.2-2) can be interpreted as total internal reflection. In this case no light beam emerges on the

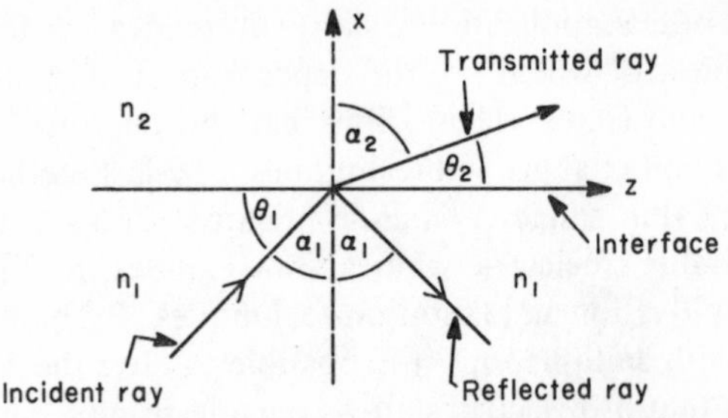

Fig. 1.2.1 *Reflection and refraction of a plane wave at a dielectric interface.*

opposite side of the dielectric interface, so that all the light is totally reflected inside medium 1. The critical angle for total internal reflection is defined by

$$\cos\theta_{1c} = n_2/n_1 \tag{1.2-3}$$

A straight light ray can be associated with a plane wave. Ray optics does not include the description of the phase of the light field. However, the notion of the optical path length, which is defined by the actual length times the refractive index of the medium, makes it easy to attach a phase to the light ray. A ray of length s has traveled an optical path length $n_1 s$ in medium 1. Its phase relative to the starting point at $s = 0$ can be defined as $\phi = -n_1 ks$ (the minus sign is necessary since we assume time dependence $e^{i\omega t}$) with the free-space propagation constant of plane waves,

$$k = 2\pi/\lambda \tag{1.2-4}$$

However, a plane wave or a ray accumulates phase shift not only by traveling in space but also by reflection from a dielectric interface. The reflection coefficient of a plane wave at a dielectric interface, which is polarized so that its electric vector is parallel to the interface, follows from Eqs. (1.6-10), (1.6-11), and (1.6-14) of [Me1]:

$$r = \frac{B}{A} = \frac{(n_1{}^2k^2 - \beta^2)^{1/2} - (n_2{}^2k^2 - \beta^2)^{1/2}}{(n_1{}^2k^2 - \beta^2)^{1/2} + (n_2{}^2k^2 - \beta^2)^{1/2}} \tag{1.2-5}$$

The parameter β (which was called k_{1z} in [Me1]) can be expressed in terms of the angle θ_1:

$$\beta = n_1 k \cos\theta_1 \tag{1.2-6}$$

For $\theta_1 > \theta_{1c}$, the reflection coefficient r, given by Eq. (1.2-5), is real and positive, so that no additional phase change occurs on reflection from the medium with index n_2. For $\theta_1 < \theta_{1c}$, total internal reflection occurs at the dielectric interface. In this case r is complex because the second square root in numerator and denominator of Eq. (1.2-5) is negative imaginary. The negative sign is necessary since a decaying instead of a growing wave must result in medium 2. Under conditions of total internal reflection, a wave that is polarized with its electric vector parallel to the interface suffers a phase shift

$$\phi = -2 \arctan[(\beta^2 - n_2{}^2k^2)^{1/2}/(n_1{}^2k^2 - \beta^2)^{1/2}] \tag{1.2-7}$$

For a wave polarized so that its magnetic vector is parallel to the interface, we obtain from Eq. (1.6-56) of [Me1] the phase shift

$$\phi = -2 \arctan[(n_1{}^2/n_2{}^2)(\beta^2 - n_2{}^2k^2)^{1/2}/(n_1{}^2k^2 - \beta^2)^{1/2}] \tag{1.2-8}$$

Having collected these few facts from ray optics and the theory of plane wave reflection at dielectric interfaces enables us to discuss mode guidance in the slab waveguide and derive the eigenvalue equation for the propagation constants of the guided modes. The rays, which belong to the mode field, propagate inside the core region with refractive index n_1. They are confined to this region by suffering total internal reflection at the dielectric interfaces. Figure 1.2.2 shows the trajectory of such a ray.

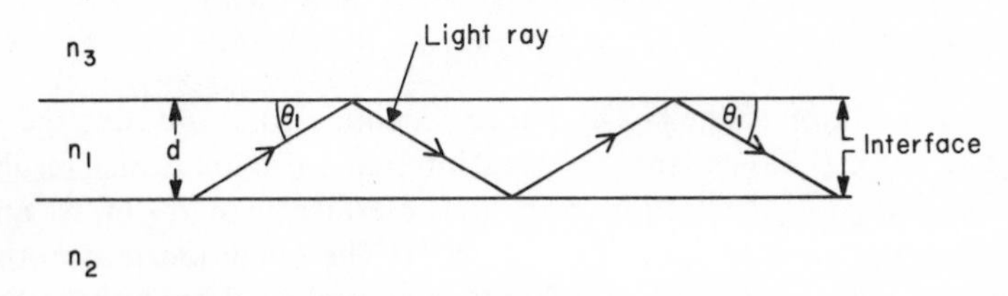

Fig. 1.2.2 *Ray trajectory of a guided wave in the slab waveguide.*

So far we have accounted for the mechanism of mode confinement and have indicated that the angle θ_1 must not exceed the critical angle of Eq. (1.2-3). One might reach the conclusion that rays with arbitrary angles can propagate in the slab waveguide provided the critical angle is not exceeded. However, this is an erroneous assumption. In order to understand the reason for the occurrence of discrete angles for rays that are associated with guided modes, we must introduce our knowledge of the phase of the ray into the picture. The rays shown in Fig. 1.2.2 are only a convenient description of the plane waves that are associated with them. Figure 1.2.3 shows the phase fronts of the plane waves as dashed lines. Two points, *A* and *B*, are marked in the figure. The phase fronts that go through these two points both belong to the same plane wave. The phase fronts of the reflected wave traveling downward have been omitted in order to keep the picture as simple as possible. The ray from point *A* to point *B* (ray *AB*) is assumed to have suffered no reflection. The longer ray from point *C* to point *D* (ray *CD*) belonging to the reflected wave has

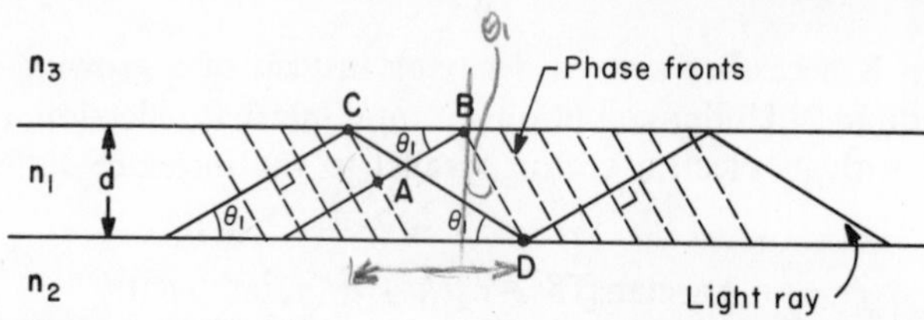

Fig. 1.2.3 *Illustration of the phase condition that leads to the eigenvalue equation. All rays that travel in the same direction belong to the same plane wave.*

suffered two total internal reflections as it travels from the phase front through A to the phase front through point B. Since all points on the same phase front of a plane wave must be in phase, we must require that the optical path length of the ray AB differ from that of the ray CD by a multiple of 2π. The distance between points B and C is $(d/\tan\theta_1) - d\tan\theta_1$. The distance between points A and B is, therefore,

$$s_1 = [(1/\tan\theta_1) - \tan\theta_1]d\cos\theta_1 = (\cos^2\theta_1 - \sin^2\theta_1)d/\sin\theta_1 \tag{1.2-9}$$

The length of the long ray through points C and D is

$$s_2 = d/\sin\theta_1 \tag{1.2-10}$$

However, in addition to the phase change accumulated in traveling the actual distance s_2, ray CD has suffered two total internal reflections which resulted in the phase change ϕ_3 (reflection from the interface with region 3) and ϕ_2 (reflection from the interface with region 2). The condition that both rays contribute to the same plane waves can thus be expressed by the relation

$$n_1(s_2 - s_1)k + \phi_2 + \phi_3 = 2N\pi \tag{1.2-11}$$

where N is an integer number. Equation (1.2-11) is a condition that determines the allowed ray angles or, through relation (1.2-6), the value of the parameter (eigenvalue) β. Such a relation is known as an eigenvalue equation. Since the phase shifts ϕ_2 and ϕ_3 are different for the two possible polarizations of the waves, we obtain two different eigenvalue equations. We designate the waves with electric vectors parallel to the interface as TE waves and obtain from Eqs. (1.2-7), (1.2-9), (1.2-10), and (1.2-11) the eigenvalue equation

$$\arctan(\gamma/\kappa) + \arctan(\delta/\kappa) = \kappa d - N\pi \tag{1.2-12}$$

with the abbreviations

$$\kappa = (n_1{}^2k^2 - \beta^2)^{1/2} = n_1 k \sin\theta_1 \tag{1.2-13}$$

$$\gamma = (\beta^2 - n_2{}^2k^2)^{1/2} = [(n_1{}^2 - n_2{}^2)k^2 - \kappa^2]^{1/2} \tag{1.2-14}$$

and

$$\delta = (\beta^2 - n_3{}^2k^2)^{1/2} = [(n_1{}^2 - n_3{}^2)k^2 - \kappa^2]^{1/2} \tag{1.2-15}$$

Taking the tangent of Eq. (1.2-12) transforms the eigenvalue equation for the TE waves into the form

$$\tan \kappa d = \kappa(\gamma+\delta)/(\kappa^2-\gamma\delta) \tag{1.2-16}$$

The corresponding procedure for TM waves results in the eigenvalue equation

$$\tan \kappa d = n_1{}^2\kappa^2(n_3{}^2\gamma+n_2{}^2\delta)/(n_2{}^2 n_3{}^2\kappa^2-n_1{}^4\gamma\delta) \tag{1.2-17}$$

We shall rederive these equations again in the next section by starting from Maxwell's equations and using the boundary conditions at the dielectric interfaces. The purpose of the present discussion was to give an intuitive explanation for the process of mode guidance in slab waveguides and to demonstrate how the mode conditions (1.2-16) and (1.2-17) can be obtained with the help of simple principles obtained from ray optics and from the properties of plane waves [Ti1].

1.3 Guided Modes of the Asymmetric Slab Waveguide

In order to obtain a complete description of the modes of dielectric waveguides, Maxwell's equations must be solved [MK1, NM1]. With the help of the operator ($\mathbf{e}_x$, $\mathbf{e}_y$, and $\mathbf{e}_z$ are unit vectors in x, y, z direction)

$$\nabla = \mathbf{e}_x \frac{\partial}{\partial x} + \mathbf{e}_y \frac{\partial}{\partial y} + \mathbf{e}_z \frac{\partial}{\partial z} \tag{1.3-1}$$

Maxwell's equations can be written in the form

$$\nabla \times \mathbf{H} = \varepsilon_0 n^2\, \partial \mathbf{E}/\partial t \tag{1.3-2}$$

and

$$\nabla \times \mathbf{E} = -\mu_0\, \partial \mathbf{H}/\partial t \tag{1.3-3}$$

The cross indicates a vector product, $\mathbf{H}$ and $\mathbf{E}$ are the magnetic and electric field vectors, and ε_0 and μ_0 are the dielectric permittivity and magnetic permeability of vacuum. We do not consider magnetic materials in this book so that the use of the vacuum constant μ_0 is sufficient. The index of refraction of the medium is designated by n, and t is the time variable.

We simplify the description of the slab waveguide by assuming that there is no variation in y direction, which we express symbolically by the equation

$$\partial/\partial y = 0 \tag{1.3-4}$$

Condition (1.3-4) is actually no restriction on the generality of the mode description since it is always possible to rotate the coordinate system in the yz plane until this condition is satisfied for any given mode. The modes of the slab waveguide can be classified as TE and TM modes. TE or transverse electric modes do not have a component of the electric field in the direction of wave propagation, while TM or transverse magnetic modes do not have a longitudinal magnetic field component. We consider TE and TM modes separately. The fields of guided modes must vanish at $x = \pm\infty$.

Guided TE Modes

TE modes have only three field components: E_y, H_x, and H_z. The position of the coordinate system relative to the slab is shown in Fig. 1.1.1. We assume that the slab is infinitely extended in the yz plane. We consider only strictly time harmonic fields whose time dependence, in complex notation, can be expressed as

$$e^{i\omega t} \tag{1.3-5}$$

The radian frequency ω is related to the actual frequency f by

$$\omega = 2\pi f \tag{1.3-6}$$

Since we are interested in obtaining the normal modes of the slab waveguide we assume that the z dependence of the mode fields is given by the function

$$e^{-i\beta z} \tag{1.3-7}$$

By combining the two factors (1.3-5) and (1.3-7) we obtain

$$e^{i(\omega t - \beta z)} \tag{1.3-8}$$

Function (1.3-8) describes a wave traveling in positive z direction with phase velocity

$$v = \omega/\beta \tag{1.3-9}$$

The eigenvalue β is identical to the quantity introduced in Eq. (1.2-6). Factor (1.3-8) is common to all field quantities and shall be omitted for brevity. With $E_x = 0$, $E_z = 0$, and $H_y = 0$, we obtain from Maxwell's Eqs. (1.3-2), (1.3-3) with the help of Eq. (1.3-4), and with the time and z dependence given by Eq. (1.3-8),

$$-i\beta H_x - (\partial H_z/\partial x) = i\omega\varepsilon_0 n^2 E_y \tag{1.3-10}$$

$$i\beta E_y = -i\omega\mu_0 H_x \tag{1.3-11}$$

$$\partial E_y/\partial x = -i\omega\mu_0 H_z \tag{1.3-12}$$

We thus obtain the H components in terms of the E_y component

$$H_x = (-i/\omega\mu_0)\,\partial E_y/\partial z = -(\beta/\omega\mu_0)E_y \tag{1.3-13}$$

and

$$H_z = (i/\omega\mu_0)\,\partial E_y/\partial x \tag{1.3-14}$$

Substitution of these two equations into Eq. (1.3-10) yields the one-dimensional reduced wave equation for the E_y component

$$(\partial^2 E_y/\partial x^2) + (n^2 k^2 - \beta^2) E_y = 0 \tag{1.3-15}$$

with $k^2 = \omega^2 \varepsilon_0 \mu_0 = (2\pi/\lambda)^2$. The problem of finding the TE modes of the slab waveguide has thus become very simple. We only need to find solutions of the one-dimensional reduced wave Eq. (1.3-15) and obtain the magnetic field components directly from Eqs. (1.3-13) and (1.3-14). The only remaining complication is the requirement that the solutions must satisfy the boundary conditions at the two dielectric interfaces at $x = 0$ and $x = -d$. The boundary conditions require that the tangential E and H fields be continuous at the dielectric discontinuities. We thus must require that E_y and H_z are continuous at $x = 0$ and $x = -d$. Solutions that satisfy these conditions for the E_y component and vanish at $x = \pm\infty$ are

$$E_y = Ae^{-\delta x}, \qquad \text{for} \quad x \geqslant 0 \tag{1.3-16}$$

$$= A \cos \kappa x + B \sin \kappa x, \qquad \text{for} \quad 0 \geqslant x \geqslant -d \tag{1.3-17}$$

$$= (A \cos \kappa d - B \sin \kappa d) e^{\gamma(x+d)}, \qquad \text{for} \quad x \leqslant -d \tag{1.3-18}$$

The refractive index n in Eq. (1.3-15) assumes the value n_3 in region 3 (see Fig. 1.1.1), for $x > 0$; n_1 in region 1, for $0 > x > -d$; and n_2 in region 2, for $x < -d$. The parameters κ, γ, and δ are defined by Eqs. (1.2-13)–(1.2-15). The E_y component shown by these three equations satisfies the reduced wave Eq. (1.3-15) and is continuous at the two dielectric interfaces. We do not need the H_x component for the moment. The H_z component is obtained from Eq. (1.3-14):

$$H_z = (-i\delta/\omega\mu_0) Ae^{-\delta x}, \qquad \text{for} \quad x \geqslant 0 \tag{1.3-19}$$

$$= (-i\kappa/\omega\mu_0)(A \sin \kappa x - B \cos \kappa x), \qquad \text{for} \quad 0 \geqslant x \geqslant -d \tag{1.3-20}$$

$$= (i\gamma/\omega\mu_0)(A \cos \kappa d - B \sin \kappa d) e^{\gamma(x+d)}, \qquad \text{for} \quad x \leqslant -d \tag{1.3-21}$$

The H_z component does not immediately satisfy the boundary conditions. The requirement of continuity of H_z at $x = 0$ and $x = -d$ leads to the following system of equations:

$$\delta A + \kappa B = 0 \tag{1.3-22}$$

$$(\kappa \sin \kappa d - \gamma \cos \kappa d) A + (\kappa \cos \kappa d + \gamma \sin \kappa d) B = 0 \tag{1.3-23}$$

This homogeneous equation system has a solution only if the system determinant vanishes. We thus obtain the determinantal or eigenvalue equation

$$\delta(\kappa \cos \kappa d + \gamma \sin \kappa d) - \kappa(\kappa \sin \kappa d - \gamma \cos \kappa d) = 0 \tag{1.3-24}$$

From Eq. (1.3-22) we obtain

$$B/A = -\delta/\kappa \tag{1.3-25}$$

The eigenvalue equation can be written in a different form:

$$\tan \kappa d = \kappa(\gamma+\delta)/(\kappa^2-\gamma\delta) \tag{1.3-26}$$

We have thus rederived the eigenvalue Eq. (1.2-16) by the precise methods of this section. The agreement of Eq. (1.3-26) with Eq. (1.2-16) justifies the heuristic method that was used in the previous section.

For future use it is convenient to express the eigenvalue equation in two alternate forms:

$$\cos \kappa d = \pm(\kappa^2-\gamma\delta)/[(\kappa^2+\gamma^2)(\kappa^2+\delta^2)]^{1/2} \tag{1.3-27}$$

and

$$\sin \kappa d = \pm\kappa(\gamma+\delta)/[(\kappa^2+\gamma^2)(\kappa^2+\delta^2)]^{1/2} \tag{1.3-28}$$

The sign of the square root must be the same in both equations. The eigenvalue equation determines the allowed values of the propagation constant β that enters Eq. (1.3-26), (1.3-27), or (1.3-28) through Eqs. (1.2-13)–(1.2-15). Equation (1.2-6) defines for each guided mode a mode angle, which is the direction of the ray or plane wave that travels inside the core region of the waveguide with refractive index n_1.

The field treatment of the guided wave problem also provides the justification for the assumption used in the preceding section that the field inside the core can be regarded as a superposition of two plane waves. Expression (1.3-17) for the E_y component of the field in the core consists of sine and cosine functions. Each of these functions can be decomposed into exponential functions of the form $\exp(\pm i\kappa x)$. If we reinstate the omitted factor (1.3-8), we see that the field in the core is formed by the superposition of plane waves of the form

$$\exp i(\omega t \mp \kappa x - \beta z) = \exp i(\omega t - \mathbf{K}\cdot\mathbf{r}) \tag{1.3-29}$$

The propagation vector $\mathbf{K}$ can be expressed as

$$\mathbf{K} = \pm\kappa\mathbf{e}_x + 0\mathbf{e}_y + \beta\mathbf{e}_z \tag{1.3-30}$$

with $\mathbf{e}_x$, $\mathbf{e}_y$, and $\mathbf{e}_z$ indicating unit vectors in x, y, and z directions. The vector $\mathbf{r}$ points from the coordinate origin to the point at which the field is being considered:

$$\mathbf{r} = x\mathbf{e}_x + y\mathbf{e}_y + z\mathbf{e}_z \tag{1.3-31}$$

According to Eq. (1.2-13) we have

$$K^2 = \kappa^2 + \beta^2 = n_1{}^2k^2 \tag{1.3-32}$$

with K indicating the magnitude of the vector **K**. This discussion makes it clear that the eigenvalue β—the propagation constant of the guided slab waveguide modes—is the z component of the propagation vector of plane waves traveling in the core. Because of condition (1.1-1) the parameter γ of Eq. (1.2-14) becomes imaginary as β becomes larger than $n_2 k$. At the point

$$\beta = n_2 k \qquad (1.3\text{-}33)$$

we have

$$\gamma = 0 \qquad (1.3\text{-}34)$$

We see from the field expression (1.3-18) that the field extends undiminished to infinite distances below the waveguide core, if $\gamma = 0$. When γ becomes imaginary the evanescent field in the substrate region turns into a radiation field, and the wave is no longer guided by the dielectric waveguide. We say that the wave reaches its cutoff point or simply that the wave is cutoff when Eq. (1.3-34) or (1.3-33) is reached. The cutoff condition (1.3-33) can easily be shown to be identical to the condition for the loss of total internal reflection from the dielectric interface at $x = -d$. By eliminating β from Eqs. (1.2-6) and (1.3-33) we obtain the condition for the critical angle, Eq. (1.2-3). This proves that cutoff is identical with the loss of total internal reflection. However, this statement is true only for the slab waveguide and does not necessarily hold for the round optical fiber to be discussed later.

We can obtain some rough information about the solutions of the eigenvalue equation by considering Fig. 1.3.1. The solid lines in this figure are the branches of the tangent as a function of κd. The dashed lines represent the

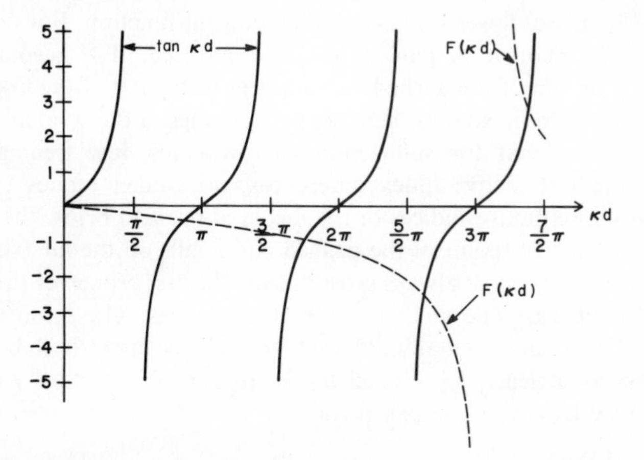

Fig. 1.3.1 *Graphical solution of the eigenvalue Eq. (1.3-26). The crossing points of the solid and dashed lines correspond to solutions.*

function $F(\kappa d)$ that represents the right-hand side of the eigenvalue Eq. (1.3–26). From Eqs. (1.2-13)–(1.2-15) we obtain

$$F(\kappa d) = \frac{\kappa d(\gamma d+\delta d)}{(\kappa d)^2-(\gamma d)(\delta d)}$$

$$= \frac{\kappa d\{[(n_1{}^2-n_2{}^2)(kd)^2-(\kappa d)^2]^{1/2}+[(n_1{}^2-n_3{}^2)(kd)^2-(\kappa d)^2]^{1/2}\}}{(\kappa d)^2-[(n_1{}^2-n_2{}^2)(kd)^2-(\kappa d)^2]^{1/2}[(n_1{}^2-n_3{}^2)(kd)^2-(\kappa d)^2]^{1/2}} \tag{1.3-35}$$

The figure was drawn for $(n_1{}^2-n_2{}^2)^{1/2}kd = 11$, and $(n_1{}^2-n_3{}^2)^{1/2}kd = 24$. The pole in the $F(\kappa d)$ curve occurs at the point where the denominator of Eq. (1.3-35) vanishes. The $F(\kappa d)$ curve ends at the point

$$(n_1{}^2-n_2{}^2)^{1/2}kd = \kappa d \tag{1.3-36}$$

since one of the square root expressions in Eq. (1.3-35) becomes imaginary as κd exceeds the value given by Eq. (1.3-36). The κd coordinates of the crossing points of the solid and dashed curves represent solutions of the eigenvalue Eq. (1.3-26). Each solution gives one TE mode of the slab waveguide. For the conditions that were used to draw Fig. 1.3.1 we obtain four guided modes. We define a parameter that combines the difference of the squares of the refractive indices of core and medium 2 with information about the operating wavelength and the width of the core:

$$V = (n_1{}^2-n_2{}^2)^{1/2}kd \tag{1.3-37}$$

As the value of V decreases, the endpoint of the dashed curve moves to the left, so that it crosses fewer branches of the tangent function. For decreasing values of V the number of guided modes is reduced. If V becomes small enough the endpoint of the dashed curve moves to the left of the first branch of the tangent function, so that there are no crossings of the solid and dashed lines. This means that for sufficiently narrow cores, low frequencies, or sufficiently small refractive index differences, no guided modes can exist. However, if the refractive indices of the media above and below the core are equal, $n_3 = n_2$, the endpoint of the dashed curve falls on the κd axis. In this case the dashed curve must always cross at least the first branch of the tangent curve so that at least one guided mode always exists. The symmetric slab waveguide is thus fundamentally different from the asymmetric slab in that it always supports at least one guided mode. If $n_3 = n_2$, we have $\gamma = \delta$, and Eq. (1.3-26) reduces to the simpler form

$$\tan(2\kappa d/2) = \frac{2\tan\kappa d/2}{1-\tan^2\kappa d/2} = \frac{2\gamma/\kappa}{1-(\gamma/\kappa)^2} \tag{1.3-38}$$

This is a second-order equation in $\tan \kappa d/2$ with the solutions

$$\tan \kappa d/2 = \gamma/\kappa \tag{1.3-39}$$

and

$$\tan \kappa d/2 = -\kappa/\gamma \tag{1.3-40}$$

Equation (1.3-39) is the well-known eigenvalue equation for the even modes of the symmetric slab waveguide, while Eq. (1.3-40) provides the propagation constants for the odd modes [Me1]. The fact that even and odd modes result can be verified from Eq. (1.3-17) by shifting the coordinate origin to the center of the core of the symmetric slab. Descriptions of the symmetric slab waveguide usually use the symbol d for the halfwidth of the core, which explains why $d/2$ appears in Eqs. (1.3-39) and (1.3-40) instead of the customary d.

The cutoff value of V for each guided TE mode can be obtained from Eq. (1.3-26). As we pointed out earlier [see Eq. (1.3-34)] $\gamma = 0$ is the cutoff point. We have at the cutoff point of every mode the relation

$$V_c = (\kappa d)_c \tag{1.3-41}$$

We thus obtain from Eqs. (1.2-13), (1.2-15), (1.3-33), and (1.3-26)

$$V_c = \arctan[(n_2{}^2 - n_3{}^2)^{1/2}/(n_1{}^2 - n_2{}^2)^{1/2}] + \nu\pi \tag{1.3-42}$$

if the arctangent function is restricted to the range 0–$\pi/2$.† For the symmetric slab waveguide we find immediately

$$V_c = \nu\pi \tag{1.3-43}$$

or

$$(\kappa d/2)_c = \nu\pi/2 \tag{1.3-44}$$

for integer values of $\nu = 0, 1, 2, \ldots$.

It remains to relate the amplitude coefficient A of the electromagnetic field to the power carried by the mode. The power is obtained by integrating the z component of the power flow vector (Poynting vector):

$$S_z = \tfrac{1}{2}\,\mathrm{Re}(\mathbf{E} \times \mathbf{H}^*) \cdot \mathbf{e}_z \tag{1.3-45}$$

over the infinite transverse cross section of the waveguide. This notation implies that S_z is a time averaged quantity. We thus obtain

$$(\beta/|\beta|)\,P = -\frac{1}{2}\int_{-\infty}^{\infty} E_y H_x{}^* \, dx = (\beta/2\omega\mu_0)\int_{-\infty}^{\infty} |E_y|^2 \, dx \tag{1.3-46}$$

where P is a real, positive quantity. The asterisk indicates complex conjugation. For the slab with its infinite extension in y direction, P is actually the power per

† Note that we assume $n_2 \geqslant n_3$.

unit length (unit length in y direction). The straightforward evaluation of Eq. (1.3-46) results in

$$A^2 = 4\kappa^2 \omega\mu_0 P/|\beta|[d+(1/\gamma)+(1/\delta)](\kappa^2+\delta^2) \tag{1.3-47}$$

Equations (1.3-27) and (1.3-28) were used to express the amplitude coefficient in this simple form. Throughout this book we normalize the modes so that the mode amplitudes are real quantities.

Guided TM Modes

Transverse magnetic, or TM, modes have the field components H_y, E_x, and E_z. Assuming again that the time and z dependence of the modes is given by the factor (1.3-8) we obtain from Maxwell's Eqs. (1.3-2) and (1.3-3)

$$i\beta H_y = i\omega\varepsilon_0 n^2 E_x \tag{1.3-48}$$

$$\frac{\partial H_y}{\partial x} = i\omega\varepsilon_0 n^2 E_z \tag{1.3-49}$$

$$i\beta E_x + \frac{\partial E_z}{\partial x} = i\omega\mu_0 H_y \tag{1.3-50}$$

The electric field components can now be expressed in terms of the H_y component:

$$E_x = (i/n^2\omega\varepsilon_0)\,\partial H_y/\partial z = (\beta/n^2\omega\varepsilon_0)\,H_y \tag{1.3-51}$$

and

$$E_z = (-i/n^2\omega\varepsilon_0)\,\partial H_y/\partial x \tag{1.3-52}$$

Substitution of these two equations into Eq. (1.3-50) yields the one-dimensional reduced wave equation for the H_y component:

$$(\partial^2 H_y/\partial x^2) + (n^2k^2 - \beta^2)\,H_y = 0 \tag{1.3-53}$$

The boundary conditions now require that the H_y and E_z components be continuous at the dielectric interfaces $x = 0$ and $x = -d$. The solution of Eq. (1.3-53) that is continuous at the two dielectric interfaces and vanishes at $x = \pm\infty$ is given by

$$H_y = (\beta/|\beta|)\,Ce^{-\delta x}, \qquad \text{for} \quad x \geqslant 0 \tag{1.3-54}$$

$$= (\beta/|\beta|)\,(C\cos\kappa x + D\sin\kappa x), \qquad \text{for} \quad 0 \geqslant x \geqslant -d \tag{1.3-55}$$

$$= (\beta/|\beta|)\,(C\cos\kappa d - D\sin\kappa d)\,e^{\gamma(x+d)}, \qquad \text{for} \quad x \leqslant -d \tag{1.3-56}$$

The factor $\beta/|\beta|$ is incorporated into the field amplitude to ensure that the transverse magnetic field changes its sign when the propagation direction is reversed [compare Eq. (3.3-11)]. The E_z component is obtained from Eq. (1.3-52):

$$E_z = (i\delta/n_3{}^2\omega\varepsilon_0)(\beta/|\beta|)Ce^{-\delta x}, \qquad \text{for} \quad x \geqslant 0 \tag{1.3-57}$$

$$= (i\kappa/n_1{}^2\omega\varepsilon_0)(\beta/|\beta|)(C\sin\kappa x - D\cos\kappa x), \qquad \text{for} \quad 0 \geqslant x \geqslant -d \tag{1.3-58}$$

$$= (-i\gamma/n_2{}^2\omega\varepsilon_0)(\beta/|\beta|)(C\cos\kappa d - D\sin\kappa d)e^{\gamma(x+d)}, \qquad \text{for} \quad x \leqslant -d \tag{1.3-59}$$

The E_z component does not satisfy the boundary conditions at the two dielectric interfaces for arbitrary values of C and D. The requirement of continuity of E_z at $x = 0$ and $x = -d$ leads to the equation system

$$(\delta/n_3{}^2)C + (\kappa/n_1{}^2)D = 0 \tag{1.3-60a}$$

$$[(\kappa/n_1{}^2)\sin\kappa d - (\gamma/n_2{}^2)\cos\kappa d]C + [(\kappa/n_1{}^2)\cos\kappa d + (\gamma/n_2{}^2)\sin\kappa d]D = 0 \tag{1.3-60b}$$

The first equation provides a relation between the two amplitude constants

$$D/C = -(n_1{}^2/n_3{}^2)\delta/\kappa \tag{1.3-61}$$

The requirement that the system determinant must vanish leads to the eigenvalue equation

$$(\delta/n_3{}^2)[(\kappa/n_1{}^2)\cos\kappa d + (\gamma/n_2{}^2)\sin\kappa d] - (\kappa/n_1{}^2)[(\kappa/n_1{}^2)\sin\kappa d - (\gamma/n_2{}^2)\cos\kappa d] = 0 \tag{1.3-62}$$

If we divide this equation by $\cos\kappa d$ and group the terms differently, we obtain again the eigenvalue Eq. (1.2-17):

$$\tan\kappa d = n_1{}^2\kappa(n_3{}^2\gamma + n_2{}^2\delta)/(n_2{}^2n_3{}^2\kappa^2 - n_1{}^4\gamma\delta) \tag{1.3-63}$$

We can use this equation to express the cosine and sine functions of κd in terms of the mode parameters κ, γ, and δ:

$$\cos\kappa d = \pm(n_2{}^2n_3{}^2\kappa^2 - n_1{}^4\gamma\delta)/[(n_2{}^4\kappa^2 + n_1{}^4\gamma^2)(n_3{}^4\kappa^2 + n_1{}^4\delta^2)]^{1/2} \tag{1.3-64}$$

and

$$\sin\kappa d = \pm n_1{}^2\kappa(n_3{}^2\gamma + n_2{}^2\delta)/[(n_2{}^4\kappa^2 + n_1{}^4\gamma^2)(n_3{}^4\kappa^2 + n_1{}^4\delta^2)]^{1/2} \tag{1.3-65}$$

Equations (1.3-63)–(1.3-65) are alternate versions of the eigenvalue equation for TM modes of the slab waveguide.

The solutions of Eq. (1.3-63) can be visualized with the help of Fig. 1.3.1. The dashed curve is slightly shifted for the TM mode case, but the principal features of the diagram are the same. For the special case of the symmetric slab we obtain from Eq. (1.3-63), with $n_2 = n_3$ for the even modes,

$$\tan \kappa d/2 = (n_1^2/n_2^2)\gamma/\kappa \tag{1.3-66}$$

and for the odd modes

$$\tan \kappa d/2 = -(n_2^2/n_1^2)\kappa/\gamma \tag{1.3-67}$$

At cutoff the relation (1.3-41) applies. However, the cutoff condition for TM modes is different from that of TE modes. Instead of Eq. (1.3-42) we now obtain from Eqs. (1.3-33), (1.3-34), and (1.3-63)

$$V_c = \arctan[(n_1^2/n_3^2)(n_2^2 - n_3^2)^{1/2}/(n_1^2 - n_2^2)^{1/2}] + \nu\pi \tag{1.3-68}$$

The arctangent function is again restricted to the range 0–$\pi/2$, ν is either zero or an integer number, and $n_2 \geqslant n_3$ has been assumed. For the symmetric slab, cutoff conditions (1.3-43) and (1.3-44) apply also to TM modes.

Cutoff conditions (1.3-42) and (1.3-68) can be used to calculate the total number of modes that can propagate on the slab waveguide. For any given value of the parameter V defined by Eq. (1.3-37), we must require $V < V_{c,N+1}$, with $V_{c,N+1}$ indicating the cutoff value of V for mode $N+1$, which is the first mode that is no longer guided. Since $\nu = 0$ corresponds to the first mode, $\nu = N$ corresponds to mode $N+1$. For TE modes we thus obtain from Eq. (1.3-42) the inequality

$$V < N\pi + \arctan[(n_2^2 - n_3^2)^{1/2}/(n_1^2 - n_2^2)^{1/2}] \tag{1.3-69}$$

For the total number of TE or TM modes we have

$$N = [(1/\pi)\{V - \arctan[\eta(n_2^2 - n_3^2)^{1/2}/(n_1^2 - n_2^2)^{1/2}]\}]_{\text{int}} \tag{1.3-70}$$

The symbol $[\;]_{\text{int}}$ indicates that the integer, whose value is just larger than the value of the number in brackets, must be taken. The parameter η is defined as

$$\eta = \begin{cases} 1, & \text{for TE modes} \\ n_1^2/n_3^2, & \text{for TM modes} \end{cases} \tag{1.3-71}$$

The total number of TE and TM modes is usually twice the number N given by Eq. (1.3-70) for TE or TM modes. However, it is possible that the number of TE modes is larger than the number of TM modes, because we always have $n_1/n_3 > 1$. In that case the total number of modes is the sum of the number of TE modes plus the number of TM modes that are calculated from Eq. (1.3-70).

The amplitude coefficient C can again be related to the total power that is carried by the mode. From Eq. (1.3-45) we obtain, with the help of Eq. (1.3-51),

$$\frac{\beta}{|\beta|}P = \frac{1}{2}\int_{-\infty}^{\infty} E_x H_y^* \, dx = (\beta/2\omega\varepsilon_0)\int_{-\infty}^{\infty} [1/n^2(x)]|H_y|^2 \, dx \quad (1.3\text{-}72)$$

The notation $n(x)$ serves as a reminder that the refractive index is different in the three sections of the waveguide. With the field expressions (1.3-54)–(1.3-56) we obtain from Eq. (1.3-72), with the help of Eqs. (1.3-64) and (1.3-65),

$$C^2 = \frac{4\omega\varepsilon_0 P}{|\beta|} n_1{}^2 n_3{}^4 \kappa^2 \cdot \left\{(n_3{}^4\kappa^2 + n_1{}^4\delta^2)\left[d + \frac{n_1{}^2 n_2{}^2}{\gamma}\frac{\kappa^2+\gamma^2}{n_2{}^4\kappa^2+n_1{}^4\gamma^2} + \frac{n_1{}^2 n_3{}^2}{\delta}\frac{\kappa^2+\delta^2}{n_3{}^4\kappa^2+n_1{}^4\delta^2}\right]\right\}^{-1} \quad (1.3\text{-}73)$$

A comparison of the expressions for TE and TM modes shows that the equations describing TM modes are somewhat more cumbersome than the corresponding equations for TE modes.

Approximate Solutions of the Eigenvalue Equations

The eigenvalues κ, γ, or β (only one of the three is independent of the others) must be obtained as solutions of Eq. (1.3-26) or (1.3-63). Near cutoff and far from cutoff it is possible to give approximate solutions in closed form.

We begin by listing near cutoff approximations and assume that $n_2 > n_3$, causing γ to become very small near cutoff, while δ remains finite. Right at cutoff we have $V = V_c$, with

$$V_c = \kappa_c d = (n_1{}^2 - n_2{}^2)^{1/2} k_c d \quad (1.3\text{-}74)$$

which follows from Eqs. (1.2-13) and (1.3-37), with $\beta = n_2 k$. k_c is the free-space propagation constant for the cutoff frequency of the guided mode. The cutoff value V_c is given by Eq. (1.3-42) for the TE-type eigenvalue equations and by Eq. (1.3-68) for TM-type eigenvalue equations. In order to be able to cover both cases simultaneously, we introduce the notation

$$m_j = \begin{cases} n_j, & \text{for TM case} \\ 1, & \text{for TE case} \end{cases} \quad (1.3\text{-}75)$$

The cutoff condition thus becomes

$$V_c = \nu\pi + \arctan[(m_1{}^2/m_3{}^2)(n_2{}^2 - n_3{}^2)^{1/2}/(n_1{}^2 - n_2{}^2)^{1/2}] \quad (1.3\text{-}76)$$

The eigenvalue Eqs. (1.3-26) and (1.3-63) are combined by writing

$$\tan \kappa d = m_1{}^2 \kappa (m_3{}^2 \gamma + m_2{}^2 \delta)/(m_3{}^2 m_2{}^2 \kappa^2 - m_1{}^4 \gamma\delta) \qquad (1.3\text{-}77)$$

We now assume that the waveguide is operated close to cutoff of the mode of interest and use

$$V = V_c + v \qquad (1.3\text{-}78)$$

with

$$v \ll 1 \qquad (1.3\text{-}79)$$

The assumption that the mode is near cutoff also allows us to use

$$\gamma d \ll V_c \qquad (1.3\text{-}80)$$

From Eq. (1.2-13) we obtain approximately

$$\kappa d = V_c + v - (\gamma^2 d^2/2V_c) \qquad (1.3\text{-}81)$$

and from Eq. (1.2-15) we find

$$\delta d = \delta_c d - (V_c v/\delta_c d) + (\gamma^2 d/2\delta_c) \qquad (1.3\text{-}82)$$

with the abbreviation

$$\delta_c d = [(n_2{}^2 - n_3{}^2)^{1/2}/(n_1{}^2 - n_2{}^2)^{1/2}] V_c \qquad (1.3\text{-}83)$$

A first-order perturbation solution of Eq. (1.3-77) yields the near cutoff approximation

$$\gamma d = \left[1 + \frac{m_1{}^2 m_3{}^2 (V_c{}^2 + \delta_c{}^2 d^2)}{\delta_c d (m_3{}^4 V_c{}^2 + m_1{}^4 \delta_c{}^2 d^2)}\right] \frac{m_2{}^2}{m_1{}^2} V_c v \qquad (1.3\text{-}84)$$

This approximate solution is valid only for asymmetric waveguides, so that $\delta \gg \gamma$ near cutoff. For symmetric guides a different approximation is needed. If $n_2 = n_3$, we find for the lowest-order solution, $v = 0$, of Eq. (1.3-77),

$$\gamma d = (m_1{}^2/m_2{}^2)\{[(m_2{}^4/m_1{}^4) V^2 + 1]^{1/2} - 1\} \qquad (1.3\text{-}85)$$

The cutoff value is $V_c = 0$ in this case, so that we have $V = v$. For higher-order modes we have for symmetric waveguides near cutoff

$$\gamma d = \tfrac{1}{2}(m_2{}^2/m_1{}^2) V_c v \qquad (1.3\text{-}86)$$

It is apparent that Eq. (1.3-85) also assumes the form Eq. (1.3-86) for very small values of $V_c \approx V = v$, but it is advantageous to use the more complicated form Eq. (1.3-85) for the lowest-order mode since it gives more accurate results.

Far from cutoff we use the fact that

$$\kappa \ll \gamma, \qquad \text{and} \qquad \kappa \ll \delta \qquad (1.3\text{-}87)$$

We write

$$\kappa d = \kappa_\infty d/(1+\varepsilon) \tag{1.3-88}$$

with (see Fig. 1.3.1)

$$\kappa_\infty d = (\nu+1)\pi \tag{1.3-89}$$

and with the assumption

$$\varepsilon \ll 1 \tag{1.3-90}$$

From Eq. (1.3-77) we find the approximation for ε:

$$\varepsilon = (m_3{}^2\gamma_\infty + m_2{}^2\delta_\infty)/m_1{}^2\gamma_\infty\delta_\infty d \tag{1.3-91}$$

The notation γ_∞ and δ_∞ is used to indicate that we use $\kappa = \kappa_\infty$ of Eq. (1.3-89) to obtain the values of γ and δ from Eqs. (1.2-14) and (1.2-15).

This approximate solution for κd is remarkably accurate. We could, of course, have written $1-\varepsilon$ instead of $(1+\varepsilon)^{-1}$, but comparison with the exact solutions of the eigenvalue equations shows that better results are obtained with the form Eq. (1.3-88). We can go one step further and use a first approximation for κd obtained from Eq. (1.3-88) to calculate improved values for γ and δ from Eqs. (1.2-14) and (1.2-15) and use these to calculate improved values of ε from Eq. (1.3-91). The far from cutoff approximations are compared to the exact solutions in Sect. 1.7 describing the rectangular dielectric waveguide.

The propagation constant β is obtained either from

$$\beta = (n_1{}^2k^2 - \kappa^2)^{1/2} \tag{1.3-92}$$

with the help of Eq. (1.3-88) or, for the near cutoff approximation, from Eqs. (1.3-84)–(1.3-86) and

$$\beta = (n_2{}^2k^2 + \gamma^2)^{1/2} \tag{1.3-93}$$

1.4 Radiation Modes of the Asymmetric Slab Waveguide

A slab waveguide can support guided modes if expression (1.3-70) for the number of TE or TM modes is larger than zero. However, the number of guided modes is always finite so that there must be other solutions of Maxwell's equations that satisfy the boundary conditions in order to provide a complete set of orthogonal modes. Such additional modes do indeed exist. It is helpful to resort to a physical argument to illustrate the nature of the various modes of the waveguide.

In Sect. 1.2 we discussed the mechanism of mode guidance on the basis of geometrical optics. There it was assumed that a ray or plane wave already exists inside the waveguide core. Once such a ray was postulated it could be shown that it would travel in the waveguide without loss of power, provided that losses in the dielectric material were ignored. The source of the guided mode field must be presumed to be inside the waveguide core and, if the waveguide is infinitely long, the source must be located at minus infinity. The fact that a source inside the core does not necessarily excite only guided modes shall not concern us for the moment.

Let us now assume that we place a source of radiation outside the waveguide core. It is easy to see that such a source can not contribute to guided modes if the waveguide structure is perfect. The waves emitted by the source outside the core reflect and refract at the core boundary, but none of its energy is trapped and travels inside the core as a guided wave. It would require imaginary angles in order to inject a ray from the outside into the core in such a way that it travels inside with less than the critical angle for total internal reflection. According to Snell's law, Eq. (1.2-2), the ray angle inside the core (whose refractive index is larger than that of the surrounding medium) is larger than the angle on the outside. A ray that can be refracted into the core must thus always exceed the critical angle. An imaginary ray angle corresponds to an evanescent field. Such fields can tunnel into the core from the outside but are themselves evanescent field tails resulting from total internal reflection and cannot result from the radiation field of a source outside the waveguide core.

If we visualize a plane wave impinging on the waveguide core from an infinite distance we know that a portion of this wave is reflected at the core boundary, while the remaining energy penetrates through the core and emerges on the other side as a plane wave. The direction of these plane waves can be obtained by applying Snell's law repeatedly. A plane wave impinging on the core from above thus results in a reflected wave above the core and a transmitted wave below the core. The total field in the region above the core is thus a standing wave. The resulting radiation field must, of course, be a solution of Maxwell's equations and it must also satisfy the boundary conditions. In addition, the t and z dependence of the field can again be described by factor (1.3-8). The radiation field thus qualifies in all respects as a mode, except that it is not confined to the waveguide core but reaches undiminished to infinite distances in x direction normal to the core. We call modes of this type radiation modes. Their propagation constants β are not constrained to a discrete set of values, since they are related to the angle of the incident plane wave which can be chosen arbitrarily. The values of the propagation constant thus form a continuum, so that we also speak of the radiation modes as modes of the continuum.

The radiation modes are necessary to describe radiation phenomena in the region around the waveguide core. We shall see in later chapters that waveguide imperfections cause some of the guided mode power to radiate away into the space outside the core. However, radiation that originates inside the core results in traveling wave fields in the outside regions, while we have seen that the radiation modes must be standing waves, at least on one side of the waveguide core. This has caused considerable confusion and makes it difficult to understand the mechanism by which the standing wave radiation modes (standing waves in transverse direction, the modes travel along the z direction) can contribute to strictly outgoing radiation. Let us first reiterate that there are no normalizable solutions of Maxwell's equations satisfying the boundary conditions at the core interface that form only traveling waves outside the core regions. (Leaky waves do not form standing wave patterns but grow exponentially in x direction.) This fact can be shown mathematically, but it is obvious from our physical argument since an incident traveling wave is always partially reflected at the core boundary resulting in a standing wave. The resolution of the seeming paradox is obtained when we consider that the radiation modes form a continuum. It is impossible to excite one single radiation mode with a source inside the core of a waveguide. In fact, any mechanism that excites a continuum mode always simultaneously excites infinitely many continuum modes in its vicinity (vicinity in the sense of β space). Only an infinitely extended source at infinity could excite a pure radiation mode. However, such a source is physically impossible. If radiation is excited by an imperfection of the waveguide it always excites infinitely many radiation modes, which superimpose themselves in such a way that the incoming parts of the standing wave are eliminated by destructive interference. It is not easy to show this rigorously, but the approximate solution of the resulting integrals by the method of stationary phase shows clearly that only the outgoing waves contribute in proper phase, while the incoming wave components of the radiation modes fail to interfere constructively.

After these introductory remarks we proceed to derive the mathematical expressions of the radiation mode fields from Maxwell's equations.

TE Radiation Modes

The analysis proceeds in close analogy to the derivation of the guided modes. The number of nonvanishing field components is the same for both types of modes. The H_x and H_z field components can again be obtained from Eqs. (1.3-13) and (1.3-14) in terms of the E_y component. The E_y component is obtained from the reduced wave Eq. (1.3-15).

We can convince ourselves easily that asymmetric slab waveguides have two types of radiation modes. As always, we assume that the refractive index n_3 of the region above the core is smaller than the index n_2 of the region below the core. A wave impinging on the core from below can thus suffer total internal reflection at the interface of regions 1 and 3. In this case we obtain an evanescent field in region 3 and a standing wave in the core and in region 2. The range of β values that belongs to this type of radiation modes follows directly from Snell's law (1.2-2). The smallest angle of the incident ray is $\theta_2 = 0$ on the outside but becomes θ_1 inside the core, so that we have

$$\beta = n_1 k \cos\theta_1 = n_2 k \tag{1.4-1}$$

The largest β value of the radiation modes thus coincides with the cutoff value (1.3-33) of the guided modes. The smallest β value which still results in total internal reflection at the upper core boundary follows again from Snell's law. This time we must require that the angle θ_3 of the emerging ray in medium 3 vanishes, so that we have

$$\beta = n_1 k \cos\theta_1' = n_3 k \tag{1.4-2}$$

The range of β values for radiation modes, which have exponentially decaying fields in region 3, is thus

$$n_3 k \leqslant |\beta| < n_2 k \tag{1.4-3}$$

This type of radiation mode with exponentially decaying fields on one side of the core is peculiar to the asymmetric slab waveguide. We see from Eq. (1.4-3) that its range shrinks to zero when $n_2 = n_3$, that is, when the waveguide is of the symmetric type. Radiation modes with β values in the range (1.4-3) are responsible for radiation phenomena with power escaping only into the substrate (that is region 2) but not into the space above the core. The electric field component E_y of radiation modes in the range (1.4-3) is described by the following expressions:

$$E_y = A_r e^{i\Delta x}, \qquad \text{for} \quad x \geqslant 0 \tag{1.4-4}$$

$$= A_r \cos\sigma x + B_r \sin\sigma x, \qquad \text{for} \quad 0 \geqslant x \geqslant -d \tag{1.4-5}$$

$$= (A_r \cos\sigma d - B_r \sin\sigma d)\cos\rho(x+d) + \bar{C}_r \sin\rho(x+d), \qquad \text{for} \quad x \leqslant -d \tag{1.4-6}$$

The constants appearing in these equations are adjusted to assure continuity of the E_y component at $x = 0$ and $x = -d$. The parameters Δ, σ, and ρ are defined by the equations

$$\Delta = (n_3^2 k^2 - \beta^2)^{1/2} \tag{1.4-7}$$

$$\sigma = (n_1^2 k^2 - \beta^2)^{1/2} \tag{1.4-8}$$

and

$$\rho = (n_2^2 k^2 - \beta^2)^{1/2} \tag{1.4-9}$$

Note that Δ is positive imaginary in the range (1.4-3). The E_y component thus decays exponentially in region 3 for $x > 0$. This notation was chosen so that we can keep the same symbol Δ also for the modes outside the range (1.4-3). The field in space 2, the substrate region, is a standing wave in accordance with our physical discussion of the origin of radiation modes. The three amplitude coefficients are necessary to keep the field expressions general. The H_z component is obtained from the E_y component with the help of Eq. (1.3-14):

$$H_z = (-\Delta/\omega\mu_0)A_r e^{i\Delta x}, \qquad \text{for} \quad x \geqslant 0 \tag{1.4-10}$$

$$= (-i\sigma/\omega\mu_0)(A_r \sin\sigma x - B_r \cos\sigma x), \qquad \text{for} \quad 0 \geqslant x \geqslant -d \tag{1.4-11}$$

$$= (-i\rho/\omega\mu_0)[(A_r \cos\sigma d - B_r \sin\sigma d)\sin\rho(x+d) - \bar{C}_r \cos\rho(x+d)], \qquad \text{for} \quad x \leqslant -d \tag{1.4-12}$$

The E_y component already satisfies the boundary conditions at $x = 0$ and $x = -d$. In order to force the H_z component to satisfy the boundary condition, requiring continuity at the two interfaces, we must satisfy the following two equations:

$$-i\sigma B_r = \Delta A_r \tag{1.4-13}$$

$$\sigma \cos\sigma d\, B_r - \rho\bar{C}_r = -\sigma \sin\sigma d\, A_r \tag{1.4-14}$$

When we determined the guided modes of the asymmetric slab waveguide we found that the boundary conditions led to a determinantal condition which furnished the eigenvalue equation from which the allowed values of the propagation constant β could be determined. Equations (1.4-13) and (1.4-14) contain three coefficients. We may regard one of them, A_r for example, as given and then consider the equation system as a set of two inhomogeneous equations from which B_r and $\bar{C}_r$ can be determined. However, the system determinant must now be nonvanishing so that no eigenvalue equation results. The values of the propagation constant β thus remain arbitrary and form a continuum in the range (1.4-3). The two amplitude coefficients can be expressed in terms of A_r:

$$B_r = (i\Delta/\rho)A_r \tag{1.4-15}$$

$$\bar{C}_r = [(\sigma/\rho)\sin\sigma d + (i\Delta/\rho)\cos\sigma d]A_r \tag{1.4-16}$$

Radiation modes cannot be normalized with respect to a finite amount of power. If we calculate the integral of the power expression (1.3-46) for one radiation mode as given by Eqs. (1.4-4)–(1.4-6), we find that the integral diverges. However, the normalization of continuum modes can be accomplished with the help of the Dirac delta function $\delta(x)$. Instead of Eq. (1.3-46) we require for radiation modes

$$\frac{1}{2}\int_{-\infty}^{\infty} [\hat{\mathbf{E}}(\rho)\times\hat{\mathbf{H}}^*(\rho')]\cdot\mathbf{e}_z\,dx = s_\rho(\beta^*/|\beta|)\,P\,\delta(\rho-\rho') \tag{1.4-17}$$

where P is always real and positive.

This equation has several new features compared to Eq. (1.3-46). The two field expressions under the integral sign belong to different radiation modes. The delta function on the right-hand side states that the integral vanishes if the two modes are different, but that it becomes infinitely large if both modes are identical. Several features of the expression were introduced to allow for the fact that the propagation constant β can become imaginary, as we shall see later. The caret on top of the field quantities states that the propagation factor (1.3-7) must not be included in the field expressions. It has been our practice to omit this factor from the equations for simplicity of notation. However, we now require specifically that this factor be absent from the field expressions. For real values of β it would not make any difference whether we include Eq. (1.3-7) in the field expressions, since the complex conjugation causes this factor to cancel out for $\rho=\rho'$, and for $\rho\neq\rho'$ the integral vanishes. But for imaginary β values the term (1.3-7) would not cancel, so that the orthogonality expression would become a function of z.

The ratio $\beta^*/|\beta|$ causes the right-hand side of the equation to become negative for waves traveling in negative z direction, so that this factor assures us that P is always positive. For imaginary β values expression (1.4-17) becomes imaginary. Since the real part of the expression on the left-hand side expresses the average power flow, imaginary values of β cause no power to flow along the z axis. Finally we must explain the factor s_ρ. For real values of β we always have $s_\rho = 1$. However, for imaginary β we may have to require $s_\rho = -1$ in order to keep P positive. For TE modes we always have $s_\rho = 1$ for all possible values of β. For TE modes Eq. (1.4-17) can be written as

$$(\beta^*/2\omega\mu_0)\int_{-\infty}^{\infty} \hat{E}_y(\rho)\,\hat{E}_y{}^*(\rho')\,dx = (\beta^*/|\beta|)\,P\,\delta(\rho-\rho') \tag{1.4-18}$$

The arguments ρ and ρ' label two different radiation modes. Equations (1.4-17) and (1.4-18) establish not only the normalization of the radiation modes but also their orthogonality. It can be shown by direct calculation that relation (1.4-18) is indeed true. This relation is used to express the amplitude coefficient A_r of the radiation mode in terms of the factor P appearing in the normalization and orthogonality condition (1.4-18). The actual calculation requires some care, since it is necessary to recognize the delta function in the expressions that result from substitution of Eqs. (1.4-4)–(1.4-6) into Eq. (1.4-18). The details of such a calculation were shown in [Mel, pp. 316–317]. The calculation results in the following expression for the amplitude coefficient A_r:

$$A_r^2 = \frac{4\omega\mu_0\rho^2\sigma^2 P}{\pi|\beta|[\rho^2(\sigma\cos\sigma d - i\Delta\sin\sigma d)^2 + \sigma^2(\sigma\sin\sigma d + i\Delta\cos\sigma d)^2]} \tag{1.4-19}$$

Next we derive the radiation modes for β values in the range

$$-n_3 k \leqslant \beta \leqslant n_3 k \tag{1.4-20}$$

These modes correspond to plane waves impinging on the core at such angles that no total internal reflection results at the upper core interface. However, instead of using only a single plane wave incident from above or below, we assume that two sources send waves toward the core, one from above and the other from below. The reason for this choice is not so obvious in the present case of the asymmetric slab waveguide. However, this procedure results in even and odd radiation modes in the symmetric case provided the waves are properly phased. We adjust our radiation modes such that even and odd modes result in the limit $n_2 = n_3$. The following field expressions satisfy the wave Eq. (1.3-15) and the boundary conditions:

$$E_y = C_r[\cos\Delta x + (\sigma/\Delta) F_i \sin\Delta x], \qquad \text{for} \quad x \geqslant 0 \tag{1.4-21}$$

$$= C_r(\cos\sigma x + F_i \sin\sigma x), \qquad \text{for} \quad 0 \geqslant x \geqslant -d \tag{1.4-22}$$

$$= C_r[(\cos\sigma d - F_i \sin\sigma d)\cos\rho(x+d) + (\sigma/\rho)(\sin\sigma d + F_i\cos\sigma d)\sin\rho(x+d)], \qquad \text{for} \quad x \leqslant -d \tag{1.4-23}$$

These field expressions have already been adjusted so that the H_z component, which follows from Eq. (1.3-14), is also continuous at the two interfaces. For simplicity, the detailed expressions for H_z are not given. The parameters Δ, σ, and ρ are defined by Eqs. (1.4-7)–(1.4-9). Note however, that in the range (1.4-20) Δ is a real parameter.

The field expressions for the radiation modes, Eqs. (1.4-21)–(1.4-23), contain two undetermined constants, C_r and F_i. One of them, C_r, can again be related to the parameter P of the normalizing expression (1.4-18). However, the constant F_i remains completely arbitrary. We are free to choose F_i according to our own convenience. If we use a certain value F_1, we obtain one set of radiation modes. A second choice F_2 results in another set of radiation modes. We thus see that we obviously have obtained two independent types of radiation modes. The freedom of choice of F_i is related to the arbitrary phases of the two plane waves that generate the radiation modes.

For future applications it is most important to choose these two types of modes to be mutually orthogonal. However, even this requirement does not completely determine both F_1 and F_2. In the case of the symmetric slab waveguide this problem was avoided by choosing even and odd radiation modes from the start. In this case no undetermined parameters occur in the equations. We find it convenient to choose F_1 and F_2 in our present situation in such a way that even and odd radiation modes result in the limit $n_2 = n_3$. This choice is not dictated by necessity but only by convenience.

The requirement that even and odd radiation modes result in the limit of a symmetric slab yields the following explicity expressions for F_1 and F_2:

$$\begin{aligned} F_{1,2} = {} & [(\sigma^2 - \rho^2) \sin 2\sigma d]^{-1} \{(\sigma^2 - \rho^2) \cos 2\sigma d + (\rho/\Delta)(\sigma^2 - \Delta^2) \\ & \pm [(\sigma^2 - \rho^2)^2 + 2(\rho/\Delta)(\sigma^2 - \rho^2)(\sigma^2 - \Delta^2) \cos 2\sigma d \\ & + (\rho^2/\Delta^2)(\sigma^2 - \Delta^2)^2]^{1/2}\} \end{aligned} \tag{1.4-24}$$

Both signs of the square root are used to determine F_1 or F_2. The plus sign belongs to the odd modes, while the minus sign belongs to the even modes, in the limit $n_2 = n_3$. The two types of radiation modes are obtained by using either F_1 or F_2 in the field expressions (1.4-21)–(1.4-23). It can be checked that the following relation is valid:

$$F_1 F_2 = -1 \tag{1.4-25}$$

The amplitude coefficient C_r of the radiation modes can again be related to the factor P by means of expression (1.4-18):

$$\begin{aligned} C_r^2 = {} & \frac{4\omega\mu_0 P}{\pi|\beta|} \\ & \cdot \left[(\cos\sigma d - F_i \sin\sigma d)^2 + \frac{\sigma^2}{\rho^2}(\sin\sigma d + F_i \cos\sigma d)^2 + \left(1 + \frac{\sigma^2}{\Delta^2} F_i^2\right)\frac{\Delta}{\rho}\right]^{-1} \end{aligned} \tag{1.4-26}$$

Actually C_r also needs the label $i = 1$ or $i = 2$, since both values of i are used to label F_i in its denominator. We omit these additional labels in order to keep the notation simpler.

The parameter ρ is used to label the radiation modes [see Eq. (1.4-18)]. It is allowed to assume all values from 0 to ∞. If

$$0 \leqslant \rho \leqslant (n_2{}^2 - n_3{}^2)^{1/2} k \tag{1.4-27}$$

β covers the range of values given by Eq. (1.4-3), so that this range of the parameter ρ belongs to the radiation modes (1.4-4)–(1.4-6). As ρ covers the range

$$(n_2{}^2 - n_3{}^2)^{1/2} k \leqslant \rho \leqslant n_2 k \tag{1.4-28}$$

the corresponding β values lie in the range (1.4-20) belonging to the modes (1.4-21)–(1.4-23). However, ρ is also allowed to fall in the range

$$n_2 k \leqslant \rho < \infty \tag{1.4-29}$$

It is apparent from Eq. (1.4-9) that the corresponding β values are imaginary. The radiation modes that correspond to imaginary β values are also described by Eqs. (1.4-21)–(1.4-23). These modes decay exponentially along the z axis and are necessary to describe the fine structure of the field in the immediate vicinity of a waveguide imperfection. Radiation modes of this kind cannot be generated by a plane wave source at infinity. We have thus found an extension of the simple intuitive range of radiation modes whose generation could easily be visualized by physical arguments. Radiation modes with evanescent fields along the z direction are not very important for most practical applications. However, they are necessary to form a complete set of orthogonal modes which is capable of expressing any field distribution satisfying Maxwell's equations. Evanescent waves of this type are familiar from the theory of hollow metallic waveguides, where they are associated with waves beyond cutoff. Cutoff has a different meaning in dielectric waveguides and is not associated with evanescent waves but with fields that lose power continuously by radiation. Such leaky waves will be discussed in Sect. 1.5. However, we can convince ourselves that the evanescent waves of our dielectric waveguide do have a close relationship to the cutoff waves in metallic waveguides. Let us assume that we enclose the slab waveguide with plane metallic plates on both sides. The two metal surfaces are assumed to be perfect conductors. The new structure is a metallic waveguide with a dielectric insert, all the modes of which belong to a discrete spectrum of β values. A finite number of them can propagate, while an infinite number is beyond cutoff in the usual waveguide sense. These cutoff modes decay exponentially in z direction. As we allow the metal surfaces to move further away from the core of the dielectric slab, but keep the region between core and metal plates filled with the two media with refractive indices n_2 and n_3, we find that the modes of the structure assume two different features. We again obtain the usual guided modes of the dielectric slab waveguide which are unaffected by the presence of the metal surfaces, provided these are sufficiently far away so that the exponentially decaying fields (in transverse direction) reach them with practically zero intensity. In addition to these surface modes of the dielectric slab, we find modes of the metallic waveguide that fill the entire volume between the two reflectors. The β values of these latter modes become closer spaced as the metal surfaces recede more from the core region and become the continuum of radiation modes in the limit of infinitely far reflectors. The cutoff modes of the metallic waveguide remain cutoff even if the reflectors are infinitely far spaced. (This is true only if we move the metal plates in discrete jumps so that they are placed at successive zeros of the mode we wish to discuss.) The modes are cutoff in the sense of metallic waveguides if their transverse nodes follow each other at distances that are spaced closer than half the plane wave wavelength in the medium filling the guide. The nature of the evanescent radiation modes is thus identical to the cutoff modes of metallic waveguides. They are evanescent because of their rapid transverse variation.

TM Radiation Modes

The TM radiation modes have the same features that we encountered in discussing TE radiation modes. With restriction (1.3-4) we have only three nonvanishing field components H_y, E_x, and E_z. The H_y component is obtained from the reduced wave Eq. (1.3-53), and the components E_x and E_z can be calculated from Eqs. (1.3-51) and (1.3-52). We list only the H_y component and find radiation modes in the range (1.4-3) [for a definition of Δ, σ, and ρ see Eqs. (1.4-7)–(1.4-9)]:

$$H_y = (\beta/|\beta|)\, D_r e^{i\Delta x}, \qquad \text{for} \quad x \geqslant 0 \tag{1.4-30}$$

$$= (\beta/|\beta|)(D_r \cos\sigma x + G_r \sin\sigma x), \qquad \text{for} \quad 0 \geqslant x \geqslant -d \tag{1.4-31}$$

$$= (\beta/|\beta|)[(D_r \cos\sigma d - G_r \sin\sigma d)\cos\rho(x+d) + K_r \sin\rho(x+d)], \qquad \text{for} \quad x \leqslant -d \tag{1.4-32}$$

The H_y component has been adjusted to satisfy the boundary conditions at $x = 0$ and $x = -d$. The requirement that E_z also satisfy the boundary conditions leads to the determination of the coefficients G_r and K_r:

$$G_r = (n_1^2/n_3^2)(i\Delta/\sigma)\, D_r \tag{1.4-33}$$

and

$$K_r = [(n_2^2/n_1^2)(\sigma/\rho)\sin\sigma d + (n_2^2/n_3^2)(i\Delta/\rho)\cos\sigma d]\, D_r \tag{1.4-34}$$

From Eq. (1.4-17) we obtain for TM modes, with the help of Eq. (1.3-51),

$$(\beta/2\omega\varepsilon_0)\int_{-\infty}^{\infty} [1/n^2(x)]\, H_y(\rho)\, H_y^*(\rho')\, dx = s_\rho(\beta^*/|\beta|)\, P\delta(\rho-\rho') \tag{1.4-35}$$

The function $n(x)$ assumes the constant values n_1, n_2, and n_3 in the three regions of the slab. The normalization condition and orthogonality relation (1.4-35) allows us to express D_r in terms of P:

$$D_r^2 = (4\omega\varepsilon_0 n_1^4 n_2^2 n_3^4 \rho^2 \sigma^2 P/\pi|\beta|)(\beta^*/\beta)\, s_\rho \cdot [n_1^4\rho^2(n_3^2\sigma\cos\sigma d - n_1^2 i\Delta\sin\sigma d)^2 + n_2^4\sigma^2(n_3^2\sigma\sin\sigma d + n_1^2 i\Delta\cos\sigma d)^2]^{-1} \tag{1.4-36}$$

(The factor s_ρ must be chosen $+1$ or -1 to keep D_r^2 positive.) The TM radiation modes of this first type decay exponentially in region 3, for $x > 0$, since $i\Delta$ is a real negative quantity for β values in the range (1.4-3).

The second type of radiation modes consists of standing waves above and below the core region. In complete analogy to the TE modes of this type, we now have

$$H_y = (\beta/|\beta|)\, S_r[\cos \Delta x + (n_3^2/n_1^2)(\sigma/\Delta)\, R_i \sin \Delta x] \qquad \text{for} \quad x \geqslant 0 \tag{1.4-37}$$

$$= (\beta/|\beta|)\, S_r(\cos \sigma x + R_i \sin \sigma x), \qquad \text{for} \quad 0 \geqslant x \geqslant -d \tag{1.4-38}$$

$$\begin{aligned} &= (\beta/|\beta|)\, S_r[(\cos \sigma d - R_i \sin \sigma d) \cos \rho(x+d) \\ &\quad + (n_2^2/n_1^2)(\sigma/\rho)(\sin \sigma d + R_i \cos \sigma d) \sin \rho(x+d)], \qquad \text{for} \quad x \leqslant -d \end{aligned} \tag{1.4-39}$$

The parameters Δ, σ, and ρ are defined by Eqs. (1.4-7)–(1.4-9), and Δ is a real quantity in the range (1.4-28) and (1.4-29) which belongs to radiation modes of this type.

We choose R_i so that modes 1 and 2 are orthogonal and even and odd radiation modes result in the limit $n_2 = n_3$ of the symmetric slab:

$$\begin{aligned} R_{1,2} = {} & [(n_2^4\sigma^2 - n_1^4\rho^2) \sin 2\sigma d]^{-1} \\ & \cdot \{(n_2^4\sigma^2 - n_1^4\rho^2) \cos 2\sigma d + (n_2^2/n_3^2)(\rho/\Delta)(n_3^4\sigma^2 - n_1^4\Delta^2) \\ & \pm [(n_2^4\sigma^2 - n_1^4\rho^2)^2 + (n_2^4/n_3^4)(\rho^2/\Delta^2)(n_3^4\sigma^2 - n_1^4\Delta^2)^2 \\ & + 2(n_2^2/n_3^2)(\rho/\Delta)(n_2^4\sigma^2 - n_1^4\rho^2)(n_3^4\sigma^2 - n_1^4\Delta^2) \cos 2\sigma d]^{1/2}\} \end{aligned} \tag{1.4-40}$$

R_1 is obtained for the positive sign of the square root, while R_2 belongs to the negative sign. The plus sign leads to odd modes, and the minus sign to even modes in the limit of a symmetric slab, $n_2 = n_3$. It must be remembered that even and odd field distributions are referred to the center of the core. In order to see that even and odd modes do indeed result from R_2 and R_1, it is necessary to transform Eqs. (1.4-37)–(1.4-39) to a coordinate system that is centered in the middle of the core.

The amplitude coefficients of the modes are obtained by substituting the field expressions (1.4-37)–(1.4-39) into Eq. (1.4-35). The tedious calculation results in

$$\begin{aligned} S_r = {} & (4\omega\varepsilon_0 P/\pi|\beta|)(\beta^*/\beta)s_\rho \\ & \cdot \left[\frac{1}{n_2^2}(\cos \sigma d - R_i \sin \sigma d)^2 + \frac{n_2^2}{n_1^4}\frac{\sigma^2}{\rho^2}(\sin \sigma d + R_i \cos \sigma d)^2 \right. \\ & \left. + \left(\frac{1}{n_3^2} + \frac{n_3^2}{n_1^4}\frac{\sigma^2}{\Delta^2} R_i^2\right)\frac{\Delta}{\rho}\right]^{-1} \end{aligned} \tag{1.4-41}$$

We must choose $s_\rho = +1$ or -1 to keep S_r^2 positive. The radiation modes will be needed in later sections to calculate radiation losses caused by waveguide irregularities.

Mode Orthogonality

All waveguide modes are mutually orthogonal to each other. Orthogonality is defined in the sense of Eq. (1.4-17). If we use any guided or radiation mode for the field labeled $E(\rho)$ and any other guided or radiation mode for $H(\rho')$, we find that the integral over the z component of the vector product of the two mode fields vanishes. The labels ρ and ρ' are here used to distinguish between two different modes. These labels may indicate either guided modes, in which case they are discrete quantities, or radiation modes, in which case they form a continuum. Only if $E(\rho)$ and $H(\rho')$ belong to the same mode is the result of the integration different from zero. For guided modes we can normalize the field to make the right-hand side equal to the power P carried by the mode. We always normalize our normal modes so that they carry the same amount of power, for example 1 W. The function $\delta(\rho-\rho')$ on the right-hand side of Eq. (1.4-17) indicates a Dirac delta function when both ρ and ρ' label continuum modes. For guided modes we must interpret this symbol as the Kronecker delta, which is unity when both indices are equal and zero otherwise. The orthogonality relation (1.4-17) has the physical meaning that the power carried by the waveguide field is simply the sum of the powers carried by all the modes. If we consider the superposition of several modes and form the power of the mode field by calculating the integral of the Poynting vector over the infinite cross section, we can use Eq. (1.4-17) to convert the expression for the total power to the sum of the powers carried by each mode. This is not a trivial statement. In fact, mode orthogonality with respect to average power as expressed by Eq. (1.4-17) applies only for lossless dielectric waveguides. If the refractive indices n_1, n_2, and n_3 are complex quantities, relation (1.4-17) does not hold. It will still hold to a good approximation for only slightly lossy waveguides, but it is no longer strictly true. However, it can be shown that another orthogonality relation applies, which follows from Eq. (1.4-17) simply by omitting the complex conjugation from the H field. This mode orthogonality is more general than the power orthogonality implied by Eq. (1.4-17). Without the complex conjugation the orthogonality relation holds for lossy as well as lossless waveguides, but the simple interpretation of power orthogonality is lost in this case. The quantity $\mathbf{E}\times\mathbf{H}$ without the complex conjugate for $\mathbf{H}$ does not have a physical meaning. This is the reason we prefer the form (1.4-17) of the orthogonality relation. The validity of Eq. (1.4-17) has been proven in many places (see, for example, Sect. 8.5 of [Me1]). Both types of orthogonality relations can, of course, be verified by direct calculation using the field expressions for the modes.

1.5 Leaky Waves

The guided and radiation modes, whose field expressions we derived in the last two sections, form a complete orthogonal set of modes. Any field distribution that obeys the restriction (1.3-4) can be expressed by series expansion into these modes. However, there are other solutions of the eigenvalue Eqs. (1.3-26) and (1.3-63) that are not part of the complete set of modes [Sol]. These additional solutions do not belong to proper field expressions, since the fields associated with them diverge at $x = \pm\infty$. The integrals in Eqs. (1.3-46) and (1.3-72) diverge and, since these modes have a discrete spectrum, they cannot even be normalized with the help of the Dirac delta function. However, if we do not try to compute the total power carried by these modes and if we ignore the normalization problem, we can interpret them as representing guided modes beyond the cutoff point.

It is helpful to use physical intuition to understand leaky waves, as these additional solutions of the eigenvalue problem are called. Using the geometrical optics argument developed in Sect. 1.2 we can obtain an eigenvalue equation for leaky waves. Since we are now considering waves that are beyond cutoff from the point of view of the usual guided modes, we must restrict ourselves to propagation constants in the range

$$-n_2 k < \beta < n_2 k \tag{1.5-1}$$

This range of β values belongs to the domain of radiation modes. The leaky waves to be considered here have thus the same propagation constants as the radiation modes, except that they form a discrete set instead of a continuum. In addition, leaky waves are always lossy so that their propagation constants have complex values. We are here about to derive an approximation to the real part of the complex propagation constants of leaky waves.

In the β range (1.5-1) both square roots appearing in Eq. (1.2-5) are real, so that there is no phase shift on reflection from the dielectric interface with region 2. With $\phi_2 = 0$ we obtain from Eq. (1.2-11) with the help of Eqs. (1.2-9), (1.2-10), (1.2-13), and (1.3-75) the eigenvalue condition

$$\kappa d = \begin{cases} N\pi + \arctan[(m_1{}^2/m_3{}^2)(\delta/\kappa)], & \text{for} \quad \delta^2 > 0 \\ N\pi, & \text{for} \quad \delta^2 < 0 \end{cases} \tag{1.5-2}$$

The real part of the propagation constant is then obtained from Eqs. (1.2-13) and (1.5-2):

$$\beta = (n_1{}^2 k^2 - \kappa^2)^{1/2} \tag{1.5-3}$$

This result holds for TE (with $m_i = 1$) and for TM modes (with $m_i = n_i$).

We obtain the imaginary part of the propagation constant by considering the reflection coefficient of plane waves that superimpose themselves on each other to form the leaky wave in the core. If we use the notation

$$\kappa = (n_1{}^2 k^2 - \beta^2)^{1/2} \tag{1.5-4}$$

$$\rho = (n_2{}^2 k^2 - \beta^2)^{1/2} \tag{1.5-5}$$

and

$$\Delta = (n_3{}^2 k^2 - \beta^2)^{1/2} \tag{1.5-6}$$

we obtain the power reflection coefficient from (1.2-5):

$$R_2 = |r|^2 = [(\kappa - \rho)/(\kappa + \rho)]^2 \tag{1.5-7}$$

for reflection from medium 2 and†

$$R_3 = |(\kappa - \Delta)/(\kappa + \Delta)|^2 \tag{1.5-8}$$

for reflection from medium 3. Concentrating on one interface for the moment, we obtain the power ΔP, that is lost per unit length, from the following equation:

$$\Delta P_2 = S(1 - R_2) \sin\theta \tag{1.5-9}$$

where S is the magnitude of the Poynting vector of the plane wave and indicates the amount of power that is flowing through the unit area in the direction of plane wave propagation. $1 - R$ is the ratio of the power flowing normal to the surface out of region 1 divided by the amount of power that is carried normal to that surface. Multiplication of $1 - R$ by the x component, $S_x = S \sin\theta$, of the Poynting vector yields the power that is lost per unit length of the waveguide (see Fig. 1.5.1). Similarly we find the power loss per unit length on the other interface

$$\Delta P_3 = S(1 - R_3) \sin\theta \tag{1.5-10}$$

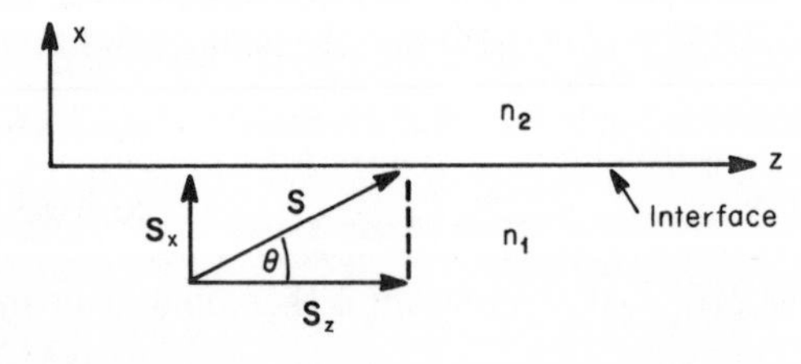

Fig. 1.5.1 *Illustration of the decomposition of the energy flow vector into components parallel and perpendicular to the dielectric interface.*

† Δ may be real or imaginary, but ρ is real in the interval (1.5-1).

The power loss coefficient of the leaky wave is defined as the amount of power that is lost per unit length divided by the power carried in the guide. The power carried by the plane wave inside the waveguide is

$$P = S_z d = Sd \cos\theta \tag{1.5-11}$$

The power loss coefficient 2α is thus

$$2\alpha = (\Delta P_2 + \Delta P_3)/2P \tag{1.5-12}$$

The factor 2 in the denominator is needed to adjust for the fact that the combined length over which the power $\Delta P_2 + \Delta P_3$ is dissipated is twice the unit length. We define α as the amplitude loss coefficient, so that 2α is the power loss. From Eqs. (1.5-7)–(1.5-12) we finally obtain, with the help of Eqs. (1.2-6) and (1.2-13), the power loss coefficient for leaky TE waves:

$$2\alpha = \frac{2}{d}\left[\frac{\rho}{(\kappa+\rho)^2} + \frac{\Delta+\Delta^*}{2|\kappa+\Delta|^2}\right]\frac{\kappa^2}{\beta} \tag{1.5-13}$$

The corresponding formula for leaky TM modes is

$$2\alpha = \frac{2}{d}\left[\frac{n_2{}^2\rho}{(n_2{}^2\kappa+n_1{}^2\rho)^2} + \frac{n_3{}^2(\Delta+\Delta^*)}{2|n_3{}^2\kappa+n_1{}^2\Delta|^2}\right]\frac{n_1{}^2\kappa^2}{\beta} \tag{1.5-14}$$

These equations can be expected to be reasonably accurate for relatively low losses. For very lossy waves the plane wave concept does not work very well.

It is important to realize that the leaky waves described so far are not modes of the structure. By definition a mode maintains its shape at all points along the cross section of the waveguide except for a phase change and except for an attenuation term of the form $e^{-\alpha z}$ in case of lossy modes. The lossy waves, whose properties we have studied by means of geometrical optics, are transient phenomena. If we connect a lossless waveguide to a guide whose dimensions are designed so that it is cutoff for the field distribution traveling in the first guide, we excite a field that behaves approximately as described by the ray optics theory of this section. However, this wave changes its shape throughout the cross section. This change of the field shape is caused by the radiation that is leaking out of the core region. Modes of dielectric waveguides must be considered throughout the infinite cross section. Outside the waveguide core we find more radiation at increasing distances from the core as we follow the leaky wave in the cutoff waveguide. We know that the true modes with propagation constants in the range (1.5-1) are radiation modes discussed in Sect. 1.4. Our leaky waves can be represented as series expansions in terms of radiation modes. Each radiation mode is lossless by itself. However, the total field must also be lossless because the power is simply redistributed from inside the waveguide core to the core region and the region outside the core. The loss

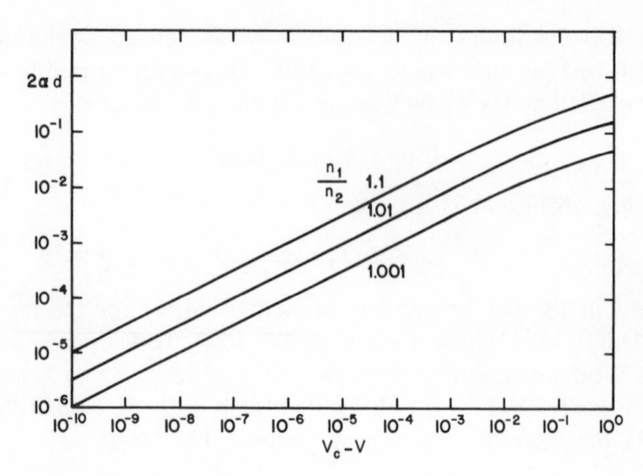

Fig. 1.5.2 *Power loss 2α of a leaky wave as a function of the frequency parameter V.*

formulas (1.5-13) and (1.5-14) account approximately for the power loss from the core region, regarding the amount of power found outside the core as lost.

Figure 1.5-2 shows a plot of the power loss coefficient (1.5-13), normalized with respect to the core width, as a function of the V parameter defined by Eq. (1.3-37). We have assumed $n_2 = n_3$ for simplicity. The curve applies to the second even TE mode, whose cutoff value is $V = V_c = 2\pi$. Since the loss increases enormously for V values just beyond cutoff, we have plotted $V_c - V$ on the horizontal axis. In order to have a feeling for the actual magnitude of the loss let us assume that the core is 5 μm wide. This corresponds approximately to a symmetric waveguide with $n_1/n_2 = 1.01$ and a vacuum wavelength of $\lambda = 1$ μm. In order to find the loss per centimeter expressed in decibels, we must divide the numbers given in Fig. 1.5.2 by the core width, $d = 5 \times 10^{-4}$ cm, and multiply the result by 4.34 to convert the absolute units to decibels. For $V_c - V = 10^{-8}$, we find for $n_1/n_2 = 1.01$ the loss value $2\alpha d = 3 \times 10^{-5}$ from Fig. 1.5.2. For our example of a 5-μm core, we thus have an actual loss of 0.26 dB/cm. This loss value increases by one order of magnitude for every two orders of magnitude increase of $V_c - V$. This means that for $V_c - V = 10^{-4}$, we already have 26 dB/cm loss. For $V_c - V = 0.1$, the loss has reached 651 dB/cm. The increase in loss levels off slightly for large values of $V_c - V$. This example dramatizes the very high losses of the leaky waves just beyond the cutoff point of the corresponding guided mode. The power that is carried in the core region escapes very fast once total internal reflection is lost.

After obtaining some insight into leaky waves by means of a geometrical optics approach, we continue their study from the point of view of the boundary value approach. We have defined a guided mode as a solution of Maxwell's equations satisfying the boundary conditions at the core boundary and vanishing at $x = \pm\infty$. By omitting our insistence on vanishing field amplitudes at infinity, we obtained the radiation mode solutions of our boundary value problem. The radiation exhibited a continuous spectrum of allowed values of their propagation constants since we arrived at an inhomogeneous equations system for the determination of the amplitude coefficients of the field functions. This fundamental difference between guided and radiation modes was achieved by allowing incoming as well as outgoing waves to appear outside the waveguide core. The incoming waves add their own amplitude coefficients to the equation system, so that we end up with more undetermined coefficients than equations to determine them. This results in an inhomogeneous equation system if we arbitrarily assume that one of the amplitude coefficients is known. The equation system can then always be solved without the need for requiring the vanishing of the system determinant. There is, therefore, no determinantal or eigenvalue equation to restrict the possible values of the propagation constant.

However, it is possible to drop the requirement of vanishing field amplitudes at infinity without introducing additional fields corresponding to incoming waves, as in the case of radiation modes. We could, for example, search for solutions of the eigenvalue equation for negative instead of positive values of the parameters γ and δ which determine the amount of field decay outside the core region. If such solutions did exist, we would not expect them to correspond to physically realizable fields, since their field amplitudes would grow exponentially outside the core region. It turns out that solutions of the eigenvalue equations with complex values of γ and δ do indeed exist. For simplicity we study such solutions for the case of the symmetric slab waveguide. The even TE modes are determined by the eigenvalue equation

$$\tan u = v/u \tag{1.5-15}$$

and the odd modes by

$$\tan u = -u/v \tag{1.5-16}$$

These equations are identical with Eqs. (1.3-39) and (1.3-40). However, we are now using the abbreviations

$$V_s = \tfrac{1}{2}V = (n_1{}^2 - n_2{}^2)^{1/2}(kd/2) \tag{1.5-17}$$

$$u = \kappa d/2 \tag{1.5-18}$$

and

$$v = \gamma d/2 = (V_s{}^2 - u^2)^{1/2} \tag{1.5-19}$$

We discuss the leaky wave solutions of the even mode eigenvalue Eq. (1.5-15) in detail and state the corresponding results for the odd modes.

We begin by exploring the possibility of finding solutions of Eq. (1.5-15) for negative values of v. We set

$$v = -w, \qquad \text{with} \quad w > 0 \tag{1.5-20}$$

and search for solutions of the transcendental equation

$$\tan u = -w/u \tag{1.5-21}$$

Figure 1.5.3 gives an indication of the possible solutions for real positive values of w. The curve $-w/u$ was drawn under the assumption that V_s is slightly less than π, so that the second even mode is just beyond cutoff, according to Eqs. (1.3-43) and (1.5-17). The construction for the usual guided

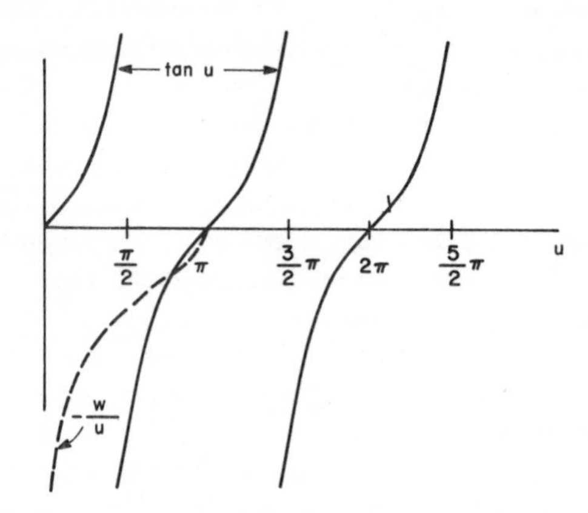

Fig. 1.5.3 *Graphical illustration of the unphysical solutions of the eigenvalue Eq. (1.5–21).*

modes differs from Fig. 1.5.3 by the fact that the dashed line representing the function v/u has the opposite sign and is folded upward to the first quadrant of the diagram. It is apparent from Fig. 1.5.3 that the dashed line crosses the tangent curves twice, provided that V_s is slightly smaller than π. There is, of course, also one crossing of the dashed curve with the second branch of the tangent curve for values $V_s > \pi$. Here we are in the regime where guided mode solutions of the second even mode also exist. The solutions of Eq. (1.5-21) for $V_s > \pi$ represent unphysical waves whose amplitudes grow with increasing distance from the core. The leaky wave solutions, that are apparent from the construction of Fig. 1.5.3, are also unphysical but merge continuously into the complex leaky wave solutions for which some physical rationale can be given. We are interested primarily in the solution that results from the crossover of the solid and dashed curve nearest the u axis. We see immediately that this unphysical solution starts at $u = \pi$ when $V_s = \pi$, that is, just at the cutoff point of the second guided even TE mode. For decreasing values of V_s the crossover point moves slowly away from the u axis and yields decreasing values of the solution u. Then comes the point where the dashed curve becomes tangential to the tangent curve touching it in only one point. For even smaller values of V_s there are no more real solutions. We shall see shortly that Eq. (1.5-21) has complex solutions for values of V_s that are too small to yield a crossing of the solid and dashed curves of Fig. 1.5.3.

The value at which the real unphysical solution of Eq. (1.5-21) just ceases to exist is of interest. We obtain the solutions at this point by requiring that the tangents of the functions $\tan u$ and $-w/u$ are parallel and that Eq. (1.5-21) is still satisfied. The requirement of parallel tangents leads to the equation

$$1/\cos^2 u = V_s^2/wu^2 \tag{1.5-22}$$

From Eq. (1.5-21) we derive

$$\cos u = -u/V_s \tag{1.5-23}$$

The negative sign of the right-hand side of Eq. (1.5-23) follows from inspection of the sign of the cosine function for $\pi/2 < u < \pi$ and the fact that u and V_s must both be positive. Equations (1.5-22) and (1.5-23) must both be satisfied simultaneously. This is possible only if

$$w = 1 \tag{1.5-24}$$

The corresponding value of V_s is obtained by solving the transcendental Eq. (1.5-21) with $w = 1$:

$$\tan u = -1/u, \qquad \text{for even modes} \tag{1.5-25}$$

We have thus found the point where real solutions of Eq. (1.5-21) cease to exist. When we lower the value of V_s below the limit that is obtained from solutions of Eq. (1.5-25) and the relation

$$u = (V_s^2 - 1)^{1/2} \tag{1.5-25a}$$

the solutions of Eq. (1.5-21) or (1.5-15), if they exist at all, must be complex. We expect that the solutions move continuously from real to complex values, so that the imaginary parts of u and v must be small. We set

$$u = u_r + iu_i \tag{1.5-26}$$

and

$$v = v_r + iv_i \tag{1.5-27}$$

and require that

$$u_i \ll u_r, \qquad v_i \ll v_r \tag{1.5-28}$$

Substitution of Eqs. (1.5-26) and (1.5-27) into Eq. (1.5-15) leads to two simultaneous equations which assume the form

$$\tan u_r = v_r/u_r \tag{1.5-29}$$

and

$$u_i V_s^2/u_r^2 = v_i/u_r - u_i v_r/u_r^2 \tag{1.5-30}$$

if we use Eq. (1.5-28) and neglect terms of higher than second order in the small imaginary quantities. Second-order terms cancel from Eq. (1.5-29) if we use the fact that v_r can deviate from -1 only by a quantity of first order. Even though second-order terms are absent from these equations, they are indeed accurate to the second order of approximation. We need two more equations to be able to determine the four unknown quantities u_r, u_i, v_r, and v_i. These additional equations are obtained from Eq. (1.5-19). We find without any approximation

$$u_r^2 + v_r^2 - u_i^2 - v_i^2 = V_s^2 \tag{1.5-31}$$

and

$$u_r u_i + v_r v_i = 0 \tag{1.5-32}$$

When we solve Eq. (1.5-32) for v_i and substitute this value into Eq. (1.5-30) we obtain the solution

$$v_r = -1 \tag{1.5-33}$$

in agreement with Eq. (1.5-24). This means that the value v_r remains constant to first order of approximation and does not change from its value $v_r = -1$ which it assumes at the point where the solution becomes complex. Equation (1.5-29) results again in Eq. (1.5-25), which now applies to the real part u_r. Since u_r and v_r have been determined, we find the imaginary part v_i from Eqs. (1.5-31) and (1.5-32):

$$v_i = (u_r/V_s)(u_r^2+1-V_s^2)^{1/2} \tag{1.5-34}$$

and also

$$u_i = (1/V_s)(u_r^2+1-V_s^2)^{1/2} \tag{1.5-35}$$

We have thus found approximate leaky wave solutions of the eigenvalue Eq. (1.5-15) for the even TE modes of the symmetric slab waveguide. A complete solution must be obtained numerically with the aid of a computer. It may be helpful to point out that computer solutions tend to fail in the region where our approximate solutions are valid. The reason for this difficulty becomes apparent when we inspect the approximate forms (1.5-29) and (1.5-30) of the complex Eq. (1.5-15). The real part of the eigenvalue equation contains the zero-order terms u_r and v_r, while the imaginary part leads to Eq.

(1.5-30), which is small of first order. A computer solution of Eq. (1.5-15), which treats the imaginary and real parts of this complex equation on an equal footing, falls into the trap of being satisfied with small values of the imaginary part, regarding them as a sign that a solution is reached. In actuality, the imaginary part is always small near the point where the complex solution merges into the real solution, so that considerable inaccuracies result.

Figure 1.5.4 shows the locus of the solution points in the complex v plane. The parameter V_s is changed continuously. At the point $v_r = -1$, $v_i = 0$, we have V_s given as the solution of Eqs. (1.5-25) and (1.5-25a). As we move away from this point, V_s decreases in value and reaches zero when v_r reaches $-\infty$. The solution curves show the trajectories of the second even and odd modes. The first even mode does not have a cutoff value, so that it never becomes a leaky wave. The first odd mode, which does go through a cutoff at $V_s = \pi/2$, never leaves the negative real v axis and does not become a proper leaky wave. It simply does not have a physical interpretation beyond its cutoff point.

Even though our derivation was based on the eigenvalue Eq. (1.5-15) of the even TE modes, it can be shown that the approximate solutions (1.5-33)–

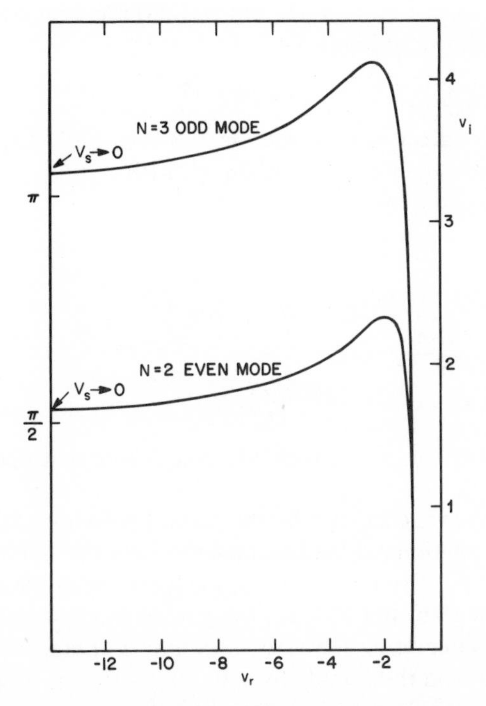

Fig. 1.5.4 *Trajectory of the leaky wave solutions in the complex v plane. The solution point moves as a function of V_s.*

(1.5-35) apply also to the odd TE modes, whose eigenvalue equation is Eq. (1.5-16). However, the real part of u is now obtained as the solution of the equation

$$\tan u_r = u_r, \qquad \text{for odd modes} \tag{1.5-36}$$

A similar analysis near the point $V_s = 0$ yields the approximate equations

$$\tan u_r = -u_i/u_r, \qquad \text{for even modes} \tag{1.5-37}$$

and

$$\tan u_r = u_r/u_i, \qquad \text{for odd modes} \tag{1.5-38}$$

and the equation

$$\exp(-u_i) = V_s/2(u_r^2 + u_i^2)^{1/2} \tag{1.5-39}$$

which holds for even and odd modes. Equation (1.5-39) shows that we must have

$$u_i \to \infty, \qquad \text{as} \quad V_s \to 0 \tag{1.5-40}$$

so that we find from Eqs. (1.5-37) and (1.5-38) the asymptotic end points of the trajectories shown in Fig. 1.5.4

$$u_r = \nu\pi/2 \tag{1.5-41}$$

with odd integer values of ν for even modes and even integer values of ν for odd modes. For $V_s = 0$ we find from Eq. (1.5-19)

$$v_i = u_r \tag{1.5-42}$$

and

$$v_r = -u_i \tag{1.5-43}$$

so that we also have

$$v_i = \nu\pi/2 \tag{1.5-43a}$$

This equation explains the asymptotic end points that are marked in Fig. 1.5.4.

Figure 1.5.5 shows schematically the general behavior of the trajectories of the leaky wave solutions as the parameter V_s runs through the values from 0 to ∞. Along the positive real v axis are the locations of the guided TE modes of the symmetric slab. For $V_s \to \infty$, the guided mode solutions move out to $v_r = \infty$. As the value of V_s decreases, each mode approaches its cutoff point at $v_r = 0$. One should think that any solution that exists past the cutoff point would immediately belong to a lossy wave with a positive imaginary part v_i. Positive values of v_i are needed since the wave should travel outward, away

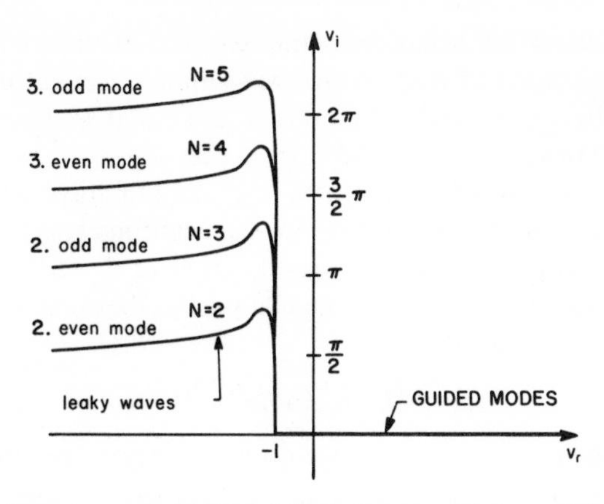

Fig. 1.5.5 *Schematic representation of the leaky wave solutions of the first four modes in the complex v plane.*

from the core. However, as we have discussed earlier and as is shown in Fig. 1.5.5, the locus of the solution of the eigenvalue equation first moves along the negative real v axis to the point $v_r = -1$. Starting at this point the solution moves out into the complex plane and can now be interpreted as a leaky mode corresponding to a wave that travels in the waveguide beyond cutoff. However, the field amplitude grows exponentially as $\exp[-(2v_r x/d)]$ for increasing values of x. We may interpret this behavior as the result of an accumulated radiation field that builds up from radiation escaping from the waveguide. Since the field must be visualized as arriving from $z = -\infty$, it appears plausible that the accumulated radiation would reach infinite values far from the core at $x = \pm\infty$, particularily since it requires an infinitely strong field at $z = -\infty$ if the field is to reach finite z values with a nonvanishing amplitude. It is thus possible to associate a certain physical reality with leaky wave solutions with $v_i > 0$. However, there are also complex solutions of the eigenvalue equation with $v_i < 0$ which correspond to waves that move inward toward the core and grow as they travel along the positive z axis. These waves and the solutions with $v_r < 0$ and $v_i = 0$ are of no particular practical value.

For small values of v_i, we can easily calculate the loss coefficient of the leaky wave. We consider a complex propagation constant

$$\beta = \beta_r - i\alpha \tag{1.5-44}$$

and obtain from Eqs. (1.2-14) and (1.5-19)

$$\beta_r = [n_2^2 k^2 + (2v_r/d)^2]^{1/2}, \qquad \text{for} \quad v_i \ll 1 \tag{1.5-45}$$

$$\alpha = -4v_r v_i/\beta_r d^2, \qquad \text{for} \quad v_i \ll 1 \tag{1.5-46}$$

We observe an interesting behavior of the real part of the propagation constant β of the leaky mode solution. As the wave approaches cutoff, v_r decreases in value and reaches $v_r = 0$ right at cutoff. At this point we have, not surprisingly, $\beta_r = n_2 k$. However, we should now expect that $\beta_r < n_2 k$ as the wave goes beyond cutoff. Instead, v_r becomes negative but stays real, so that we have again $\beta_r > n_2 k$ quite contrary to our expectations and to the behavior of the leaky wave transient fields discussed earlier in this section. In fact, as long as v_i remains small, we have for the real part of the propagation constant of the leaky mode solution

$$\beta_r = [n_2{}^2 k^2 + (4/d^2)]^{1/2} \tag{1.5-47}$$

independent of the mode number. If we do not neglect v_i we obtain

$$\beta_r = [n_2{}^2 k^2 + (v_r{}^2 - v_i{}^2)\, 4/d^2]^{1/2} \tag{1.5-48}$$

Using Eqs. (1.5-17) and (1.5-31) we can write this expression also in the form

$$\beta_r = [n_1{}^2 k^2 + (u_i{}^2 - u_r{}^2)\, 4/d^2]^{1/2} \tag{1.5-49}$$

For a mode of high order we have $u_r \gg 1$ and obtain from Eq. (1.5-29) with $v_r = -1$ and from Eq. (1.5-36) the approximation

$$u_r = N\pi/2 \tag{1.5-50}$$

Neglecting u_i we have from Eq. (1.5-49)

$$\beta_r = [n_1{}^2 k^2 - (N\pi/d)^2]^{1/2} \tag{1.5-51}$$

with even integer N for even modes and odd N for odd modes. This expression agrees with the β value given by Eqs. (1.5-2) and (1.5-3) which we deduced from our intuitive treatment of leaky waves.

The loss coefficient of Eq. (1.5-46) can be written in the following form if we use $v_r = -1$ and set $u_r \approx V_s \gg 1$ in Eq. (1.5-34):

$$\alpha = (4/\beta d^2)[(N\pi/2)^2 - V_s{}^2]^{1/2} \tag{1.5-52}$$

This approximation also holds only for high-order modes, so that we could replace u_r with $N\pi/2$ according to Eq. (1.5-50). In order to be able to compare the loss coefficient (1.5-52) with our earlier formula (1.5-13) we must first introduce the specialization to a symmetric slab waveguide. Setting $\Delta = \rho$ and using the fact that for a high-order, relatively low-loss mode we must have $\kappa \gg \rho$, we obtain from Eq. (1.5-13)

$$\alpha = 2\rho/\beta d \tag{1.5-53}$$

Next we replace β in Eq. (1.5-5) with expression (1.5-3) and use Eq. (1.5-17) with the result

$$\alpha = (4/\beta d^2)[(N\pi/2)^2 - V_s{}^2]^{1/2} \tag{1.5-54}$$

We thus see once more that for high-order modes our intuitive theory agrees with the result obtained from the leaky mode solutions of the eigenvalue equation. This agreement justifies our interpretation of the leaky mode solutions as belonging to waves that propagate beyond cutoff. It does seem surprising that we get this agreement. The leaky mode solutions are actually quite unphysical since their field amplitudes grow with increasing distance from the waveguide core. As we discussed earlier, power that is radiated from the core is actually not lost since it is simply redistributed from the core into the outside region. One wonders why this redistribution of power should appear as loss in a theory that treats the field throughout all infinite space as belonging to one and the same mode. It would not have been surprising if we had had no loss at all. In fact, our leaky mode does not show any loss just beyond the cutoff point. Only some distance (in V_s) beyond cutoff does the propagation constant assume complex values. The loss must thus indicate power outflow towards infinity in transverse direction. That the loss values of the leaky mode solution indeed agree with the loss calculation, which directly consideres the power outflow from the core, is not a trivial result. The conclusion to be reached from this discussion is that one should regard the loss coefficient obtained from the complex solutions of the eigenvalue equation with some reservation, since it is not entirely clear what constitutes loss in this theory. The intuitive formula (1.5-13) was obtained from the idea of counting all energy as lost once it escapes from the core. The intuitive theory does not actually describe a mode but a transient field condition. However, this is the situation one encounters as a wave enters a waveguide section where it is cutoff. The formula (1.5-13) must thus be considered more reliable for such applications.

1.6 Hollow Dielectric Waveguides

The leaky waves and leaky mode solutions discussed in the previous section occurred for waves that went beyond cutoff in a waveguide capable of supporting guided modes. We now consider a dielectric slab waveguide that does not support any guided mode in the proper sense of the word. So far we have always assumed that the slab waveguide of Fig. 1.1.1 had a core whose refractive index n_1 was larger than that of the surrounding media. If we invert the situation and assume that

$$n_1 < n_3 \leqslant n_2 \tag{1.6-1}$$

we obtain a structure that we call a hollow dielectric waveguide [MS1]. The name suggests that the core region may be air or vacuum with $n_1 = 1$. We thus think of a hollow channel in a dielectric medium. It is, of course, not necessary that $n_1 = 1$; the type of leaky waveguide to be discussed here requires only that relationship (1.6-1) hold. According to Snell's law, Eq. (1.2-2), there can be no total internal reflection in this case. Every ray leaves the core with an increased angle after penetrating the core boundary.

Hollow dielectric waveguides have been successfully used in laser resonator structures [Sh1]. In this application a gas is confined inside a glass or quartz tube and is induced to exhibit gain by means of an electrical discharge. Gas lasers show increasing gain for decreasing tube diameter. The gain depends on the mechanism that depopulates the lower energy level of the atomic or molecular system that is used in the laser. This depopulation of the lower laser level is accomplished by collisions of the molecules with the wall. In a narrow tube these collisions occur more frequently, thus increasing the gain. However, as the tube diameter is decreased, the walls of the tube begin to interact with the electromagnetic field in the cavity. In a typical laser, the cavity is formed by external mirrors. The laser tube only has the purpose of confining the gas mixture. As the tube becomes too narrow its walls take part in shaping the mode field of the laser. This effect could be harmful if we think in terms of an open laser resonator defined solely by the mirrors. However, the tube walls can be used to advantage to guide the electromagnetic wave and thus confine not only the gas mixture but also the fields. The fact that a hollow dielectric waveguide does not actually have guided modes does not preclude its usefulness as a waveguide in this laser application. We have seen in the previous section that leaky waves propagate in dielectric waveguides for some distance. They may suffer high losses, but the amount of loss depends on the particular geometry and operating conditions that are being used.

The field of a hollow dielectric waveguide is also described by Eqs. (1.3-16)–(1.3-26) for TE modes and by Eqs. (1.3-54)–(1.3-63) for TM modes. The eigenvalue equations have the same form: Eq. (1.3-26) for TE modes and Eq. (1.3-63) for TM modes. The only difference is that now γ and δ defined by Eqs. (1.2-14) and (1.2-15) are imaginary for real values of β. There are thus no real solutions of the eigenvalue equations.

We obtain a very useful description of the leaky waves in the hollow dielectric waveguide by the geometrical optics approach that was used to describe leaky waves in the usual dielectric waveguide. If we ignore the eigenvalue equation and determine the real and imaginary parts of the propagation constants simply from the laws of ray optics and plane wave refraction and reflection at a dielectric interface, we obtain again Eqs. (1.5-2) and (1.5-3) for the real part of the propagation constant and Eq. (1.5-13) for twice its imaginary part, which is the power loss coefficient.

However, there is a difference in the behavior of the loss of leaky waves of ordinary waveguides compared to the leaky waves of hollow dielectric waveguides. In ordinary dielectric waveguides with $n_1 > n_2 \geqslant n_3$, we can assume that either ρ or Δ of Eqs. (1.5-5) and (1.5-6) is very nearly zero, so that we have $\kappa \gg \rho$. For symmetric waveguides this assumption leads to the useful approximate form (1.5-53) for the loss coefficient. In low-loss hollow dielectric waveguides, we must assume that the ray impinges on the core boundary with very small angles, so that κ, given by Eq. (1.2-13), is very nearly zero. On the other side of the core boundary the angles are larger, so that we have

$$\kappa \ll \rho \tag{1.6-2}$$

and

$$\kappa \ll \Delta \tag{1.6-3}$$

A useful formula for the power loss coefficient 2α is thus obtained from Eq. (1.5-13):

$$2\alpha = (2\kappa^2/\beta d)(\rho^{-1}+\Delta^{-1}) \tag{1.6-4}$$

The allowed values of κ are given by Eq. (1.5-2):

$$\kappa = N\pi/d \tag{1.6-5}$$

so that we obtain from Eq. (1.5-5)

$$\rho = [(W/d)^2+(N\pi/d)^2]^{1/2} \tag{1.6-6}$$

and from Eq. (1.5-6)

$$\Delta = [(n_3{}^2-n_2{}^2)k^2+(W/d)^2+(N\pi/d)^2]^{1/2} \tag{1.6-7}$$

with the abbreviation

$$W = (n_2{}^2-n_1{}^2)^{1/2}kd \tag{1.6-8}$$

Finally, we have for the propagation constant

$$\beta = [n_1{}^2k^2-(N\pi/d)^2]^{1/2} \tag{1.6-9}$$

In order to achieve low leaky wave loss in the hollow dielectric waveguide we must choose the mode number N small and the core width d large. In addition, the refractive index difference n_3-n_1 must be made as large as possible. Low-loss propagation in a hollow dielectric waveguide is thus possible only for low-order modes in a guide that is capable of supporting very many leaky modes.

Let us consider an example: We use the free space wavelength $\lambda = 1\ \mu$m, $d = 50\ \mu$m, $n_1 = 1$, $n_2 = n_3 = 1.5$, and consider the lowest-order mode with $N = 1$. With these values, we obtain from Eq. (1.6-4)

$$2\alpha = 4.77\times 10^{-2}\quad \text{cm}^{-1} = 0.207\quad \text{dB/cm} \tag{1.6-10}$$

This loss is moderately low. It is far too high for a waveguide for long distance communications. But it is low enough to be useful for short laser cavities or integrated optics applications. The loss figure can, of course, be reduced to arbitrarily low values if we use larger values of d. However, we pay a penalty for trying to reduce the mode loss by increasing the core width. The waveguide becomes more vulnerable to losses caused by waveguide curvature. This effect is discussed in [MS1].

The leaky modes of the hollow dielectric waveguide do not have cutoff conditions since they are not properly guided in the usual sense. We obtain the total number of guided modes traveling in the forward direction by counting all the modes with positive propagation constant β. We use the symbol M to indicate the total number of forward traveling modes (we are talking only of TE modes for simplicity) and obtain $N = M$ from Eq. (1.6-9) by requiring that $\beta = 0$:

$$M = n_1 kd/\pi \tag{1.6-11}$$

For the condition used in our loss example we find from Eq. (1.6-11) that $M = 150$.

We conclude our discussion of hollow dielectric waveguides by considering the solutions of the eigenvalue Eq. (1.3-39)

$$\tan(\kappa d/2) = \gamma/\kappa \tag{1.6-12}$$

for the even TE modes of the symmetric hollow slab waveguide with $n_2 = n_3$. The x dependence of the field is given by [see Eqs. (1.3-16) and (1.3-18) with $\gamma = \delta$]

$$e^{-\gamma x} \tag{1.6-13}$$

We must require that the imaginary part of γ is positive in order to assure that the waves move outward, away from the core. The difference between our present solution and the solution discussed in the previous section is that, according to Eq. (1.6-2), the imaginary part of γ must now be much larger than its real part because the imaginary part corresponds to ρ, while the real part is even smaller than κ. We again set

$$\kappa d/2 = u_r + iu_i \tag{1.6-14}$$

and

$$\gamma d/2 = v_r + iv_i \tag{1.6-15}$$

and require that

$$u_r \gg u_i \tag{1.6-16}$$

and

$$v_i \gg v_r \tag{1.6-17}$$

Substitution of Eqs. (1.6-14) and (1.6-15) into Eq. (1.6-12) and use of the inequalities (1.6-15) and (1.6-17) yields the two simultaneous equations

$$[\tan u_r(1-\tanh^2 u_i)]/(1+\tan^2 u_r \tanh^2 u_i) = v_i u_i/u_r^2 \qquad (1.6\text{-}18)$$

and

$$[\tanh u_i(1+\tan^2 u_r)]/(1+\tan^2 u_r \tanh^2 u_i) = v_i/u_r \qquad (1.6\text{-}19)$$

The right-hand side of Eq. (1.6-19) is very large. If we assume that $\tan u_r \gg 1$ and $\tan u_r \tanh u_i \gg 1$, we obtain from Eq. (1.6-19)

$$1/u_i = v_i/u_r \qquad (1.6\text{-}20)$$

provided that u_i itself is small. With the same assumptions we can write Eq. (1.6-18) approximately in the form

$$1/u_i^2 \tan u_r = v_i u_i/u_r^2 \qquad (1.6\text{-}21)$$

Next we set (N = odd integer)

$$u_r = N\pi/2 - \eta \qquad (1.6\text{-}22)$$

and obtain from Eq. (1.6-21)

$$\eta/u_i^2 = v_i u_i/(N\pi/2)^2 \qquad (1.6\text{-}23)$$

and from Eq. (1.6-20)

$$u_i v_i = N\pi/2 \qquad (1.6\text{-}24)$$

The relation

$$\kappa^2 + \gamma^2 = (n_1^2 - n_2^2)k^2 \qquad (1.6\text{-}25)$$

that follows from Eqs. (1.2-13) and (1.2-14) leads, with our present assumptions, to the two relations

$$v_i = [W_s^2 + (N\pi/2)^2]^{1/2} \qquad (1.6\text{-}26)$$

and

$$v_r = -(u_i/v_i)\, N\pi/2 \qquad (1.6\text{-}27)$$

where we have used the abbreviation

$$W_s = [n_2^2 - n_1^2]^{1/2} kd/2 \qquad (1.6\text{-}28)$$

We now obtain immediately from Eq. (1.6-24)

$$u_i = \frac{N\pi/2}{[W_s^2 + (N\pi/2)^2]^{1/2}} \qquad (1.6\text{-}29)$$

and from Eqs. (1.6-23), (1.6-24), and (1.6-29)

$$\eta = \frac{N\pi/2}{W_s^2 + (N\pi/2)^2} \tag{1.6-30}$$

Equation (1.6-27) finally yields

$$v_r = \frac{-(N\pi/2)^2}{W_s^2 + (N\pi/2)^2} \tag{1.6-31}$$

Our approximation is apparently valid for $W_s \gg 1$.

It is noteworthy that this approximate solution implies that the leaky waves of the hollow dielectric waveguide also grow in transverse direction away from the core since v_r is negative, so that Eq. (1.6-13) is an exponentially growing function.

We obtain the propagation constant from the equation

$$\beta_r - i\alpha = (n_1^2k^2 - \kappa^2)^{1/2} \tag{1.6-32}$$

so that we have for its real part

$$\beta_r = [n_1^2k^2 - (N\pi/d)^2]^{1/2} \tag{1.6-33}$$

and for its imaginary part, which is the amplitude attenuation constant,

$$\alpha = \frac{u_r u_i}{\beta_r} \frac{4}{d^2} = \frac{4(N\pi/2)^2}{\beta_r[W_s^2 + (N\pi/2)^2]^{1/2} d^2} \tag{1.6-34}$$

Equation (1.6-33) is identical with Eq. (1.6-9) and the equation for the attenuation constant (1.6-34) is identical with (1.6-4) if we remember that for the symmetrical slab we must set $\rho = \Delta$ and $W = 2W_s$.

The agreement of the solution of the eigenvalue equation with the result of the geometrical optics treatment is much better for the hollow dielectric waveguide than it was for the leaky waves of the ordinary dielectric waveguide. In our present case we did not have to assume that we were dealing with a high-order mode. The two methods of approximate analysis agree for any mode, provided that the losses are low. Our results, Eqs. (1.6-26)–(1.6-34), apply also to the odd modes. It is interesting that we must now associate odd values of N with even modes and even N values with odd modes, contrary to the assignment that is listed below Eq. (1.5-51). The difference is caused by the fact that at the core boundary the fields of low-loss modes in the hollow dielectric waveguide are very small, while they are strong for the modes of the ordinary waveguide just below cutoff.

The trajectory of the solution points of the complex v values for $0 < W_s < \infty$ is shown schematically in Fig. 1.6.1. This figure was drawn with the help of numerical computer solutions of the eigenvalue equations for even

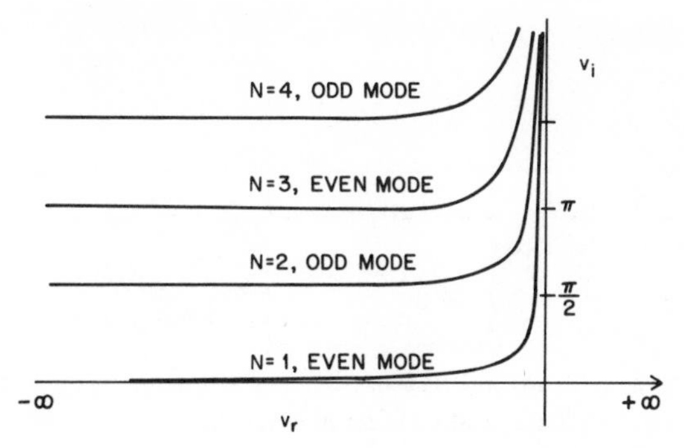

Fig. 1.6.1 *Leaky wave trajectories in the complex v plane for hollow dielectric waveguides.*

and odd modes of the symmetric hollow dielectric waveguide. For very large values of W_s, the trajectory of solutions points moves up along the positive imaginary v axis reaching $v_r = 0$ and $v_i = \infty$ at $W_s = \infty$. For decreasing values of W_s, the values of v_r become increasingly more negative, while v_i decreases. In the limit of $W_s = 0$, we have $v_r = -\infty$ and

$$v_i = \nu\pi/2, \qquad \text{for} \quad W_s = 0 \tag{1.6-35}$$

with even integer values of ν for even modes and odd ν values for odd modes. The assignment of even and odd integers to even and odd modes is again reversed compared to the corresponding situation for ordinary dielectric waveguides.

1.7 Rectangular Dielectric Waveguides

The dielectric slab waveguide is a useful model for more complicated waveguide structures. Its simplicity allows us to study the properties of wave propagation in dielectric waveguides without the encumberance of tedious mathematical expressions. However, in most practical applications more complicated waveguides are used. The waveguides used in integrated optics are usually rectangular strips of dielectric material that are embedded in other

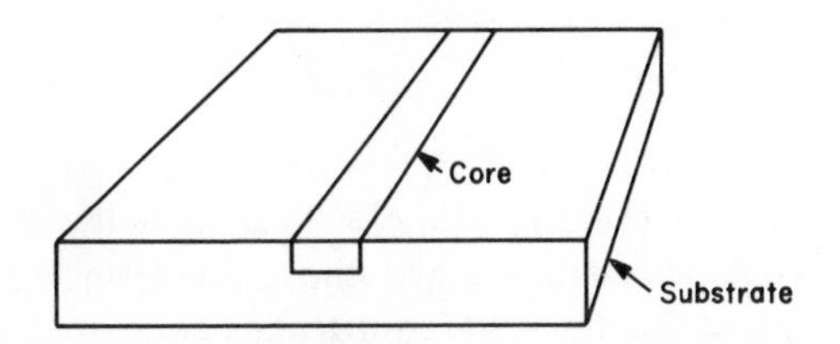

Fig. 1.7.1 *Rectangular dielectric waveguide in an integrated optics application.*

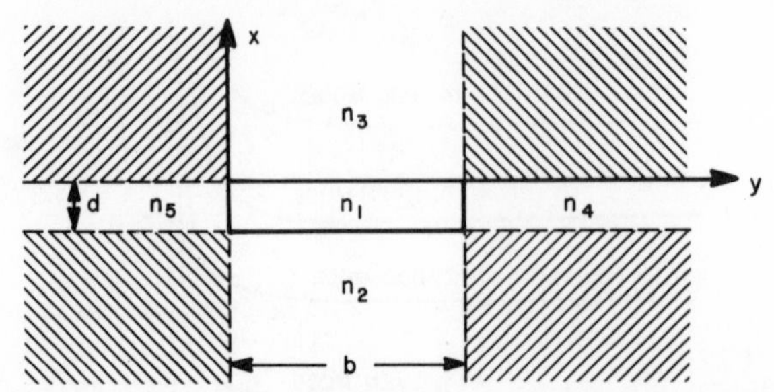

Fig. 1.7.2 *Schematic of the five dielectric regions of a rectangular dielectric waveguide. The field in the shaded regions is ignored in the approximation.*

dielectrics. Figure 1.7.1 shows the geometry of a rectangular dielectric waveguide. The rectangular strip is embedded in the material of the substrate of the integrated circuit. We analyze a structure that is more general: Instead of assuming that the waveguide core is embedded in the material of a substrate we allow the materials on all four sides of the rectangular core to be different. This geometry is shown in Fig. 1.7.2. We have chosen the labeling of the refractive indices to coincide as nearly as possible with Fig. 1.1.1.

An exact analytical treatment of this problem is not possible. Approximate solutions by numerical methods have been obtained that can be made as accurate as desired [GL1]. However, here we are following an approximate analytical approach that was developed by Marcatili [Mi1]. This method works only for modes far from cutoff. If the mode is not very close to cutoff, its field is confined almost exclusively to the region of the core, and only very little field energy is carried in the surrounding media. In particular, there is very little field energy in the shaded regions of Fig. 1.7.2, so that we can simplify the analysis by ignoring these regions completely. We can verify the statement that most of the field is contained in the core by examining the solutions of the slab waveguide. The ray angle θ_1 is small for well-guided modes, so that we see from Eq. (1.2-6) that the propagation constant approaches

$$\beta \approx n_1 k \tag{1.7-1}$$

Because k is a large quantity, we thus have from Eqs. (1.2-13)–(1.2-15)

$$\kappa \ll \gamma \tag{1.7-2}$$

and

$$\kappa \ll \delta \tag{1.7-3}$$

According to Eq. (1.3-25) we have $|B/A| \gg 1$, so that the sine term in the field expression (1.3-17) predominates over the cosine term. At the upper core boundary, $x = 0$, we find that the field is quite weak compared to its maximum value inside the core. Inequalities (1.7-2) and (1.7-3) also cause $\sin \kappa d$ of

Eq. (1.3-28) to be small. We thus see from Eq. (1.3-17) that the field is also weak at the lower core boundary, $x = -d$. The E_y field amplitude just outside the core is equal to the value on the inside and drops off with increasing distance from the core. It is thus clear that the field in the core is much stronger than the field outside as long as the mode is not too close to cutoff. The same considerations hold also for the TM mode [Eq. (1.3-55)].

It is possible to express the transverse field components in terms of the longitudinal components. With the time and z dependence, Eq. (1.3-8), we obtain from Maxwell's Eqs. (1.3-2) and (1.3-3) (compare p. 13 of [Me1])

$$E_x = -(i/K_j^2)[(\beta\, \partial E_z/\partial x) + (\omega\mu_0\, \partial H_z/\partial y)] \tag{1.7-4}$$

$$E_y = -(i/K_j^2)[(\beta\, \partial E_z/\partial y) - (\omega\mu_0\, \partial H_z/\partial x)] \tag{1.7-5}$$

$$H_x = -(i/K_j^2)[(\beta\, \partial H_z/\partial x) - (\omega n_j^2 \varepsilon_0\, \partial E_z/\partial y)] \tag{1.7-6}$$

$$H_y = -(i/K_j^2)[(\beta\, \partial H_z/\partial y) + (\omega n_j^2 \varepsilon_0\, \partial E_z/\partial x)] \tag{1.7-7}$$

The parameter K_j is defined by

$$K_j = (n_j^2 k^2 - \beta^2)^{1/2} \tag{1.7-8}$$

The refractive index n_j assumes the values n_1, n_2, n_3, n_4, and n_5 for fields in the five regions of the waveguide. The longitudinal components E_z and H_z must satisfy the reduced wave equation

$$(\partial^2\psi/\partial x^2) + (\partial^2\psi/\partial y^2) + K_j^2\psi = 0 \tag{1.7-9}$$

There are two types of modes that the waveguide can support. One type, which we call E_{pq}^x modes, is polarized predominantly in the x direction. The other mode, E_{pq}^y, is polarized predominantly in the y direction. We can adjust the amplitude coefficients of E_z and H_z so that one of the transverse field components vanishes. The following set of field components satisfies the reduced wave equation and relations (1.7-4)–(1.7-7) and describes an E_{pq}^x mode in region 1. The factor (1.3-8) is omitted throughout:

$$E_z = A \cos\kappa_x(x+\xi) \cos\kappa_y(y+\eta) \tag{1.7-10}$$

$$H_z = -A(\varepsilon_0/\mu_0)^{1/2} n_1^2 (\kappa_y/\kappa_x)(k/\beta) \sin\kappa_x(x+\xi) \sin\kappa_y(y+\eta) \tag{1.7-11}$$

$$E_x = (iA/\kappa_x\beta)(n_1^2 k^2 - \kappa_x^2) \sin\kappa_x(x+\xi) \cos\kappa_y(y+\eta) \tag{1.7-12}$$

$$E_y = -iA(\kappa_y/\beta) \cos\kappa_x(x+\xi) \sin\kappa_y(y+\eta) \tag{1.7-13}$$

$$H_x = 0 \tag{1.7-14}$$

$$H_y = iA(\varepsilon_0/\mu_0)^{1/2} n_1^2 (k/\kappa_x) \sin\kappa_x(x+\xi) \cos\kappa_y(y+\eta) \tag{1.7-15}$$

with $k = \omega^2\varepsilon_0\mu_0$ and

$$K_1^2 = n_1^2 k^2 - \beta^2 = \kappa_x^2 + \kappa_y^2 \tag{1.7-16}$$

The longitudinal components E_z and H_z were chosen to satisfy the reduced wave Eq. (1.7-9). This requirement leads to condition (1.7-16). Except for this restriction κ_x and κ_y are arbitrary at this point. The sine and cosine functions were selected so that the contributions of E_z and H_z result in the same functional dependence of the transverse components. The phase parameters ξ and η, which are added to the arguments of the sine and cosine functions, are necessary to make the solution general. We could have used a superposition of a sine and cosine function with arbitrary amplitude coefficients instead of using a circular function with an arbitrary amplitude and phase. Both methods are, of course, equivalent. The various undetermined coefficients become fixed by the requirement that the fields satisfy the boundary conditions on the four sides of the core.

The choice of the amplitude coefficient for H_z may appear arbitrary. However, it was dictated by the desire to make the H_y component vanish. Relation (1.7-1) results in the inequalities

$$\kappa_x \ll \beta \tag{1.7-17}$$

and

$$\kappa_y \ll \beta \tag{1.7-18}$$

which follow from Eq. (1.7-16) and apply to modes far from cutoff. Owing to inequality (1.7-17) we see that $E_y \ll E_z$. Since $E_z \ll E_x$ we can neglect E_y as a quantity of second order in κ/β. To the approximation used in this analysis we have thus only four nonvanishing field components.

On the outside we must require all field components in regions 2–5 to vanish at infinite distance from the core. We are thus limited to use decaying exponential functions for those coordinates that lead away from the core. The field components in these regions are chosen similarly to the field in the core, with the additional requirement that the electric field component tangential to the particular face of the core passes continuously through the core boundary. We thus have in region 2

$$E_z = A \cos \kappa_x(\xi - d) \cos \kappa_y(y+\eta) \exp[\gamma_2(x+d)] \tag{1.7-19}$$

$$H_z = -A(\varepsilon_0/\mu_0)^{1/2} n_2{}^2 (\kappa_y/\gamma_2)(k/\beta) \cos \kappa_x(\xi - d) \sin \kappa_y(y+\eta) \exp[\gamma_2(x+d)] \tag{1.7-20}$$

$$E_x = +iA[(\gamma_2{}^2 + n_2{}^2 k^2)/\gamma_2 \beta] \cos \kappa_x(\xi - d) \cos \kappa_y(y+\eta) \exp[\gamma_2(x+d)] \tag{1.7-21}$$

$$E_y \approx 0 \tag{1.7-22}$$

$$H_x = 0 \tag{1.7-23}$$

$$H_y = iA(\varepsilon_0/\mu_0)^{1/2} n_2{}^2 (k/\gamma_2) \cos \kappa_x(\xi - d) \cos \kappa_y(y+\eta) \exp[\gamma_2(x+d)] \tag{1.7-24}$$

with

$$K_2^2 = n_2^2k^2 - \beta^2 = \kappa_y^2 - \gamma_2^2 \tag{1.7-25}$$

The amplitude coefficients of these field components were chosen to ensure continuity of the E_z component at $x = -d$ and to make $H_x = 0$. The E_y component vanishes to the same approximation as the E_y component inside the core. The y-dependent functions were chosen to coincide with the corresponding functions inside the core. Equation (1.7-25) is again a consequence of Eq. (1.7-9).

The field in region 3 is similarly

$$E_z = A \cos\kappa_x\xi \cos\kappa_y(y+\eta) \exp(-\gamma_3 x) \tag{1.7-26}$$

$$H_z = A(\varepsilon_0/\mu_0)^{1/2}n_3^2(\kappa_y/\gamma_3)(k/\beta) \cos\kappa_x\xi \sin\kappa_y(y+\eta) \exp(-\gamma_3 x) \tag{1.7-27}$$

$$E_x = -iA[(\gamma_3^2+n_2^2k^2)/\gamma_3\beta] \cos\kappa_x\xi \cos\kappa_y(y+\eta) \exp(-\gamma_3 x) \tag{1.7-28}$$

$$E_y \approx 0 \tag{1.7-29}$$

$$H_x = 0 \tag{1.7-30}$$

$$H_y = -iA(\varepsilon_0/\mu_0)^{1/2}n_3^2(k/\gamma_3) \cos\kappa_x\xi \cos\kappa_y(y+\eta) \exp(-\gamma_3 x) \tag{1.7-31}$$

with

$$K_3^2 = n_3^2k^2 - \beta^2 = \kappa_y^2 - \gamma_3^2 \tag{1.7-32}$$

In regions 4 and 5 we adjust the amplitudes so that the strong component E_x is continuous at the core boundary and obtain in region 4

$$E_z = A(n_1^2/n_4^2) \cos\kappa_y(b+\eta) \cos\kappa_x(x+\xi) \exp[-\gamma_4(y-b)] \tag{1.7-33}$$

$$H_z = -A(\varepsilon_0/\mu_0)^{1/2}n_1^2(\gamma_4/\kappa_x)(k/\beta) \cos\kappa_y(b+\eta) \sin\kappa_x(x+\xi) \exp[-\gamma_4(y-b)] \tag{1.7-34}$$

$$E_x = iA(n_1^2/n_4^2)[(n_4^2k^2-\kappa_x^2)/\kappa_x\beta] \cos\kappa_y(b+\eta) \times \sin\kappa_x(x+\xi) \exp[-\gamma_4(y-b)] \tag{1.7-35}$$

$$H_y = iA(\varepsilon_0/\mu_0)^{1/2}n_1^2(k/\kappa_x) \cos\kappa_y(b+\eta) \sin\kappa_x(x+\xi) \exp[-\gamma_4(y-b)] \tag{1.7-36}$$

with

$$K_4^2 = n_4^2k^2 - \beta^2 = \kappa_x^2 - \gamma_4^2 \tag{1.7-37}$$

In region 5 we have finally

$$E_z = A(n_1^2/n_5^2)\cos\kappa_y\eta\cos\kappa_x(x+\xi)\exp(\gamma_5 y) \tag{1.7-38}$$

$$H_z = A(\varepsilon_0/\mu_0)^{1/2}n_1^2(\gamma_5/\kappa_x)(k/\beta)\cos\kappa_y\eta\sin\kappa_x(x+\xi)\exp(\gamma_5 y) \tag{1.7-39}$$

$$E_x = iA(n_1^2/n_5^2)[(n_5^2k^2-\kappa_x^2)/\kappa_x\beta]\cos\kappa_y\eta\sin\kappa_x(x+\xi)\exp(\gamma_5 y) \tag{1.7-40}$$

$$H_y = iA(\varepsilon_0/\mu_0)^{1/2}n_1^2(k/\kappa_x)\cos\kappa_y\eta\sin\kappa_x(x+\xi)\exp(\gamma_5 y) \tag{1.7-41}$$

with

$$K_5^2 = n_5^2k^2 - \beta^2 = \kappa_x^2 - \gamma_5^2 \tag{1.7-42}$$

It is noteworthy that, in the spirit of our approximation, the E_x component is continuous only if we neglect the κ_x^2 term in the numerator of the equations compared to the term n^2k^2. The H_x component vanishes exactly, while E_y is only approximately zero in regions 4 and 5.

We complete the analysis of the modes of the rectangular dielectric waveguide by matching the remaining boundary conditions. One tangential electric field component has already been matched by proper choice of the field amplitudes. In regions 2 and 3 we require that the H_z component pass continuously through the core boundary at $x=0$ and $x=-d$. This requirement also causes H_y to be continuous as required. The remaining tangential component E_y, which would have to be continuous in an exact treatment, is neglected since it is small compared to all other field components. We thus obtain the following two equations:

$$(n_1^2/\kappa_x)\sin\kappa_x(\xi-d) - (n_2^2/\gamma_2)\cos\kappa_x(\xi-d) = 0 \tag{1.7-43}$$

and

$$(n_1^2/\kappa_x)\sin\kappa_x\xi + (n_3^2/\gamma_3)\cos\kappa_x\xi = 0 \tag{1.7-44}$$

If the sine and cosine functions in Eq. (1.7-43) are expanded by means of the addition theorems, these two equations represent a system of homogeneous simultaneous equations for the two unknowns, $\sin\kappa_x\xi$ and $\cos\kappa_x\xi$. A solution is possible only if the determinant of the equation system vanishes. We thus obtain the eigenvalue equation

$$\tan\kappa_x d = n_1^2\kappa_x(n_3^2\gamma_2+n_2^2\gamma_3)/(n_3^2n_2^2\kappa_x^2-n_1^4\gamma_2\gamma_3) \tag{1.7-45}$$

We recognize this equation as the eigenvalue equation of TM modes of the infinite slab, Eq. (1.3-63). It is clear that relative to the two core boundaries at $x=0$ and $x=-d$ the field polarization corresponds to a TM mode, because the transverse electric field is polarized normal to the core boundary.

The phase parameter ξ is determined by Eq. (1.7-44):

$$\tan \kappa_x \xi = -(n_3{}^2/n_1{}^2)(\kappa_x/\gamma_3) \tag{1.7-46}$$

which corresponds to Eq. (1.3-61). We see that the eigenvalue Eq. (1.7-45) can be used to determine κ_x, since we can express γ_2 in terms of κ_x with the help of Eqs. (1.7-16) and (1.7-25):

$$\gamma_2 = [(n_1{}^2 - n_2{}^2)k^2 - \kappa_x{}^2]^{1/2} \tag{1.7-47}$$

Similarly it is possible to express γ_3 in terms of κ_x by means of Eqs. (1.7-16) and (1.7-32):

$$\gamma_3 = [(n_1{}^2 - n_3{}^2)k^2 - \kappa_x{}^2]^{1/2} \tag{1.7-48}$$

This leaves κ_x as the only unknown quantity in Eq. (1.7-45).

At the core boundaries in regions 4 and 5 we require that H_z assume equal values on either side of the dielectric interface. The E_x component is already continuous by our choice of wave amplitudes, provided we neglect $\kappa_x{}^2$ compared to n^2k^2. We obtain the following two equations from the requirement of continuity of H_z at $y = b$ and $y = 0$:

$$\kappa_y \sin \kappa_y(b+\eta) - \gamma_4 \cos \kappa_y(b+\eta) = 0 \tag{1.7-49}$$

and

$$\kappa_y \sin \kappa_y \eta + \gamma_5 \cos \kappa_y \eta = 0 \tag{1.7-50}$$

The E_z component is not continuous at these two interfaces. However, far from cutoff condition (1.7-2) indicates that $(\mu_0/\varepsilon_0)^{1/2} H_z$ of Eq. (1.7-34) is considerably larger than E_z of Eq. (1.7-33). Matching of the H_z component is thus more important than letting E_z satisfy the boundary conditions. In addition, it is obvious that the E_z component satisfies the boundary condition in the limit $n_1 = n_4 = n_5$. The approximation thus gets better for small index differences.

From Eq. (1.7-50) we find that the phase parameter η is given by the relation

$$\tan \kappa_y \eta = -\gamma_5/\kappa_y \tag{1.7-51}$$

In Eq. (1.7-49) we expand the sine and cosine functions by means of the addition theorems and obtain an equation system for the unknown quantities $\sin \kappa_y \eta$ and $\cos \kappa_y \eta$. The condition of vanishing system determinant of this equation system yields the eigenvalue equation

$$\tan \kappa_y b = \kappa_y(\gamma_4 + \gamma_5)/(\kappa_y{}^2 - \gamma_4 \gamma_5) \tag{1.7-52}$$

which we recognize as Eq. (1.3-26) for TE modes of the slab. It is also apparent that Eq. (1.7-51) corresponds to Eq. (1.3-25). This result is reasonable. E_x is the dominant electric field component of the E^x_{pq} modes, so that the field appears as a TE mode when viewed from region 4 or 5. Equation (1.7-52) determines κ_y, since we obtain from Eqs. (1.7-16), (1.7-37), and (1.7-42)

$$\gamma_4 = [(n_1^2 - n_4^2)k^2 - \kappa_y^2]^{1/2} \tag{1.7-53}$$

and

$$\gamma_5 = [(n_1^2 - n_5^2)k^2 - \kappa_y^2]^{1/2} \tag{1.7-54}$$

Once κ_x and κ_y are determined, we obtain the propagation constant from Eq. (1.7-16):

$$\beta = [n_1^2 k^2 - (\kappa_x^2 + \kappa_y^2)]^{1/2} \tag{1.7-55}$$

This completes our determination of the E_{pq}^x mode. The integers p and q are the mode numbers which we obtain from the solutions of the two eigenvalue Eqs. (1.7-45) and (1.7-52). These numbers indicate the number of maxima of the field distribution in x and y direction. We have obtained a field description inside the waveguide core and in regions 2–5 as shown in Fig. 1.7.2. However, in the shaded regions of this figure the field remains unknown. Ignoring these areas makes it impossible to obtain an exact solution by this method. We have seen that we could not satisfy all the boundary conditions, even in the four unshaded regions, since our method is inherently not correct. However, we shall see later that our approximation compares favorably with more exact numerical solutions of the problem.

The E_{pq}^y modes are obtained in close analogy to the derivation of the E_{pq}^x modes. It is now the E_y component that is dominant, while the E_x component nearly vanishes, and $H_y = 0$. We restrict the description of the field components to stating only the E_z and H_z components in all four media. The corresponding transverse components are obtained by differentiation from Eqs. (1.7-4)–(1.7-7). We obtain in the core of the waveguide, that is, in medium 1,

$$E_z = B \cos \kappa_x(x+\xi) \cos \kappa_y(y+\eta) \tag{1.7-56}$$

$$H_z = B(\varepsilon_0/\mu_0)^{1/2} n_1^2 (\kappa_x/\kappa_y)(k/\beta) \sin \kappa_x(x+\xi) \sin \kappa_y(y+\eta) \tag{1.7-57}$$

In medium 2 we have

$$E_z = B(n_1^2/n_2^2) \cos \kappa_x(\xi - d) \cos \kappa_y(y+\eta) \exp[\gamma_2(x+d)] \tag{1.7-58}$$

$$H_z = -B(\varepsilon_0/\mu_0)^{1/2} n_1^2 (\gamma_2/\kappa_y)(k/\beta) \cos \kappa_x(\xi - d) \sin \kappa_y(y+\eta) \exp[\gamma_2(x+d)] \tag{1.7-59}$$

The following field components belong to medium 3:

$$E_z = B(n_1^2/n_3^2) \cos \kappa_x \xi \cos \kappa_y(y+\eta) \exp(-\gamma_3 x) \tag{1.7-60}$$

$$H_z = B(\varepsilon_0/\mu_0)^{1/2} n_1^2 (\gamma_3/\kappa_y)(k/\beta) \cos \kappa_x \xi \sin \kappa_y(y+\eta) \exp(-\gamma_3 x) \tag{1.7-61}$$

The field in medium 4 is described by

$$E_z = B \cos \kappa_y(b+\bar{\eta}) \cos \kappa_x(x+\bar{\xi}) \exp[-\gamma_4(y-b)] \tag{1.7-62}$$

$$H_z = -B(\varepsilon_0/\mu_0)^{1/2} n_4^2 (\kappa_x/\gamma_4)(k/\beta) \cos \kappa_y(b+\bar{\eta}) \sin \kappa_x(x+\bar{\xi}) \exp[-\gamma_4(y-b)] \tag{1.7-63}$$

Finally we have in medium 5:

$$E_z = B \cos \kappa_y \bar{\eta} \cos \kappa_x(x+\bar{\xi}) \exp(\gamma_5 y) \tag{1.7-64}$$

$$H_z = B(\varepsilon_0/\mu_0)^{1/2} n_5^2 (\kappa_x/\gamma_5)(k/\beta) \cos \kappa_y \bar{\eta} \sin \kappa_x(x+\bar{\xi}) \exp(\gamma_5 y) \tag{1.7-65}$$

Equations (1.7-16), (1.7-25), (1.7-32), (1.7-37), and (1.7-42) also apply to this case.

In regions 2 and 3 we require continuity of the H_z component at the core boundary and obtain in the usual way the eigenvalue equation for TE modes of a slab:

$$\tan \kappa_x d = \kappa_x(\gamma_2+\gamma_3)/(\kappa_x^2-\gamma_2\gamma_3) \tag{1.7-66}$$

and the following equation for $\bar{\xi}$,

$$\tan \kappa_x \bar{\xi} = \gamma_3/\kappa_x \tag{1.7-67}$$

Continuity of E_y to first order in κ_y/k at these two interfaces is assured by our choice of the amplitude coefficients of E_z and H_z. The boundary condition for E_z is not satisfied. However, the magnitude of E_z is small compared to the other electric field components and, in addition, better compliance with this boundary condition is obtained for decreasing refractive index differences.

In regions 4 and 5 we match the H_z components on either side of the core boundary and obtain the eigenvalue equation for TM modes of the asymmetric slab:

$$\tan \kappa_y b = n_1^2 \kappa_y(n_5^2\gamma_4+n_4^2\gamma_5)/(n_4^2 n_5^2 \kappa_y^2 - n_1^4 \gamma_4\gamma_5) \tag{1.7-68}$$

and the equation for $\bar{\eta}$,

$$\tan \kappa_y \bar{\eta} = (n_5^2/n_1^2)(\kappa_y/\gamma_5) \tag{1.7-69}$$

The H_x component becomes continuous after we have matched the H_z component. Continuity of the E_z component is assured by our choice of the amplitude coefficients and the E_x component is neglected, since it is small to second order in κ_y/k compared to the E_y component. The two eigenvalue equations are again used to determine κ_x and κ_y, since γ_j can be expressed in terms of κ_x and κ_y as before. The propagation constant is finally obtained from Eq. (1.7-55).

Our approximate analysis does not hold near cutoff since the fields detach themselves from the core and reach strongly into the shaded regions of Fig.

1.7.2. This breakdown of the theory near cutoff is also apparent from the propagation constant. We write Eq. (1.7-55) in the form

$$\beta d = [n_2{}^2 k^2 d^2 + V^2 - (\kappa_x{}^2 + \kappa_y{}^2)\, d^2]^{1/2} \tag{1.7-70}$$

If n_2 is larger than any of the other refractive indices—except n_1—total internal reflection will first break down at the core boundary $x = -d$. According to Eq. (1.3-74) we have right at cutoff $V = V_c$ and

$$\kappa_c d = V_c \tag{1.7-71}$$

so that the propagation constant assumes the value

$$\beta = (n_2{}^2 k^2 - \kappa_y{}^2)^{1/2} < n_2 k \tag{1.7-72}$$

Even if we assume $n_2 = n_3 = n_4 = n_5$, Eq. (1.7-72) implies that the propagation constant, at the point where the wave is no longer guided, is smaller than the plane wave propagation constant in the medium outside the waveguide core. This is an unphysical result which shows that our theory just does not work near the cutoff point, because when most of the field energy is traveling outside the core we must have $\beta = n_2 k$.

Marcatili [Mi1] has evaluated the results of the approximate field analysis and compared them to computer solutions of the problem by Goell. Of the many figures given in Marcatili's paper we reproduce two that apply to the case of a square guide and of a rectangular guide with $b = 2d$. In both cases we have $n_2 = n_3 = n_4 = n_5$. The solid curves in Fig. 1.7.3 were obtained from exact solutions of the eigenvalue Eqs. (1.7-45), (1.7-52), (1.7-66) and (1.7-68). The dashed curves show the result of the far from cutoff approximation (1.3-88) and (1.3-91) of these eigenvalue equations. However, Marcatili used the approximation

$$\delta_\infty = \gamma_\infty = (n_1{}^2 - n_2{}^2)^{1/2} k \tag{1.7-73}$$

in Eq. (1.3-91). It is apparent how remarkably well this approximation reproduces the exact solutions of the eigenvalue equations, particularly far from cutoff. Finally, Fig. 1.7.3 shows the results of Goell's computer analysis as dash–dotted curves. The agreement between the exact solutions of the approximate theory and the computer solutions of our problem is so close that the two types of curves are indistinguishable over most of their range. Only very close to the cutoff point is there a noticeable deviation. The discrepancy between the approximate analytical solutions and the computer solutions is most pronounced for the fundamental E_{11}^x and E_{11}^y modes. (Goell has produced very interesting pictures of the power distribution of the various modes which are shown in his paper [Gl1].)

Lack of space prevents us from discussing such interesting applications of our theory as directional couplers formed with two rectangular dielectric

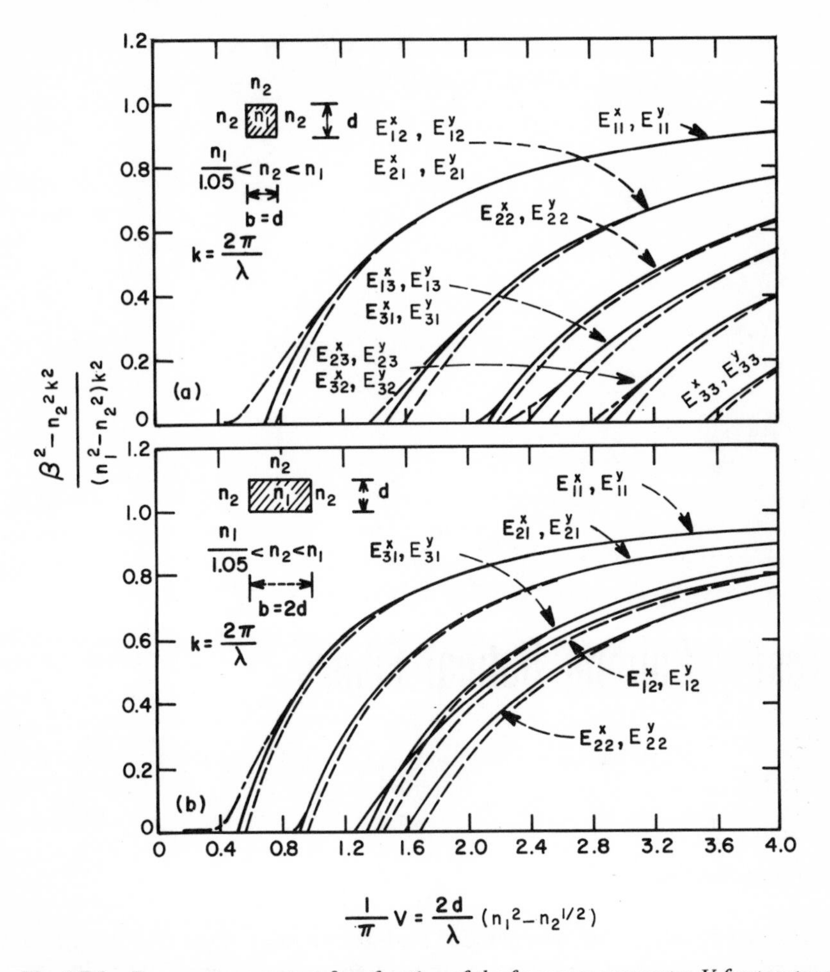

Fig. 1.7.3 *Propagation constant β as function of the frequency parameter V for two types of rectangular dielectric waveguides. The solid curves represent solutions of the eigenvalue equations of the approximate theory, the dashed curves were obtained from the far from cutoff approximations, the dash–dotted curves are Geoll's results.* (From *E. A. J. Marcatili, "Dielectric rectangular waveguide and directional coupler for integrated optics,"* Bell. Syst. Tech. J. ***48**, 2071–2102 (1969). Copyright 1969, The American Telephone and Telegraph Co., reprinted by permission.)*

waveguides. This application is also contained in Marcatili's paper. The simplest method of computing the coupling coefficient between two adjacent dielectric waveguides consists in applying the coupling formula (10.2-18) or, in case of identical waveguides, the simplified formula (10.2-22) of [Me1] to our mode field expressions.

2

Weakly Guiding Optical Fibers

2.1 Introduction

The asymmetric slab waveguides and rectangular dielectric waveguides discussed in Chap. 1 are used in integrated optics circuits. The distances bridged by these types of waveguides are usually very short, no more than a few centimeters. We now turn our attention to optical waveguides for long distance communications. Such guides may typically be a few kilometers in length. The most promising waveguide for long distance optical communications purposes is the round optical fiber, which consists of a core of a dielectric material with refractive index n_1 and a cladding of another dielectric material whose refractive index n_2 is less than n_1. Interest in optical fibers of this type soared after it could be demonstrated that low-loss fibers are indeed feasible. The glass that is routinely used in optical instruments is far too lossy to be useful for optical fibers suitable for bridging long distances. Losses on the order of 1000 dB/km are not at all unusual and are found in high quality optical glasses used for the manufacture of lenses. Losses on the order of 100 dB/km are exceptionally good for typical high-quality optical materials. However, the loss figures that are presently being discussed and that have been experimentally demonstrated are on the order of a few decibels per kilometer [KKM, KSZ]. Optical fibers with such low losses are very exciting and open up the possibility of new types of communications systems [KH1, MS2].

The theory of optical fibers is well understood and has been described in several books [Ky1, KB1, Me1]. However, the complete description of the guided and radiation modes of optical fibers is complicated, so that its application to the theory of fibers with imperfection is not very practical. It is our purpose to present a simplified description of the theory of optical fibers which results in simple expressions that can be used to handle problems of mode conversion and radiation loss phenomena caused by waveguide imperfections.

The exact description of the modes of round fibers is involved, since they are six component hybrid fields of great mathematical complexity. The simplification in the description of these modes is made possible by the realization that most fibers for practical applications use core materials whose refractive index is only very slightly higher than that of the surrounding cladding. By introducing the assumption that

$$n_1 - n_2 \ll 1 \tag{2.1-1}$$

considerable simplifications result. Instead of a six-component field only four field components need to be considered, and the field description is further simplified by use of rectangular cartesian instead of cylindrical coordinates. The eigenvalue equation of the simplified modes is far less complex than the corresponding equation of the exact mode fields. The simplified eigenvalue equation allows us to obtain simple approximate solutions for the regions close to cutoff and far from cutoff.

It was demonstrated in [Me1] that the simplified modes can be obtained by superposition of two nearly degenerate modes of the exact solution of the boundary value problem. This method has the advantage of establishing the relation of the approximate modes with the exact modes and showing clearly the limitations and the extent of the approximations. In this book we use the opposite approach and derive the simplified modes directly from Maxwell's equations. This procedure has the advantage of greater simplicity, but we lose the ability to judge the validity of the approximation.

Snyder [Sr1] first realized the simplification that results in the eigenvalue equation if condition (2.1-1) is applicable. Marcatili and Gloge [Ge1] perfected the description of the simplified modes of the round fiber. It was Gloge who first used the term "weakly guiding fiber" for a waveguide whose core and cladding have very nearly the same refractive index.

2.2 Guided Modes of the Optical Fiber

All dielectric waveguides support a finite number of guided modes in addition to the infinite continuum of radiation modes that are not guided by the structure but are, nevertheless, solutions of the same boundary value problem. We begin the approximate analytical treatment of round optical fibers by deriving the guided modes of the structure.

Marcatili realized that the description of the modes of the weakly guiding fiber becomes much simpler if the components of the field vectors are expressed in rectangular cartesian coordinates instead of the cylindrical coordinates that appear so much more suitable to the cylindrical geometry of the waveguide.

The cross section of the optical fiber is shown in Fig. 2.2.1. Region 1 with refractive index n_1 is the fiber core, region 2 with index n_2 is the cladding. In all our work we assume that the cladding is infinitely extended, in spite of the fact that it has a finite radius for practical fibers. The justification for assuming an infinitely extended cladding region comes from the fact that the guided modes have exponentially decaying fields outside the core. If the cladding radius is large enough, the guided mode fields have decayed to insignificant values at the outer boundary of the cladding. All practical fibers are designed to ensure that the guided mode field does not reach the outer boundary of the cladding. In the opposite case the fiber would suffer high radiation losses, since the outer fiber surface is never perfectly smooth on account of accumulating dust and other environmental effects.

The assumption of an infinite cladding radius is more questionable when we study the radiation modes. These solutions of the boundary value problem

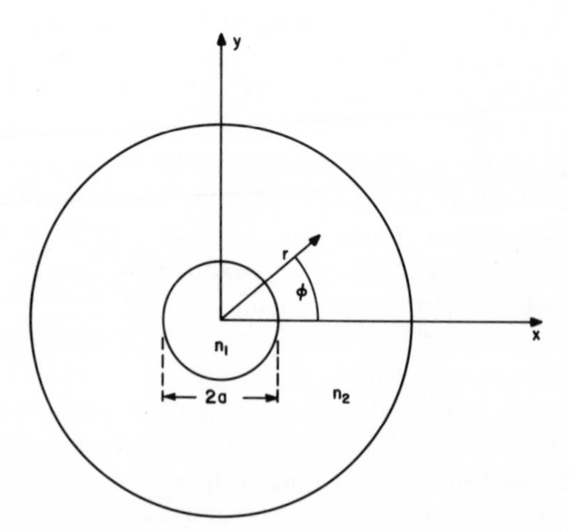

Fig. 2.2.1 *Cross section of a round optical fiber.*

reach out to infinity with undiminished strength and are certainly strongly affected by the outer cladding boundary. However, we can justify our procedure even for the radiation modes. The finite radius of the cladding has the effect of trapping some of the radiation field, causing cladding modes to appear. These modes have a discrete spectrum of allowed values of their propagation constants. But the density of these modes is much higher than the core modes, so that they form almost a continuum. When we calculate the interaction of the core modes with the radiation modes of the fiber with infinite cladding, we must keep in mind that in actuality we would have coupling of the core modes and the cladding modes. The fact that a portion of the radiation field does not actually escape freely but finds itself trapped by reflection at the outer cladding boundary does not alter our conclusions very much. In most practical cases cladding modes will be suppressed by a lossy coating on the outside of the fiber, or they will scatter out of the cladding on account of the rough outer surface. In either case, power will not endure very long in cladding modes and may be considered as being lost, just as it would be had it radiated away freely. In those cases where these conditions are not met it is necessary to study the interaction of core and cladding modes in detail.

The derivation of the simplified guided modes of the fiber uses again the longitudinal E_z and H_z components from which the transverse components are derived by means of Eqs. (1.7-4)–(1.7-7). The longitudinal components must satisfy the reduced wave Eq. (1.7-9) which we now express in cylindrical coordinates:

$$\frac{\partial^2\psi}{\partial r^2} + \frac{1}{r}\frac{\partial\psi}{\partial r} + \frac{1}{r^2}\frac{\partial^2\psi}{\partial\phi^2} + K_j^2\psi = 0 \qquad (2.2\text{-}1)$$

with

$$K_j^2 = n_j^2k^2 - \beta^2 \qquad (2.2\text{-}2)$$

and

$$k = \omega(\varepsilon_0\mu_0)^{1/2} = 2\pi/\lambda \qquad (2.2\text{-}3)$$

Equation (2.2-1) is obtained from Maxwell's equations by eliminating the transverse field components and solving Maxwell's equations for either E_z or H_z. All field components have the common factor

$$\exp[i(\omega t - \beta z)] \qquad (2.2\text{-}4)$$

which we omit from the equations for brevity.

To solve the reduced wave equation we substitute the trial solution

$$\psi = F(r)\cos\nu\phi \qquad (2.2\text{-}5)$$

into Eq. (2.2-1), with the result

$$\frac{d^2F}{dr^2} + \frac{1}{r}\frac{dF}{dr} + \left(K_j^2 - \frac{v^2}{r^2}\right)F = 0 \tag{2.2-6}$$

The same result would have been obtained had we used the sine instead of the cosine function in Eq. (2.2-5). Equation (2.2-6) is Bessel's differential equation. Its solutions are cylinder functions which we indicate collectively by the symbol

$$F(r) = Z_v(K_j r) \tag{2.2-7}$$

where $Z_v(K_j r)$ may be either a Bessel or a Neumann function. As solutions of a second-order differential equation there must be two independent types of functions. In the core region we need to identify $Z_v(K_1 r)$ with a Bessel function, $J_v(K_1 r)$, since the solution must remain finite at $r = 0$.

When we derived the approximate solutions of the modes of the rectangular waveguide we used E_z and H_z in such a way that one of the transverse field components vanished. For the same reason we are now using the following solution of Eq. (2.2-1) in the region $r < a$:

$$E_z = \frac{iA\kappa}{2\beta_v}\left[J_{v+1}(\kappa r)\begin{Bmatrix} \sin(v+1)\phi \\ -\cos(v+1)\phi \end{Bmatrix} + J_{v-1}(\kappa r)\begin{Bmatrix} \sin(v-1)\phi \\ -\cos(v-1)\phi \end{Bmatrix}\right] \tag{2.2-8}$$

$$H_z = -\frac{iA\kappa}{2k}\left(\frac{\varepsilon_0}{\mu_0}\right)^{1/2}\left[J_{v+1}(\kappa r)\begin{Bmatrix} \cos(v+1)\phi \\ \sin(v+1)\phi \end{Bmatrix} - J_{v-1}(\kappa r)\begin{Bmatrix} \cos(v-1)\phi \\ \sin(v-1)\phi \end{Bmatrix}\right] \tag{2.2-9}$$

We have combined two solutions with $v+1$ and $v-1$ in order to achieve cancellation of the E_x and H_y components. Instead of K_1 we have used our usual notation

$$\kappa = K_1 = (n_1^2k^2 - \beta_v^2)^{1/2} \tag{2.2-10}$$

The two choices of sine and cosine functions are intended to obtain a complete set of orthogonal functions. Since both sine and cosine functions are allowed as solutions of Eq. (2.2-1), both functions are needed. The upper and lower set of functions in Eqs. (2.2-8) and (2.2-9) belong together, respectively. We now obtain the transverse components from Eqs. (2.2-8) and (2.2-9) with the help of Eqs. (1.7-4)–(1.7-7). In order to be able to carry out the derivations, we need the relations

$$\frac{\partial}{\partial x} = \frac{\partial r}{\partial x}\frac{\partial}{\partial r} + \frac{\partial \phi}{\partial x}\frac{\partial}{\partial \phi} \tag{2.2-11}$$

and

$$\frac{\partial}{\partial y} = \frac{\partial r}{\partial y}\frac{\partial}{\partial r} + \frac{\partial \phi}{\partial y}\frac{\partial}{\partial \phi} \tag{2.2-12}$$

Using the transformations

$$r = (x^2 + y^2)^{1/2} \tag{2.2-13}$$

and

$$\phi = \arctan(y/x) \tag{2.2-14}$$

we obtain

$$\partial r/\partial x = x/r = \cos\phi, \tag{2.2-15}$$

$$\partial r/\partial y = y/r = \sin\phi, \tag{2.2-16}$$

$$\partial \phi/\partial x = -y/r^2 = -(1/r)\sin\phi, \tag{2.2-17}$$

and

$$\partial \phi/\partial y = x/r^2 = (1/r)\cos\phi \tag{2.2-18}$$

At this point we use condition (2.1-1) to simplify the analysis. Since the propagation constant of the guided mode must lie within the limits

$$n_2 k < \beta < n_1 k \tag{2.2-19}$$

we use approximately

$$\beta = nk \tag{2.2-20}$$

with $n \approx n_1 \approx n_2$. With the help of the functional relations of the Bessel function, we obtain the following expressions for the transverse field components

$$E_y = AJ_\nu(\kappa r)\begin{Bmatrix} \cos\nu\phi \\ \sin\nu\phi \end{Bmatrix} \tag{2.2-21}$$

and

$$H_x = -nA\frac{\beta_\nu}{|\beta_\nu|}\left(\frac{\varepsilon_0}{\mu_0}\right)^{1/2} J_\nu(\kappa r)\begin{Bmatrix} \cos\nu\phi \\ \sin\nu\phi \end{Bmatrix} \tag{2.2-22}$$

The remaining components are $E_x = 0$ and $H_y \approx 0$.

Outside the core in the cladding region at $r > a$, we obtain similarly

$$E_z = -\frac{A\gamma}{2\beta_\nu}\frac{J_\nu(\kappa a)}{H_\nu^{(1)}(i\gamma a)} \times\left[H_{\nu+1}^{(1)}(i\gamma r)\begin{Bmatrix} \sin(\nu+1)\phi \\ -\cos(\nu+1)\phi \end{Bmatrix} + H_{\nu-1}^{(1)}(i\gamma r)\begin{Bmatrix} \sin(\nu-1)\phi \\ -\cos(\nu-1)\phi \end{Bmatrix}\right] \tag{2.2-23}$$

$$H_z = \frac{A\gamma}{2k}\left(\frac{\varepsilon_0}{\mu_0}\right)^{1/2} \frac{J_\nu(\kappa a)}{H_\nu^{(1)}(i\gamma a)}$$

$$\times \left[H_{\nu+1}^{(1)}(i\gamma r) \begin{Bmatrix} \cos(\nu+1)\phi \\ \sin(\nu+1)\phi \end{Bmatrix} - H_{\nu-1}^{(1)}(i\gamma r) \begin{Bmatrix} \cos(\nu-1)\phi \\ \sin(\nu-1)\phi \end{Bmatrix} \right] \tag{2.2-24}$$

$$E_y = A \frac{J_\nu(\kappa a)}{H_\nu^{(1)}(i\gamma a)} H_\nu^{(1)}(i\gamma r) \begin{Bmatrix} \cos \nu\phi \\ \sin \nu\phi \end{Bmatrix} \tag{2.2-25}$$

$$H_x = -nA \frac{\beta_\nu}{|\beta_\nu|}\left(\frac{\varepsilon_0}{\mu_0}\right)^{1/2} \frac{J_\nu(\kappa a)}{H_\nu^{(1)}(i\gamma a)} H_\nu^{(1)}(i\gamma r) \begin{Bmatrix} \cos \nu\phi \\ \sin \nu\phi \end{Bmatrix} \tag{2.2-26}$$

The Hankel function of the first kind, $H_\nu^{(1)}$, has now replaced the Bessel function. The choice of this function is dictated by the requirement that the field must vanish at $r \to \infty$. Only if this requirement is satisfied do we obtain a guided mode. The parameter γ appearing in the argument of the Hankel function is defined by the usual expression [compare Eq. (1.2-14)]

$$\gamma = -iK_2 = (\beta_\nu{}^2 - n_2{}^2 k^2)^{1/2} \tag{2.2-27}$$

For guided modes, whose propagation constants must fall in the interval (2.2-19), γ is a real quantity. This means that the argument of the Hankel function is imaginary. The amplitude coefficient of these field expressions was chosen to ensure that E_y remains continuous at the core boundary $r = a$. Since the ratios γ/β and κ/β are both small, it is apparent that the longitudinal field components are much smaller than the transverse components. The fields are thus almost transverse. The electric field of the mode that is described by Eqs. (2.2-8), (2.2-9), and (2.2-21)–(2.2-26) is polarized in y direction. For a complete field description we also need the mode with the orthogonal polarization. For $r < a$ we obtain by the same arguments as before

$$E_z = \frac{iA\kappa}{2\beta_\nu}\left[J_{\nu+1}(\kappa r) \begin{Bmatrix} \cos(\nu+1)\phi \\ \sin(\nu+1)\phi \end{Bmatrix} - J_{\nu-1}(\kappa r) \begin{Bmatrix} \cos(\nu-1)\phi \\ \sin(\nu-1)\phi \end{Bmatrix} \right] \tag{2.2-28}$$

$$H_z = \frac{iA\kappa}{2k}\left(\frac{\varepsilon_0}{\mu_0}\right)^{1/2}$$

$$\times \left[J_{\nu+1}(\kappa r) \begin{Bmatrix} \sin(\nu+1)\phi \\ -\cos(\nu+1)\phi \end{Bmatrix} + J_{\nu-1}(\kappa r) \begin{Bmatrix} \sin(\nu-1)\phi \\ -\cos(\nu-1)\phi \end{Bmatrix} \right] \tag{2.2-29}$$

$$E_x = AJ_\nu(\kappa r) \begin{Bmatrix} \cos \nu\phi \\ \sin \nu\phi \end{Bmatrix} \tag{2.2-30}$$

$$H_y = nA\frac{\beta_\nu}{|\beta_\nu|}\left(\frac{\varepsilon_0}{\mu_0}\right)^{1/2} J_\nu(\kappa r)\begin{Bmatrix} \cos\nu\phi \\ \sin\nu\phi \end{Bmatrix} \tag{2.2-31}$$

and the field in the cladding for $r > a$ is given by

$$E_z = -\frac{A\gamma}{2\beta_\nu}\frac{J_\nu(\kappa a)}{H_\nu^{(1)}(i\gamma a)} \times\left[H_{\nu+1}^{(1)}(i\gamma r)\begin{Bmatrix} \cos(\nu+1)\phi \\ \sin(\nu+1)\phi \end{Bmatrix} - H_{\nu-1}^{(1)}(i\gamma r)\begin{Bmatrix} \cos(\nu-1)\phi \\ \sin(\nu-1)\phi \end{Bmatrix}\right] \tag{2.2-32}$$

$$H_z = -\frac{A\gamma}{2k}\left(\frac{\varepsilon_0}{\mu_0}\right)^{1/2}\frac{J_\nu(\kappa a)}{H_\nu^{(1)}(i\gamma a)} \times\left[H_{\nu+1}^{(1)}(i\gamma r)\begin{Bmatrix} \sin(\nu+1)\phi \\ -\cos(\nu+1)\phi \end{Bmatrix} + H_{\nu-1}^{(1)}(i\gamma r)\begin{Bmatrix} \sin(\nu-1)\phi \\ -\cos(\nu-1)\phi \end{Bmatrix}\right] \tag{2.2-33}$$

$$E_x = A\frac{J_\nu(\kappa a)}{H_\nu^{(1)}(i\gamma a)}H_\nu^{(1)}(i\gamma r)\begin{Bmatrix} \cos\nu\phi \\ \sin\nu\phi \end{Bmatrix} \tag{2.2-34}$$

$$H_y = nA\frac{\beta_\nu}{|\beta_\nu|}\left(\frac{\varepsilon_0}{\mu_0}\right)^{1/2}\frac{J_\nu(\kappa a)}{H_\nu^{(1)}(i\gamma a)}H_\nu^{(1)}(i\gamma r)\begin{Bmatrix} \cos\nu\phi \\ \sin\nu\phi \end{Bmatrix} \tag{2.2-35}$$

Now we have $E_y = 0$ and $H_x \approx 0$. We have collected the field expressions for two types of guided modes whose transverse fields are polarized orthogonally to each other. These field expressions are approximate solutions of Maxwell's equations, but they do not yet satisfy all the boundary conditions requiring the tangential field components to pass continuously through the core boundary. In order to be able to satisfy the boundary conditions, we need to transform the transverse field components to cylindrical coordinates. For a general vector field we obtain the r and ϕ components from the x and y components by means of the transformations

$$F_r = F_x\cos\phi + F_y\sin\phi \tag{2.2-36}$$

and

$$F_\phi = -F_x\sin\phi + F_y\cos\phi \tag{2.2-37}$$

Since either the x or y component of the electric field vanishes, it is apparent that the E_ϕ component is simply proportional to the E_x or E_y components with the same proportionality factor (the function $\cos\phi$ or $\sin\phi$) for the fields on either side of the core boundary. Since we have adjusted the field amplitudes to ensure that the E_y or E_x components have the same value on either side of the boundary, we see that continuity of the E_ϕ components is already achieved.

The H_ϕ components, if treated exactly, would be multiplied with n_1 and n_2 inside and outside the core. However, to the approximation that ignores the difference of the two refractive indices, we see that continuity of E_ϕ results in approximate continuity of H_ϕ. We thus need to consider only the E_z and H_z components.

When we equate the E_z components on either side of the core boundary, we must equate the coefficients of $\sin(\nu+1)\phi$ [or $\cos(\nu+1)\phi$] and $\sin(\nu-1)\phi$ [or $\cos(\nu-1)\phi$] separately. Starting with the coefficients of $\sin(\nu+1)\phi$ [or $\cos(\nu+1)\phi$] we obtain from Eqs. (2.2-8) and (2.2-23) [or alternately from Eqs. (2.2-28) and (2.2-32)] the following equation:

$$\kappa J_{\nu+1}(\kappa a)/J_\nu(\kappa a) = i\gamma H_{\nu+1}^{(1)}(i\gamma a)/H_\nu^{(1)}(i\gamma a) \tag{2.2-38}$$

Exactly the same equation results also from the requirement that the H_z component be continuous at the interface.

When we equate the coefficient of $\sin(\nu-1)\phi$ [or $\cos(\nu-1)\phi$], we obtain the equation

$$\kappa J_{\nu-1}(\kappa a)/J_\nu(\kappa a) = i\gamma H_{\nu-1}^{(1)}(i\gamma a)/H_\nu^{(1)}(i\gamma a) \tag{2.2-39}$$

Equations (2.2-38) and (2.2-39) are two different forms of the eigenvalue equation of the simplified modes of the weakly guiding round fiber. If these two equations were not identical, our problem would have no solution, since we cannot have two different equations for the one unknown—the propagation constant. To show the equivalence of the two forms we use the recursion relation of the cylinder functions [JE1, GR1]

$$Z_{\nu+1}(x) = (2\nu/x)\,Z_\nu(x) - Z_{\nu-1}(x) \tag{2.2-40}$$

which holds for the Bessel and Hankel functions. When this recursion relation is used in Eq. (2.2-38) to replace the functions of order $\nu+1$ with the functions of order $\nu-1$, Eq. (2.2-39) is obtained.

The eigenvalue Eq. (2.2-39) is amazingly simple compared to the complicated form of the exact eigenvalue equation. [See, for example, Eq. (8.2-49) of [Me1].] However, simplicity has to be bought at a price. The exact eigenvalue equation has twice as many solutions as the simple Eq. (2.2-39). The exact field solutions of the round optical fiber are classified as $HE_{\nu\mu}$ or $EH_{\nu\mu}$ modes [Sn1, Me1]. The propagation constants of $HE_{\nu+1,\mu}$ and $EH_{\nu-1,\mu}$ modes are almost identical. They become exactly the same in the limit $n_1 \to n_2$. We are thus faced with a case of near degeneracy. Comparison of the simplified mode solutions with the exact modes shows that the simplified modes are actually a superposition of $HE_{\nu+1,\mu}$ and $EH_{\nu-1,\mu}$ modes [Me1]. The near degeneracy of the exact theory has thus become a definite degeneracy, and the two types of modes have merged into one. However, the total number of modes is the same in both theories, because we now have a fourfold degeneracy since both polarizations and both choices of sine or cosine functions lead to the same eigenvalue equation.

The dispersion curves, representing the propagation constants as functions of frequency, are very nearly the same for the simplified and exact modes in case of weakly guiding fibers. Owing to the near degeneracy of the *HE* and *EH* modes, their dispersion curves are almost indistinguishable. The simplified description is thus able to reproduce the dispersion characteristics of the modes. This enables us to study the problem of pulse distortion with the use of the simplified eigenvalue equation. Problems of mode conversion and radiation losses can also be studied with the help of the simplified modes. Instead of determining how each *HE* or *EH* mode couples to other modes we now find how the superpositions of $HE_{\nu+1,\mu}$ modes and $EH_{\nu-1,\mu}$ modes couple to each other and to radiation. For purposes of determining the power transfer between groups of guided modes and for the study of radiation losses we can gain all the information that is required.

However, in spite of the obvious advantages of the simplified theory, it is prudent to keep in mind that the simplified modes do not represent true modes in the usual sense of the word. Even though we cannot determine this fact from the approximate analysis, comparison with the exact theory teaches us that the simplified modes must decompose as they travel along the waveguide. Because they are actually superpositions of $HE_{\nu+1,\mu}$ and $EH_{\nu-1,\mu}$ modes that travel with slightly different velocities, the simplified modes change their shape as they travel along the guide. This feature of the simplified modes becomes clear when we realize that the field shape of the superposition of two modes depends on their relative phase relationships. Because of their different phase velocities the relative phases of the $HE_{\nu+1,\mu}$ and $EH_{\nu-1,\mu}$ modes keep changing as a function of z, so that the superposition fields also change their shape. Only after a distance corresponding to one beat wavelength does the original phase relationship, and therefore the field shape, restore itself. The beat wavelength between modes 1 and 2 is defined as $\Lambda = 2\pi/(\beta_1 - \beta_2)$.

This discussion makes it clear that the labeling of the simplified modes does not coincide with the conventional mode labels. A simplified mode with the label ν corresponds to an $HE_{\nu+1,\mu}$ and an $EH_{\nu-1,\mu}$ mode. In particular, the important dominant HE_{11} mode is now characterized by the label $\nu = 0$. The label μ is used to distinguish the different solutions of the eigenvalue equation for a given value of ν.

The amplitude coefficient A of the guided modes can be related to the power P carried by the modes with the help of

$$P = \frac{1}{2}\int_0^{2\pi}\int_0^{\infty} (E_x H_y^* - E_y H_x^*) r \, dr \, d\phi \qquad (2.2\text{-}41)$$

which follows from the power density vector Eq. (1.3-45) by integration over the infinite cross section. We obtain the same relation for all four types of modes:

$$A = \left[\frac{4(\mu_0/\varepsilon_0)^{1/2}\gamma^2 P}{e_\nu \pi a^2 n(n_1{}^2 - n_2{}^2)k^2 |J_{\nu-1}(\kappa a)\,J_{\nu+1}(\kappa a)|}\right]^{1/2} \tag{2.2-42}$$

with

$$e_\nu = \begin{cases} 2, & \text{for} \quad \nu = 0 \\ 1, & \text{for} \quad \nu \neq 0 \end{cases} \tag{2.2-42a}$$

A derivation of this formula is given at the end of this section.

The eigenvalue Eq. (2.2-39) must be solved by numerical techniques. However, near cutoff and far from cutoff of each mode, approximate closed-form solutions can be worked out. Near cutoff we have the inequality

$$\gamma a \ll 1 \tag{2.2-43}$$

We can thus use the approximations of the Hankel functions for small arguments. For $\nu = 0$, we use

$$H_0^{(1)}(i\gamma a) = (2i/\pi)\ln(\Gamma\gamma a/2) \tag{2.2-44}$$

with

$$\Gamma = 1.781672 \tag{2.2-45}$$

For $\nu \neq 0$, the following approximation holds:

$$H_\nu^{(1)}(i\gamma a) = -[i(\nu-1)!/\pi](2/i\gamma a)^\nu \tag{2.2-46}$$

Substitution of these relations into Eq. (2.2-38) yields, for $\nu = 0$,

$$\gamma a = (2/\Gamma)\exp[-(1/\kappa a)\,J_0(\kappa a)/J_1(\kappa a)] \tag{2.2-47}$$

At cutoff we have $\gamma a = 0$ and $\kappa_c d = V_c$. This last relation follows from

$$(\kappa^2 + \gamma^2)a^2 = V^2 = (n_1{}^2 - n_2{}^2)k^2 a^2 \tag{2.2-48}$$

We thus obtain the cutoff condition for $\nu = 0$ modes from Eq. (2.2-47):

$$J_1(V_c) = 0 \tag{2.2-49}$$

It is apparent from Eq. (2.2-48) that near cutoff we have, to a good approximation,

$$\kappa a \approx V_c + (V - V_c) - [(\gamma a)^2/2V_c] \tag{2.2-50}$$

so that we finally find the near cutoff approximation for $\nu = 0$ modes [neglecting the $(\gamma a)^2$ term in Eq. (2.2-50)]

$$\gamma a = \frac{2}{\Gamma}\exp\left[-\frac{1}{V}\frac{J_0(V)}{J_1(V)}\right] \tag{2.2-51}$$

For the modes with $\nu = 1$ we obtain from Eqs. (2.2-39), (2.2-44), and (2.2-46),

$$\kappa a J_0(\kappa a)/J_1(\kappa a) = (\gamma a)^2 \ln(\Gamma \gamma a/2) \tag{2.2-52}$$

The cutoff condition for $\nu = 1$ modes is thus

$$J_0(V_c) = 0 \tag{2.2-53}$$

We use Eqs. (2.2-50) and (2.2-53) and expand the function $J_0(\kappa a)$ with the help of the functional relation for the derivative of Bessel functions,

$$J_\nu'(\kappa a) = -(\nu/\kappa a) J_\nu(\kappa a) + J_{\nu-1}(\kappa a) = (\nu/\kappa a) J_\nu(\kappa a) - J_{\nu+1}(\kappa a) \tag{2.2-54}$$

(where the prime indicates the derivative with respect to the whole argument) obtaining

$$J_0(\kappa a) = -[(V-V_c)-(\gamma a)^2/2V_c] J_1(V_c) \tag{2.2-55}$$

If we now replace κa with V_c on the left-hand side of Eq. (2.2-52) and substitute Eq. (2.2-55), we obtain, for modes with $\nu = 1$,

$$(\gamma a)^2 = 2V_c(V-V_c)/[1-2\ln(\Gamma\gamma a/2)] \tag{2.2-56}$$

This is still an implicit equation for γa, but it can easily be solved with a few iterations. V_c is obtained as the solution of Eq. (2.2-53).

To find the approximation for $\nu \geqslant 2$ we expand $J_{\nu-1}(\kappa a)$ with the help of Eqs. (2.2-50) and (2.2-54) and the cutoff condition

$$J_{\nu-1}(V_c) = 0, \qquad \text{for} \quad \nu = 2, 3, \ldots \tag{2.2-57}$$

and obtain

$$J_{\nu-1}(\kappa a) = -[(V-V_c)-(\gamma a)^2/2V_c] J_\nu(V_c) \tag{2.2-58}$$

Equations (2.2-39) and (2.2-50) thus yield the near cutoff approximation for all modes with $\nu \geqslant 2$

$$\gamma a = \{2[(\nu-1)/\nu] V_c(V-V_c)\}^{1/2} \tag{2.2-59}$$

V_c is obtained as the solution of Eq. (2.2-57). The roots of the Bessel functions which solve Eqs. (2.2-49), (2.2-53), and (2.2-57) can be found in tables of Bessel functions [Je1, As1]. In Eq. (2.2-51) we did not expand the Bessel function $J_1(V)$ as we did in the other equations. The reason is twofold: The usual expansion of the type (2.2-55) or (2.2-58) would not work for the $V = 0$ solution of Eq. (2.2-49), since the expression $J_1(V_c)/V_c = 0/0$ is encountered in Eq. (2.2-54) for $V_c = 0$. It is, of course, easy to determine the limit of this undetermined ratio. The easiest approach would be to use the small argument approximation

$$J_\nu(V) = (1/\nu!)(V/2)^\nu \tag{2.2-60}$$

of the Bessel functions. However, the second reason for not approximating Eq. (2.2-51) any further is its wider applicability. A comparison of the approximate expression (2.2-51) with the exact solution of Eq. (2.2-39) [Mel] shows that the approximation works well for V values as large as $V = 1$. The agreement is poorer if we use approximations for the Bessel functions in Eq. (2.2-51).

It might appear as if $V_c = 0$ were also an admissible solution of Eq. (2.2-57). However, we can show with the help of the small argument approximation (2.2-60) and the small argument approximation (2.2-46) for the Hankel functions that $\kappa_c a = V_c = 0$ is not a solution of Eq. (2.2-39). The right-hand side of the equation vanishes for $\gamma \to 0$. On the left-hand side we have

$$\lim_{V_c \to 0} V_c J_{\nu-1}(V_c)/J_\nu(V_c) = 2\nu \qquad (2.2\text{-}61)$$

This expression vanishes only if $\nu = 0$, showing that $V_c = 0$ is allowed only for $\nu = 0$ or for the cutoff condition (2.2-49). This shows that the lowest-order mode with $\nu = 0$ is theoretically not cutoff, since it is guided for arbitrarily small frequencies or arbitrarily thin cores. However, as a practical matter, we find with the help of Eq. (2.2-51), $\gamma a = 2.2 \times 10^{-10}$, for $V = 0.3$. In order to appreciate the meaning of this result we use the large argument approximation of the Hankel function

$$H_\nu^{(1)}(i\gamma r) = (2/\pi i \gamma r)^{1/2} \exp\{-i[\nu(\pi/2) + (\pi/4)]\} \exp(-\gamma r) \qquad (2.2\text{-}62)$$

We thus see that for large γr the field, Eq. (2.2-25), outside the core is proportional to $\exp(-\gamma r)$,

$$E_y \propto \exp(-\gamma r) \qquad (2.2\text{-}63)$$

The parameter γ is thus a measure for the radius r_m at which the field intensity has decayed to $1/e$ of its value at the core boundary. We can define the mode radius by the equation

$$r_m = a + \gamma^{-1} \qquad (2.2\text{-}64)$$

The mode radius for the lowest-order mode, $\nu = 0$, with $V = 0.3$ is thus 4.5×10^9 times the core radius. This means, for all practical purposes, that the field is no longer guided. It seems advisable to let r_m become no more than $10a$. This value is reached for the lowest-order mode, $\nu = 0$, if $V = 0.95$.

Far from cutoff γa becomes large. With Eq. (2.2-62) the eigenvalue Eq. (2.2-39) assumes its far from cutoff approximation

$$\kappa a J_{\nu-1}(\kappa a)/J_\nu(\kappa a) = -\gamma a \qquad (2.2\text{-}65)$$

The limiting value $\kappa = \kappa_\infty$, which is reached for $\gamma \to \infty$, is thus given as the root of the equation

$$J_\nu(\kappa_\infty a) = 0 \tag{2.2-66}$$

Using a method developed by Snyder [Sr1, Sr7] it was shown in [Me1] that an approximate solution of Eq. (2.2-65) is

$$\kappa a = \kappa_\infty a[1-(2\nu/V)]^{1/2\nu} \tag{2.2-67}$$

It can also be shown that the solution for $\nu = 0$ is indeed

$$\kappa a = \kappa_\infty a \exp(-1/V) \tag{2.2-68}$$

as one would expect by taking the limit of Eq. (2.2-67), for $\nu \to 0$.

Gloge [Ge1] introduced the notation $LP_{\nu\mu}$ modes for the approximate mode solutions discussed in this section. The first index, ν, corresponds to the integer that enters the circular functions, $\cos \nu\phi$ and $\sin \nu\phi$. The second index, μ, labels the roots of Eq. (2.2-39) for a given value of ν. The LP_{01} mode thus corresponds to the important dominant HE_{11} mode of the exact treatment of the fiber. Generally, we know that an $LP_{\nu\mu}$ mode is a superposition of $HE_{\nu+1,\mu}$ and $EH_{\nu-1,\mu}$ modes. The notation LP modes was chosen to suggest "linearly polarized modes."

For fibers supporting many modes, it is advantageous to be able to estimate the total number of modes that are possible for a given value of V. According to Eq. (2.2-67) κ is very nearly equal to κ_∞ for very large values of V. We thus obtain the total number of modes by estimating the number of roots of the equation

$$J_\nu(z_m) = 0 \tag{2.2-69}$$

This number is constrained by the condition

$$z_m \leqslant V \tag{2.2-70}$$

An approximate formula for the roots of Eq. (2.2-69) for large values of m is [AS1]

$$\kappa_\infty a = z_m = (\nu+2m-\tfrac{1}{2})\pi/2 \tag{2.2-71}$$

with $m = 1, 2, 3, \ldots$. For constant values of z_m, the integer values of ν and m satisfying Eq. (2.2-71) lie along lines parallel to the dashed line depicted in Fig. 2.2.2. The largest value of $z_m = \kappa a$ is obtained from Eq. (2.2-48) with the cutoff value $\gamma = 0$. It is $z_m = V$. Neglecting the $\frac{1}{2}$ term in Eq. (2.2-71), we see that all values of ν and m that are allowed by Eq. (2.2-71) lie inside the triangle formed by the two coordinate axes and the slanted solid line passing through the points $m = V/\pi$ and $\nu = 2V/\pi$ in Fig. 2.2.2.

Each point with integer coordinate values ν and m represents one solution of Eq. (2.2-69) and, consequently, can be associated with one mode of a given polarization and ϕ dependence. It can be seen with the help of Fig. 2.2.3 that

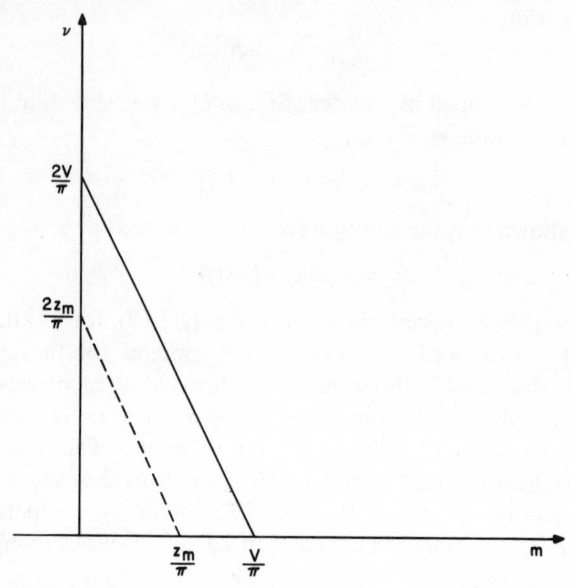

Fig. 2.2.2 *The ν, m mode number plane. The dashed curve indicates the location of the points $\nu+2m=$ const.*

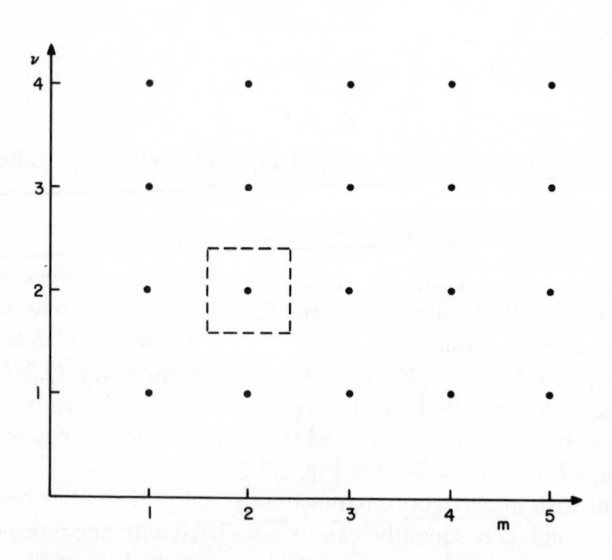

Fig. 2.2.3 *Each point represents a mode in the ν, m plane.*

each representation point occupies a square of unit area in ν, m space. The area in this space thus represents the number of modes. The total number of modes is four times the area of the triangle shown in Fig. 2.2.2, since each mode can appear in two mutually orthogonal polarizations and since we can have the sine or cosine function in Eq. (2.2-21). The area of the triangle is $(V/\pi)^2$, so that we obtain the approximate total number of modes that can exist for a given value of V from

$$N = 4V^2/\pi^2 \tag{2.2-72}$$

We found earlier [Eq. (1.3-70)] that the total number of TE and TM modes is $2V/\pi$. The total number of modes of the round fiber is thus equal to the square of the total number of slab waveguide modes for the same value of V. However, formula (2.2-72) is not very precise, since we used a number of crude approximations for its derivation. Formula (1.3-70) for the modes of the slab [with the restriction (1.3-4)] holds exactly. A better approximation of the total mode number is given by the formula [Ge1]

$$N = \tfrac{1}{2}V^2 \tag{2.2-73}$$

Finally, we study the distribution of power that is carried by the guided modes in the core and in the cladding. The amount of power that is contained inside the core is given by

$$P_{\text{core}} = \frac{1}{2}\int_0^{2\pi}\int_0^a r(E_x H_y^* - E_y H_x^*)\, dr\, d\phi \tag{2.2-74}$$

Substitution of either Eqs. (2.2-21) and (2.2-22) or (2.2-30) and (2.2-31) yields the result

$$P_{\text{core}} = e_\nu(\pi/4)\, a^2 n(\varepsilon_0/\mu_0)^{1/2} A^2 [J_\nu^{\,2}(\kappa a) - J_{\nu+1}(\kappa a)\, J_{\nu-1}(\kappa a)] \tag{2.2-75}$$

with

$$e_\nu = \begin{cases} 2, & \text{for} \quad \nu = 0 \\ 1, & \text{for} \quad \nu \neq 0 \end{cases} \tag{2.2-76}$$

For $\nu = 0$, we use the relation $J_{-1} = -J_1$. The power in the cladding is obtained similarly by integrating Eq. (2.2-74) from a to infinity:

$$\begin{aligned} P_{\text{clad}} &= e_\nu(\pi/4)\, a^2 n(\varepsilon_0/\mu_0)^{1/2} A^2 [J_\nu(\kappa a)/H_\nu^{(1)}(i\gamma a)]^2 \\ &\quad \times [H_{\nu+1}^{(1)}(i\gamma a)\, H_{\nu-1}^{(1)}(i\gamma a) - H_\nu^{(1)^2}(i\gamma a)] \end{aligned} \tag{2.2-77}$$

With the aid of the eigenvalue equation in its two forms (2.2-38) and (2.2-39), we can eliminate the Hankel functions from Eq. (2.2-77) and obtain

$$P_{\text{clad}} = -e_\nu(\pi/4)\, a^2 n(\varepsilon_0/\mu_0)^{1/2} A^2 [J_\nu^{\,2}(\kappa a) + (\kappa^2/\gamma^2)\, J_{\nu+1}(\kappa a)\, J_{\nu-1}(\kappa a)] \tag{2.2-78}$$

The sum of $P_{\text{core}}+P_{\text{clad}}$ is the total power P. By adding Eqs. (2.2-75) and (2.2-78), we obtain Eq. (2.2-42) with the help of Eq. (2.2-48). The negative sign in front of Eq. (2.2-78) looks disturbing, but for those γa values that are allowed by the eigenvalue equation, $J_{\nu+1}J_{\nu-1}$ is always negative, so that P_{clad} is always positive. The ratio of cladding power to total power P is obtained by substitution of Eq. (2.2-42) into Eq. (2.2-78):

$$P_{\text{clad}}/P = (1/V^2)[(\kappa a)^2+(\gamma a)^2 J_\nu^{\,2}(\kappa a)/J_{\nu-1}(\kappa a)\,J_{\nu+1}(\kappa a)] \tag{2.2-79}$$

Very far from cutoff we approximate Eq. (2.2-67):

$$\kappa a = \kappa_\infty a - \kappa_\infty a/V \tag{2.2-80}$$

Expanding the Bessel function with the help of Eqs. (2.2-80), (2.2-66), and (2.2-54) yields

$$J_\nu(\kappa a) = -\kappa_\infty a J_{\nu-1}(\kappa_\infty a)/V = \kappa_\infty a J_{\nu+1}(\kappa_\infty a)/V \tag{2.2-81}$$

and

$$J_{\nu\pm1}(\kappa a) = [1\pm(\nu\pm1)/V]\,J_{\nu\pm1}(\kappa_\infty a) \tag{2.2-82}$$

With these approximations we obtain from Eq. (2.2-79), with the help of Eqs. (2.2-48) and (2.2-80),

$$P_{\text{clad}}/P = (\kappa_\infty a)^4[1-(2/V)]/V^4 \tag{2.2-83}$$

As usual we obtain $\kappa_\infty a$ as the root of Eq. (2.2-66). The far from cutoff approximation (2.2-83) shows clearly that the amount of power carried in the cladding becomes smaller as V increases.

Near cutoff we have $\gamma a \ll 1$. For $\nu \neq 0$, we use Eq. (2.2-58) and express the ratio $J_\nu/J_{\nu+1}$ with the help of Eqs. (2.2-38) and (2.2-46) in the form

$$J_\nu(\kappa a)/J_{\nu+1}(\kappa a) = \kappa a/2\nu \tag{2.2-84}$$

Thus we can approximate Eq. (2.2-79) as follows:

$$P_{\text{clad}}/P = 1 - (\gamma a)^2/2\nu[V_c(V-V_c)-\tfrac{1}{2}(\gamma a)^2] \tag{2.2-85}$$

To obtain this expression κa was replaced by V_c, which is obtained as the root of Eq. (2.2-57). The parameter γa must be obtained from Eq. (2.2-51) or (2.2-59), depending on the value of ν. For $\nu \geqslant 2$, where Eq. (2.2-59) applies, we have

$$P_{\text{clad}}/P = 1/\nu, \qquad \text{for} \quad \nu \geqslant 2, \quad V = V_c \tag{2.2-86}$$

We have thus obtained the value of the power ratio at the cutoff point $V = V_c$. Actually, to this approximation, the ratio is independent of $V-V_c$. However, because of Eq. (2.2-57) the derivative of Eq. (2.2-79) taken with respect to V is negative infinite at $V = V_c$, so that the ratio drops abruptly from the value $1/\nu$ with increasing $V - V_c$. It is interesting that P_{clad}/P does not start with the value of unity at the cutoff point of the modes with $\nu \geqslant 2$. This means that a significant amount of power resides inside the core even at the cutoff point. For the modes with $\nu = 0$ and $\nu = 1$, $P_{\text{clad}}/P = 1$ at $V = V_c$.

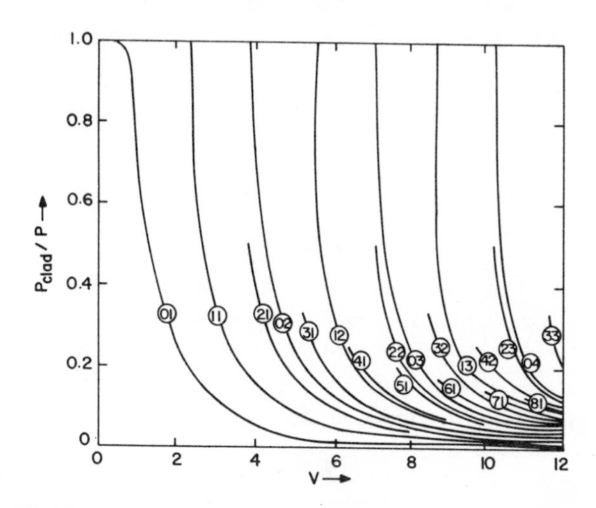

Fig. 2.2.4 *Fractional power contained in the cladding as a function of the frequency parameter V.* (From *D. Gloge, "Weakly guiding fibers,"* Appl. Opt. ***10***, *2252–2258 (1971), reproduced with permission.*)

For $\nu = 0$, we have the cutoff condition (2.2-49), so that the expansion of J_1 yields, with the help of Eqs. (2.2-50) and (2.2-54),

$$J_1(\kappa a) = [V - V_c - (\gamma a)^2/2V_c]\, J_0(\kappa a) \tag{2.2-87}$$

Using $J_{-1} = -J_1$ and Eq. (2.2-48), we obtain from Eq. (2.2-79) near cutoff of the $\nu = 0$ modes,

$$P_{\text{clad}}/P = 1 - (\gamma a)^2/[V(V - V_c) - \tfrac{1}{2}(\gamma a)^2]^2 \tag{2.2-88}$$

with γa of Eq. (2.2-47). The exponential dependence of γa ensures that the second term in Eq. (2.2-88) vanishes at $V = V_c$, even if $V_c = 0$. The power ratio is thus unity at $V = V_c$. Figure 2.2.4 shows the ratio P_{clad}/P for several modes as a function of V. The abrupt dropoff from either 1 or $1/\nu$ at the cutoff of all modes with $\nu \neq 0$ is quite apparent. Only the LP_{01} mode carries most of the power in the cladding for some distance (in terms of $V - V_c$) from its cutoff point. Figure 2.2.4 was plotted from Eq. (2.2-79) by Gloge [Ge1].

2.3 Waveguide Dispersion and Group Velocity

Having laid the theoretical groundwork for the description of the modes of the weakly guiding fiber in the preceding section we now show some numerical results.

The propagation constant β of the guided modes is obtained as a solution of the eigenvalue Eqs. (2.2-38) or (2.2-39). The propagation constant of a plane wave in a homogeneous medium with refractive index n is

$$\beta_{\mathrm{pw}} = nk \tag{2.3-1}$$

with

$$k = \omega/c = 2\pi/\lambda \tag{2.3-2}$$

representing the plane wave propagation constant in vacuum. The difference of the refractive indices of core and cladding is only slight in weakly guiding fibers. When most of the field power is carried in the core, the wave has a propagation constant that approaches Eq. (2.3-1) with $n = n_1$. Near cutoff, most of the field energy is traveling outside the waveguide core, so that the propagation constant approaches Eq. (2.3-1) with $n = n_2$. The propagation constants of all the guided modes are thus contained in the interval

$$n_2 k < \beta < n_1 k \tag{2.3-3}$$

The approximate eigenvalue equation of weakly guiding fibers, Eq. (2.2-39), does not contain the refractive indices of core and cladding explicitly. Using relation (2.2-48) it is thus clear that its solutions, $\kappa a = f(V)$ or alternately $\gamma a = g(V)$, are universal functions of V and do not depend on the geometry and index difference of any particular waveguide. This result holds only to the approximation involved in deriving Eq. (2.2-39). By comparison, we see from Eq. (1.3-39) that the solutions for the TE modes of the symmetric slab waveguide are truly general functions of V, while the exact solutions of the eigenvalue Eq. (1.3-66) for TM modes depend on the refractive index ratios.

To the approximation applicable for weakly guiding fibers, we can define the universal function of V:

$$b = (\gamma a)^2/V^2 = [(\beta/n_2 k)^2 - 1]/[(n_1/n_2)^2 - 1] \tag{2.3-4}$$

The right-hand side of Eq. (2.3-4) is obtained from Eqs. (2.2-27) and (2.2-48). Since Eq. (2.3-4) is a universal function of V only to the extent that the weakly guiding assumption (2.1-1) applies, we can write Eq. (2.3-4) with the help of Eq. (2.3-3) in the simpler approximate form

$$b = (\beta/n_2 k - 1)/(n_1/n_2 - 1) \tag{2.3-5}$$

According to Eq. (2.3-3), b varies between 0 and 1.

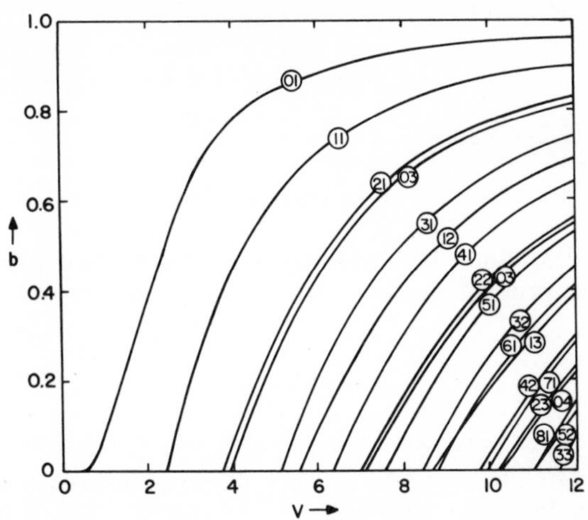

Fig. 2.3.1 *Normalized propagation constant b as function of normalized frequency V for the guided modes of the optical fiber.* (From *D. Gloge,* "*Weakly guiding fibers,*" Appl. Opt. **10**, *2252–2258* (*1971*), *reproduced with permission.*)

Figure 2.3.1 shows the normalized propagation constant b as a function of V for the first 20 guided modes of the weakly guiding fiber. The number pairs written on each curve give the values of ν and μ of the $LP_{\nu\mu}$ modes. Instead of regarding b as a normalized propagation constant, we can also use the plots of Fig. 2.3.1 to obtain the important radial decay parameter γa from the relation

$$\gamma a = V\sqrt{b} \tag{2.3-6}$$

The propagation constant β is related to the phase velocity of the guided wave (ω = radian frequency) by

$$v = \omega/\beta \tag{2.3-7}$$

A more important quantity is the group velocity

$$v_g = \frac{d\omega}{d\beta} = c\,\frac{dk}{d\beta} \tag{2.3-8}$$

The phase velocity determines the speed at which the planes of constant phase move through the waveguide. This quantity is not directly observable at optical frequencies. The group velocity gives the speed of pulse propagation and is thus an important observable quantity. Closely related to the group velocity is the group delay. It determines the transit time of a pulse traveling through the waveguide of length L. The group delay is defined as

$$\tau_g = L\,\frac{d\beta}{d\omega} = \frac{L}{c}\frac{d\beta}{dk} = \frac{LV}{ck}\frac{d\beta}{dV} \tag{2.3-9}$$

Adopting Gloge's [Ge1] notation, we define

$$\Delta = (n_1 - n_2)/n_2 \ll 1 \tag{2.3-10}$$

and write Eq. (2.2-48) approximately as

$$V = \sqrt{2}\, n_2 \sqrt{\Delta}\, ka \tag{2.3-11}$$

Equation (2.3-9) contains the derivative of β. In order to express it in terms of the normalized propagation constant b, we obtain from Eqs. (2.3-5) and (2.3-10)

$$\beta = (b\Delta + 1)\, n_2 k \tag{2.3-12}$$

The refractive index parameter Δ is small. Since the refractive index depends only slightly on frequency, its derivative is even smaller and shall be neglected. We thus use the following approach:

$$\frac{d}{dV}(bn_2 k\Delta) = \sqrt{\Delta}\,\frac{d}{dV}(bn_2 k\sqrt{\Delta}) = \frac{\sqrt{\Delta}}{\sqrt{2}\,a}\frac{d(Vb)}{dV} \tag{2.3-13}$$

Using these equations we can write the total group delay as the sum

$$\tau_g = \tau_m + \tau_w \tag{2.3-14}$$

with the delay that is characteristic of the material

$$\tau_m = (L/c)\, d(n_2 k)/dk \tag{2.3-15}$$

and the waveguide delay

$$\tau_w = (L/c)\, n_2 \Delta\, d(Vb)/dV \tag{2.3-16}$$

The derivative appearing in the formula for the waveguide delay can be eliminated with the help of the eigenvalue Eq. (2.2-39). As a first step we use Eq. (2.3-4) to obtain

$$\frac{d(Vb)}{dV} = 2\frac{\gamma a}{V}\frac{d(\gamma a)}{dV} - \frac{(\gamma a)^2}{V^2} \tag{2.3-17}$$

Next we take the V derivative of the eigenvalue Eq. (2.2-39) and obtain with the help of Eqs. (2.2-40) and (2.2-54) (which also holds for the Hankel functions)

$$\kappa a\left[\frac{J_{\nu-1}(\kappa a)\,J_{\nu+1}(\kappa a)}{J_\nu^{\,2}(\kappa a)} - 1\right]\frac{d(\kappa a)}{dV} + \gamma a\left[\frac{H_{\nu-1}^{(1)}(i\gamma a)\,H_{\nu+1}^{(1)}(i\gamma a)}{[H_\nu^{(1)}(i\gamma a)]^2} - 1\right]\frac{d(\gamma a)}{dV} = 0 \tag{2.3-18}$$

We divide Eqs. (2.2-38) and (2.2-39) by γ and multiply the resulting equations, obtaining

$$\frac{H_{\nu+1}^{(1)}(i\gamma a)\,H_{\nu-1}^{(1)}(i\gamma a)}{[H_\nu^{(1)}(i\gamma a)]^2} = -\frac{(\kappa a)^2}{(\gamma a)^2}\frac{J_{\nu+1}(\kappa a)\,J_{\nu-1}(\kappa a)}{J_\nu^{\,2}(\kappa a)} \tag{2.3-19}$$

From Eq. (2.2-48) we derive

$$\kappa a\,\frac{d(\kappa a)}{dV} + \gamma a\,\frac{d(\gamma a)}{dV} = V \tag{2.3-20}$$

Use of these three equations and (2.3-4) enables us to eliminate the derivative from Eq. (2.3-17):

$$\frac{d(Vb)}{dV} = b\left[1 - \frac{2J_\nu^{\,2}(\kappa a)}{J_{\nu-1}(\kappa a)\,J_{\nu+1}(\kappa a)}\right] \tag{2.3-21}$$

The group delay Eq. (2.3-15), which is characteristic of the material of the optical fiber, is independent of the particular mode. The waveguide delay Eq. (2.3-16) with Eq. (2.3-21) is different for every guided mode. A light pulse that is shared by many guided modes thus splits up into many pulses arriving at the end at different times, corresponding to their group delay Eq. (2.3-16). This type of pulse dispersion is called "multimode dispersion" to distinguish it from the "material dispersion" Eq. (2.3-15).

Far from cutoff b approaches unity, while $J_\nu(\kappa a)$ goes to zero according to Eq. (2.2-66). This consideration shows that for $V \to \infty$, Eq. (2.3-21)

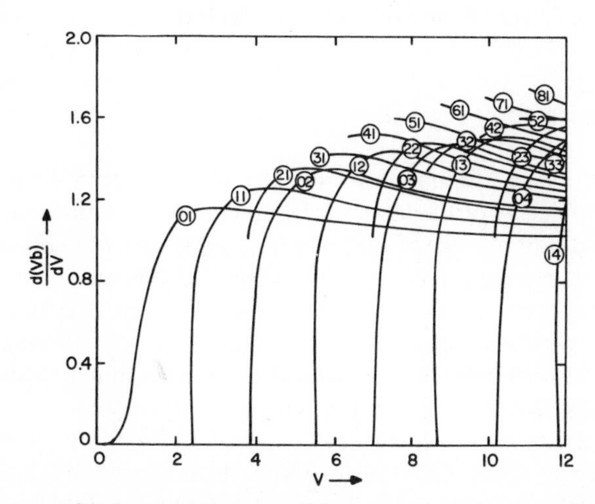

Fig. 2.3.2 *Normalized waveguide group delay as a function of V.* (From *D. Gloge,* "*Weakly guiding fibers,*" Appl. Opt. ***10****, 2252–2258 (1971), reproduced with permission.*)

approaches unity for all modes. Function (2.3-21) is plotted in Fig. 2.3.2 for the first 20 modes of the fiber. The group delay caused by waveguide dispersion is obtained from this figure with the help of Eq. (2.3-16). It is apparent from the figure that the modes with $\nu = 0$ and $\nu = 1$ have zero waveguide delay at cutoff. However, the total group delay is the sum of the material delay Eq. (2.3-15) and the waveguide delay Eq. (2.3-16), so that it always remains finite.

At cutoff $J_{\nu-1}(\kappa a) = 0$, according to Eq. (2.2-57), for all modes with $\nu \geqslant 2$. By using the approximation Eqs. (2.2-58) and (2.2-59) and expressing $J_{\nu+1}$ by J_ν with the help of Eqs. (2.2-40) and (2.2-57), we find the cutoff value of Eq. (2.3-21) for modes with $\nu \geqslant 2$,

$$d(Vb)/dV = 2(\nu - 1)/\nu \tag{2.3-22}$$

The normalized propagation constant b was replaced with its value, Eq. (2.3-4), and approximated with the use of Eq. (2.2-59).

The arrival time difference between the mode with the largest waveguide group delay and the least delay [which is given by Eq. (2.3-21) assuming the value unity] is

$$\Delta\tau = (L/c)(n_1 - n_2)[1 - (2/\nu_m)] \tag{2.3-23}$$

In this formula ν_m is the largest ν value that can occur for the V at which the waveguide operates. We see from Fig. 2.2.2 that this value is approximately

$$\nu_m = 2V/\pi. \tag{2.3-24}$$

The fastest modes with $d(Vb)/dV = 0$ have been omitted from this consideration since it is unlikely that V would be close enough to the cutoff value V_c of one of these modes. Furthermore, a mode with $\nu = 0$ or $\nu = 1$ close to its cutoff point would suffer very high losses from accidental bends or irregularities of the waveguide core, so that it would not take part in the transport of energy.

Pulse distortion in multimode waveguides is caused predominantly by the different group delays of the modes. The difference $\Delta\tau$ in the arrival times of the leading and trailing edge of the pulse is given by Eqs. (2.3-23) and (2.3-24). The pulse shape depends on the amount of power that is carried by each mode. In single-mode fibers, pulse distortion is caused by the fact that the group velocity is not constant with frequency, so that different portions of the pulse spectrum travel with different velocities. This effect is much smaller than the multimode pulse distortion and depends on the frequency content of the signal and the dispersion curve of the fiber material. In addition to the material dispersion, single-mode pulse distortion is also caused by the frequency dependence of the term (2.3-16). However, for most materials the material dispersion predominates [Ge2].

2.4 Radiation Modes of the Optical Fiber

The radiation modes of the optical fiber are analogous to the radiation modes of the slab waveguide discussed in Sect. 1.4. The radiation modes are necessary to form a complete orthogonal set together with the guided modes. A series expansion in terms of guided modes plus an integral over radiation modes is sufficient to express any arbitrary solution of Maxwell's equations. Such series expansions will be discussed in Chap. 3.

The mathematical description of the guided modes was simplified by the fact that these modes are very nearly transverse if the refractive index difference between core and cladding is small. This same simplification is not possible for all the radiation modes. Just as in the slab waveguide case, the radiation modes of the fiber have propagation constants that lie in the range

$$-n_2 k \leqslant \beta \leqslant n_2 k \tag{2.4-1}$$

In addition there are evanescent modes with a continuous spectrum of imaginary β values in the range

$$-i\infty < \beta < i\infty \tag{2.4-2}$$

The combined set of modes is labeled by means of the parameter

$$\rho = (n_2{}^2 k^2 - \beta^2)^{1/2} \tag{2.4-3}$$

that covers the range

$$0 \leqslant \rho < \infty \tag{2.4-4}$$

For small values of $n_1 - n_2$, the radiation modes with β values near $\pm n_2 k$ are very nearly transverse and can be simplified in just the same way as the guided modes. However, inside of range (2.4-1), far from the endpoints, the radiation modes are not even approximately transverse to the waveguide axis. When we discussed the slab waveguide we pointed out that radiation modes can be regarded as waves that are incident on the slab at certain angles. These incident waves are transverse waves with respect to their direction of propagation. But the electric and magnetic fields may have substantial components in z direction, so that they are not transverse with respect to the waveguide axis. We are thus not allowed to use the approximation $\beta = nk$ (using n to approximate either n_1 or n_2).

It thus appears as though the radiation modes of the fiber may be far more complicated than the guided modes. However, we can again make use of the small index difference $n_1 - n_2$ that permits waves at large angles to pass the core–cladding interface without appreciable reflection or refraction. Instead of trying to satisfy the boundary conditions at the core–cladding interface, we now use the opposite approach and ignore this boundary for those radiation modes that lie deeper inside the range (2.4-1). The radiation modes inside this range are thus "free space" modes, which means they are solutions of Maxwell's equations in a homogeneous medium. If the medium were vacuum, they would be truly free space waves. We are thus using two types of modes in the range (2.4-1). Deeper inside this range we use the modes of homogeneous space, while very near the endpoints we work out approximate solutions of Maxwells' equations that satisfy the boundary conditions at the core–cladding interface. The boundary conditions for modes with $|\beta| \approx n_2 k$ are important since they represent waves that impinge on the boundary at grazing angles suffering significant refraction and reflection [Sr5].

The derivation of the modes with $|\beta| \approx n_2 k$ is very similar to the derivation of the guided modes. We use again the approximation $\beta = nk$ and choose superpositions of E_z and H_z components that cause one of the transverse field components to vanish exactly. A second component then vanishes approximately, so that a four-component field results if the vector components are expressed in cartesian coordinates. The only difference between our present treatment and the case of the guided modes is that we now allow incoming as well as outgoing waves, thus adding one additional amplitude coefficient, with the result that no determinantal condition needs to be satisfied, so that no eigenvalue equation for β exists. The propagation constant β remains arbitrary and ranges through a continuum of values. With this introduction we simply state the field expressions.

For radiation modes whose E component is polarized predominantly in y direction and whose propagation constant is $|\beta| \approx n_2 k$, we have, inside the core at $r < a$,

$$E_z = \frac{iB\sigma}{2\beta}\left[J_{\nu+1}(\sigma r)\begin{Bmatrix} \sin(\nu+1)\phi \\ -\cos(\nu+1)\phi \end{Bmatrix} + J_{\nu-1}(\sigma r)\begin{Bmatrix} \sin(\nu-1)\phi \\ -\cos(\nu-1)\phi \end{Bmatrix}\right] \tag{2.4-5}$$

$$H_z = -\frac{iB}{2k}\left(\frac{\varepsilon_0}{\mu_0}\right)^{1/2}\left[J_{\nu+1}(\sigma r)\begin{Bmatrix} \cos(\nu+1)\phi \\ \sin(\nu+1)\phi \end{Bmatrix} - J_{\nu-1}(\sigma r)\begin{Bmatrix} \cos(\nu-1)\phi \\ \sin(\nu-1)\phi \end{Bmatrix}\right] \tag{2.4-6}$$

$$E_y = BJ_\nu(\sigma r)\begin{Bmatrix} \cos\nu\phi \\ \sin\nu\phi \end{Bmatrix} \tag{2.4-7}$$

$$H_x = -nB\frac{\beta}{|\beta|}\left(\frac{\varepsilon_0}{\mu_0}\right)^{1/2} J_\nu(\sigma r)\begin{Bmatrix} \cos\nu\phi \\ \sin\nu\phi \end{Bmatrix} \tag{2.4-8}$$

In addition, we have $E_x = 0$ and $H_y \approx 0$. The parameter σ is defined as

$$\sigma = (n_1{}^2k^2 - \beta^2)^{1/2} \tag{2.4-9}$$

In the (infinite) cladding at $r > a$, the field components are

$$E_z = \frac{iC\rho}{2\beta}\left\{[H_{\nu+1}^{(1)}(\rho r)+DH_{\nu+1}^{(2)}(\rho r)]\begin{Bmatrix} \sin(\nu+1)\phi \\ -\cos(\nu+1)\phi \end{Bmatrix}\right.$$

$$\left. + [H_{\nu-1}^{(1)}(\rho r)+DH_{\nu-1}^{(2)}(\rho r)]\begin{Bmatrix} \sin(\nu-1)\phi \\ -\cos(\nu-1)\phi \end{Bmatrix}\right\} \tag{2.4-10}$$

$$H_z = -\frac{iC\rho}{2k}\left(\frac{\varepsilon_0}{\mu_0}\right)^{1/2}\left\{[H_{\nu+1}^{(1)}(\rho r)+DH_{\nu+1}^{(2)}(\rho r)]\begin{Bmatrix} \cos(\nu+1)\phi \\ \sin(\nu+1)\phi \end{Bmatrix}\right.$$

$$\left. - [H_{\nu-1}^{(1)}(\rho r)+DH_{\nu-1}^{(2)}(\rho r)]\begin{Bmatrix} \cos(\nu-1)\phi \\ \sin(\nu-1)\phi \end{Bmatrix}\right\} \tag{2.4-11}$$

$$E_y = C[H_\nu^{(1)}(\rho r)+DH_\nu^{(2)}(\rho r)]\begin{Bmatrix} \cos\nu\phi \\ \sin\nu\phi \end{Bmatrix} \tag{2.4-12}$$

$$H_x = -nC\frac{\beta}{|\beta|}\left(\frac{\varepsilon_0}{\mu_0}\right)^{1/2}[H_\nu^{(1)}(\rho r)+DH_\nu^{(2)}(\rho r]\begin{Bmatrix} \cos\nu\phi \\ \sin\nu\phi \end{Bmatrix} \tag{2.4-13}$$

where $H_\nu^{(1)}$ and $H_\nu^{(2)}$ are the Hankel functions of the first and second kind. The parameter ρ is defined by Eq. (2.4-3). The amplitude coefficients C and D are determined from the requirement that the E_z and E_ϕ components pass continuously through the core boundary. The E_ϕ component is obtained from Eqs. (2.4-7) and (2.4-12) with the help of Eq. (2.2-37) using $E_x = 0$. Continuity of the H_z and H_ϕ components is approximately satisfied because of our assumption that $n_1 \approx n_2 = n$. The boundary conditions and use of the functional relations for cylinder functions lead to the results

$$C = (i\pi a/4)[\sigma J_{\nu+1}(\sigma a)\, H_\nu^{(2)}(\rho a) - \rho J_\nu(\sigma a)\, H_{\nu+1}^{(2)}(\rho a)]\, B \tag{2.4-14}$$

and

$$D = -\frac{\sigma J_{\nu+1}(\sigma a)\, H_\nu^{(1)}(\rho a) - \rho J_\nu(\sigma a)\, H_{\nu+1}^{(1)}(\rho a)}{\sigma J_{\nu+1}(\sigma a)\, H_\nu^{(2)}(\rho a) - \rho J_\nu(\sigma a)\, H_{\nu+1}^{(2)}(\rho a)} \tag{2.4-15}$$

These equations were obtained by comparing the coefficients of the $\cos(\nu+1)\phi$ and $\sin(\nu+1)\phi$ terms. Comparison of the coefficients of the circular functions with $(\nu-1)\phi$ leads to the same expressions. In the case of Eq. (2.4-15) this is immediately obvious. Instead of $\nu+1$ we would have $\nu-1$ in the equations. The equivalence of these two forms is established with the help of Eq. (2.2-40). In Eq. (2.4-14), replacing $J_{\nu-1}$ and $H_{\nu-1}^{(2)}$ [which we would get from the comparison of the $(\nu-1)\phi$ coefficients] with the help of Eq. (2.2-40) seems to change the sign of the expression. However, we need to know that the relation

$$\rho H_\nu^{(2)}(\rho a)\, H_{\nu+1}^{(1)}(\rho a) - \rho H_\nu^{(1)}(\rho a)\, H_{\nu+1}^{(2)}(\rho a) = 4/i\pi a \tag{2.4-16}$$

was used to obtain Eq. (2.4-14). When we derive the corresponding expression with $\nu+1$ replaced by $\nu-1$, Eq. (2.4-16) causes a sign change of the whole expression which cancels the sign change that results from application of Eq. (2.2-40) to the numerator. Once again comparisons of the coefficients of the two types of circular functions result in the same expressions.

Finally we use the orthogonality relation [compare Eq. (1.4-17)]

$$\frac{1}{2}\int_0^{2\pi}\int_0^{\infty} [\mathbf{E}(\rho)\times\mathbf{H}^*(\rho')]\cdot\mathbf{e}_z r\,dr\,d\phi = s_\rho(\beta^*/|\beta|)\,P\delta(\rho-\rho') \tag{2.4-17}$$

to relate the remaining amplitude coefficient B to P:

$$B = \frac{(\mu_0/\varepsilon_0)^{1/4}(8\rho P)^{1/2}}{(e_\nu n)^{1/2} a\pi^{3/2}\,|\sigma J_{\nu-1}(\sigma a)\, H_\nu^{(1)}(\rho a) - \rho J_\nu(\sigma a)\, H_{\nu-1}^{(1)}(\rho a)|} \tag{2.4-18}$$

where e_ν is defined by Eq. (2.2-42a).

The radiation modes with $|\beta| \approx n_2 k$ but with the $\mathbf{E}$ vector polarized predominantly in x direction have the form

$$E_z = \frac{iB\sigma}{2\beta}\left[J_{\nu+1}(\sigma r)\left\{\begin{matrix}\cos(\nu+1)\phi\\ \sin(\nu+1)\phi\end{matrix}\right\} - J_{\nu-1}(\sigma r)\left\{\begin{matrix}\cos(\nu-1)\phi\\ \sin(\nu-1)\phi\end{matrix}\right\}\right] \tag{2.4-19}$$

$$H_z = \frac{iB\sigma}{2k}\left(\frac{\varepsilon_0}{\mu_0}\right)^{1/2} \times\left[J_{\nu+1}(\sigma r)\left\{\begin{matrix}\sin(\nu+1)\phi\\ -\cos(\nu+1)\phi\end{matrix}\right\} + J_{\nu-1}(\sigma r)\left\{\begin{matrix}\sin(\nu-1)\phi\\ -\cos(\nu-1)\phi\end{matrix}\right\}\right] \tag{2.4-20}$$

$$E_x = BJ_\nu(\sigma r)\left\{\begin{matrix}\cos\nu\phi\\ \sin\nu\phi\end{matrix}\right\} \tag{2.4-21}$$

$$H_y = nB\frac{\beta}{|\beta|}\left(\frac{\varepsilon_0}{\mu_0}\right)^{1/2} J_\nu(\sigma r)\left\{\begin{matrix}\cos\nu\phi\\ \sin\nu\phi\end{matrix}\right\} \tag{2.4-22}$$

with $E_y = 0$ and $H_x \approx 0$. These expressions hold at $r < a$. The corresponding fields in the cladding at $r > a$ are obtained from the field in the core by replacing σ with ρ and the Bessel functions with combinations of Hankel functions of the first and second kind. We do not show these fields in detail, but comparison of the core and cladding fields Eqs. (2.4-5)–(2.4-13) with the above field expressions for the core field of the opposite polarization should make it clear how these cladding fields are formed. Expressions (2.4-14)–(2.4-18) also hold for the fields with the orthogonal polarization. The two types of circular functions that are indicated in all these field expressions belong to the two possible choices and are necessary to obtain a complete set of functions. The upper or lower sets of functions belong together.

The transverse field components of these radiation modes were obtained from the E_z and H_z components with the help of Eqs. (1.7-4)–(1.7-7). This derivation requires us to use the approximate expression $|\beta| \approx nk$ for the propagation constant. As indicated earlier, this procedure is allowed only for those radiation modes whose propagation constants are close to the ends of the interval (2.4-1). Deeper inside this interval we must use different approximations. Since all of our field approximations applied only to the case $n_1 - n_2 \ll 1$, it is logical to use the modes of the homogeneous medium ("free space modes") to describe radiation modes with β values that are substantially different from $n_2 k$. Of the infinite choice of forms for these "free space modes" we use a form that resembles the field expressions of the true radiation modes of the fiber that we just discussed. We thus choose

$$E_z = \frac{iF\rho}{2\beta}\left[J_{\nu+1}(\rho r)\begin{Bmatrix} \sin(\nu+1)\phi \\ -\cos(\nu+1)\phi \end{Bmatrix} + J_{\nu-1}(\rho r)\begin{Bmatrix} \sin(\nu-1)\phi \\ -\cos(\nu-1)\phi \end{Bmatrix}\right] \tag{2.4-23}$$

$$H_z = -\frac{iF\rho}{2k}\left(\frac{\varepsilon_0}{\mu_0}\right)^{1/2}\left[J_{\nu+1}(\rho r)\begin{Bmatrix} \cos(\nu+1)\phi \\ \sin(\nu+1)\phi \end{Bmatrix} - J_{\nu-1}(\rho r)\begin{Bmatrix} \cos(\nu-1)\phi \\ \sin(\nu-1)\phi \end{Bmatrix}\right] \tag{2.4-24}$$

$$E_y = FJ_\nu(\rho r)\begin{Bmatrix} \cos\nu\phi \\ \sin\nu\phi \end{Bmatrix} \tag{2.4-25}$$

$$\begin{aligned} H_x = -\frac{F}{4}\left(\frac{\varepsilon_0}{\mu_0}\right)^{1/2}\Bigg\{ & 2\left(\frac{\beta}{k}+\frac{n^2 k}{\beta}\right)J_\nu(\rho r)\begin{Bmatrix} \cos\nu\phi \\ \sin\nu\phi \end{Bmatrix} \\ & + \left(\frac{n^2 k}{\beta}-\frac{\beta}{k}\right)\left[J_{\nu+2}(\rho r)\begin{Bmatrix} \cos(\nu+2)\phi \\ \sin(\nu+2)\phi \end{Bmatrix} + J_{\nu-2}(\rho r)\begin{Bmatrix} \cos(\nu-2)\phi \\ \sin(\nu-2)\phi \end{Bmatrix}\right]\Bigg\} \end{aligned} \tag{2.4-26}$$

$$\begin{aligned} H_y = & \frac{F}{4}\left(\frac{\varepsilon_0}{\mu_0}\right)^{1/2}\left(\frac{\beta}{k}-\frac{n^2 k}{\beta}\right) \\ & \times\left[J_{\nu+2}(\rho r)\begin{Bmatrix} \sin(\nu+2)\phi \\ -\cos(\nu+2)\phi \end{Bmatrix} - J_{\nu-2}(\rho r)\begin{Bmatrix} \sin(\nu-2)\phi \\ -\cos(\nu-2)\phi \end{Bmatrix}\right] \end{aligned} \tag{2.4-27}$$

The E_x component vanishes precisely. These fields hold throughout all of space. The modes with the orthogonal polarization are

$$E_z = \frac{iF\rho}{2\beta}\left[J_{\nu+1}(\rho r)\begin{Bmatrix}\cos(\nu+1)\phi \\ \sin(\nu+1)\phi\end{Bmatrix} - J_{\nu-1}(\rho r)\begin{Bmatrix}\cos(\nu-1)\phi \\ \sin(\nu-1)\phi\end{Bmatrix}\right] \quad (2.4\text{-}28)$$

$$H_z = \frac{iF\rho}{2k}\left(\frac{\varepsilon_0}{\mu_0}\right)^{1/2} \times\left[J_{\nu+1}(\rho r)\begin{Bmatrix}\sin(\nu+1)\phi \\ -\cos(\nu+1)\phi\end{Bmatrix} + J_{\nu-1}(\rho r)\begin{Bmatrix}\sin(\nu-1)\phi \\ -\cos(\nu-1)\phi\end{Bmatrix}\right] \quad (2.4\text{-}29)$$

$$E_x = FJ_\nu(\rho r)\begin{Bmatrix}\cos\nu\phi \\ \sin\nu\phi\end{Bmatrix} \quad (2.4\text{-}30)$$

$$H_x = -\frac{F}{4}\left(\frac{\varepsilon_0}{\mu_0}\right)^{1/2}\left(\frac{\beta}{k} - \frac{n^2k}{\beta}\right) \times\left[J_{\nu+2}(\rho r)\begin{Bmatrix}\sin(\nu+2)\phi \\ -\cos(\nu+2)\phi\end{Bmatrix} - J_{\nu-2}(\rho r)\begin{Bmatrix}\sin(\nu-2)\phi \\ -\cos(\nu-2)\phi\end{Bmatrix}\right] \quad (2.4\text{-}31)$$

$$H_y = \frac{F}{4}\left(\frac{\varepsilon_0}{\mu_0}\right)^{1/2}\left\{2\left(\frac{\beta}{k} + \frac{n^2k}{\beta}\right)J_\nu(\rho r)\begin{Bmatrix}\cos\nu\phi \\ \sin\nu\phi\end{Bmatrix} + \left(\frac{\beta}{k} - \frac{n^2k}{\beta}\right)\left[J_{\nu+2}(\rho r)\begin{Bmatrix}\cos(\nu+2)\phi \\ \sin(\nu+2)\phi\end{Bmatrix} + J_{\nu-2}(\rho r)\begin{Bmatrix}\cos(\nu-2) \\ \sin(\nu-2)\end{Bmatrix}\right]\right\} \quad (2.4\text{-}32)$$

It is now the E_y component that vanishes identically. The amplitude coefficient of both polarizations is obtained with the help of Eq. (2.4-17):

$$F = \left[\frac{4s_\rho(\mu_0/\varepsilon_0)^{1/2}k\rho\beta^{*2}P}{e_\nu\pi|\beta|(\beta^{*2}+n^2k^2)}\right]^{1/2} \quad (2.4\text{-}33)$$

with e_ν defined by Eq. (2.2-42a) and $s_\rho = +1$ or -1 so that $F^2 > 0$.

The radiation modes of homogeneous space of refractive index n, Eqs. (2.4-23)–(2.4-32), are exact. However, they are, of course, only approximate solutions of the waveguide problem with $n_1 \neq n_2$. These field solutions approach the radiation modes of the fiber, Eqs. (2.4-5)–(2.4-22), in the limit $|\beta| = nk$ and $n_1 \to n_2 = n$. In order to perform this transition, we need the relations

$$J_\nu(x)\,H_{\nu+1}^{(1)}(x) - J_{\nu+1}(x)\,H_\nu^{(1)}(x) = 2/i\pi x \quad (2.4\text{-}34)$$

and

$$J_\nu(x)\,H_{\nu+1}^{(2)}(x) - J_{\nu+1}(x)\,H_\nu^{(2)}(x) = -2/i\pi x \quad (2.4\text{-}35)$$

2.5 Cutoff and Total Internal Reflection

In Sect. 1.3 we discussed the cutoff condition for modes of the slab waveguide and pointed out its connection with the critical angle of total internal reflection. In fact, the cutoff condition for slab waveguide modes can be derived from the critical angle at which total internal reflection is lost at a plane dielectric interface. The mode fields in round optical fibers can locally be expressed as quasi-plane waves. Extending our knowledge of the cutoff condition in slab waveguides, we may be tempted to derive the cutoff condition for fiber modes by applying the idea of the critical angle of total internal reflection to the quasi-plane waves of the fiber modes. However, this procedure leads to the wrong results. For most modes, cutoff in optical fibers is not related to the loss of total internal reflection. Instead we show in this section that cutoff in optical fibers is related to the phenomenon of radiation in bent dielectric waveguides. The two explanations of cutoff—loss of total internal reflection in slabs and radiation caused by bent dielectric interfaces—do not contradict each other. The critical angle of total internal reflection is a property of plane dielectric interfaces. It is not surprising that other phenomena take over if the plane interface is being curved. The feeling of surprise is associated with our habit of thinking in terms of geometrical optics, of which total internal reflection and its critical angle are integral parts.

We base our discussion on the form (2.2-21) of the guided mode field inside the fiber core. The Bessel function is expressed by its approximation for large argument and any order number $\nu < \kappa r$ [GR1]:

$$J_\nu(\kappa r) \approx \frac{(e^{i\psi}+e^{-i\psi})}{(2\pi)^{1/2}[(\kappa r)^2-\nu^2]^{1/4}} \tag{2.5-1}$$

with

$$\psi = [(\kappa r)^2-\nu^2]^{1/2} - \nu \arccos(\nu/\kappa r) - (\pi/4) \tag{2.5-1a}$$

We express the argument ψ of the exponential functions by the first two terms of a Taylor series expansion in the vicinity of the point $r = a$:

$$\psi(r) \approx \psi(a) + [\kappa^2-(\nu/a)^2]^{1/2} r + \text{const.} \tag{2.5-2}$$

Reinstating the omitted factor (2.2-4), we approximate Eq. (2.2-21) with the help of Eq. (2.5-1):

$$\begin{aligned} E_y &= A J_\nu(\kappa r) \cos \nu\phi \, e^{i(\omega t-\beta z)} \\ &\approx \frac{1}{2} \frac{(e^{i\psi}+e^{-i\psi})(e^{i\nu\phi}+e^{-i\nu\phi})\, e^{-i\beta z} e^{i\omega t}}{(2\pi)^{1/2}[(\kappa r)^2-\nu^2]^{1/4}} \end{aligned} \tag{2.5-3}$$

This approximation shows that the mode field consists of the superposition of four plane waves. Instead of traveling parallel to the waveguide axis, these quasi-plane waves spiral around the axis. The rays associated with these waves are called skew rays. The four quasi-plane waves belong to two skew rays following the path of a right- and left-handed screw. Each ray, in turn, is decomposed into a component approaching and receding from the core boundary. For our purposes it is sufficient to consider one of the two skew rays approaching the dielectric interface. The corresponding quasi-plane wave follows from Eqs. (2.5-2) and (2.5-3):

$$\exp\{-i\mathbf{K}_c \cdot \mathbf{r}\} = \exp\{-i\{[\kappa^2-(\nu/a)^2]^{1/2} r + (\nu/a)(a\phi) + \beta z\}\} \tag{2.5-4}$$

With the unit vectors $\mathbf{e}_r$, $\mathbf{e}_\phi$, and $\mathbf{e}_z$ in r, ϕ, and z directions, we can express the vector $\mathbf{r}$ as

$$\mathbf{r} = r\mathbf{e}_r + (a\phi)\,\mathbf{e}_\phi + z\mathbf{e}_z \tag{2.5-5}$$

and write the propagation vector of the quasi-plane wave in the form

$$\mathbf{K}_c = [\kappa^2-(\nu/a)^2]^{1/2}\,\mathbf{e}_r + (\nu/a)\,\mathbf{e}_\phi + \beta\mathbf{e}_z \tag{2.5-6}$$

The cosine of the angle between the propagation vector $\mathbf{K}_c$ and the direction normal to the core boundary is given by [Eq. (2.2-10) is used to simplify the denominator]

$$\cos\alpha = (1/|\mathbf{K}_c|)\,\mathbf{K}_c \cdot \mathbf{e}_r = [\kappa^2-(\nu/a)^2]^{1/2}/n_1 k \tag{2.5-7}$$

The sine of this angle is correspondingly

$$\sin\alpha = (1-\cos^2\alpha)^{1/2} = [(n_1 k)^2-\kappa^2+(\nu/a)^2]^{1/2}/n_1 k \tag{2.5-7a}$$

With the help of Eq. (2.2-10) we obtain

$$n_1 \sin\alpha = [\beta^2+(\nu/a)^2]^{1/2}/k \tag{2.5-8}$$

The cutoff condition for guided modes of the fiber is

$$\gamma = 0 \tag{2.5-9}$$

with γ defined by Eq. (2.2-27). Equation (2.5-9) leads to

$$\beta = n_2 k \tag{2.5-10}$$

That Eq. (2.5-9) or (2.5-10) is indeed the proper cutoff condition for guided fiber modes follows from the form of the mode fields outside the fiber core. Equations (2.2-23)–(2.2-26) show that the field outside the core is expressed by Hankel functions with the argument $i\gamma r$. For imaginary argument values, the Hankel function decays exponentially at infinite distance from the fiber core. However, as γ becomes imaginary the argument of the Hankel function becomes real. It now describes a radiating cylindrical wave at infinite distances from the fiber core. The mode is no longer guided when it loses power by radiation. It is thus clear that the mode description forces us to accept Eq. (2.5-9) or (2.5-10) as the cutoff condition.

Let us now look at the angle α right at cutoff. Substitution of Eq. (2.5-10) into Eq. (2.5-8) results in

$$n_1 \sin\alpha = n_2[1+(\nu/n_2 ka)^2]^{1/2} \tag{2.5-11}$$

Equation (2.5-11) leads to the inequality

$$n_1 \sin\alpha > n_2, \qquad \text{for} \quad \nu \neq 0 \tag{2.5-12}$$

By definition, the critical angle for total internal reflection follows from Eq. (1.2-1), with $\alpha_2 = 0$,

$$n_1 \sin\alpha_{1c} = n_2 \tag{2.5-13}$$

We are thus forced to the conclusion that, right at the cutoff point of the guided mode, its direction of propagation with respect to the nomal to the core–cladding interface obeys the inequality

$$\alpha > \alpha_{1c} \tag{2.5-14}$$

At the cutoff point of the mode field, the rays (associated with this field) impinge on the dielectric interface with an angle that is larger than the critical angle, so that from the point of view of ray optics total internal reflection is still taking place. Only if $\nu = 0$ do we have $\alpha = \alpha_{1c}$, so that cutoff coincides with the loss of total internal reflection.

In order to understand the physical reason for the cutoff in fibers, we take a look at curved slab waveguides. It is well known that lossless transmission through curved slab waveguides is impossible [Me1, Mi2]. Consider a guided mode traveling in a dielectric slab that is bent into a large hollow cylinder. The wave is traveling in a direction perpendicular to the axis of the cylinder. If the radius of curvature is large, the field near the curved slab is very similar to the field in the straight slab. However, far from the slab, in the direction away from the cylinder, the exponential field decay is changing into a radiation field. Since a guided mode is defined as a wave whose field tends to zero at infinity and does not lose power by radiation, we see that the wave in the curved slab waveguide is no longer a guided mode in the strict sense. In fact, in accordance with the discussion in Sect. 1.5 we may call it a leaky wave. Leaving the outer radius of the cylinder fixed but shrinking the inner radius results in a solid dielectric rod in the limit of vanishing inner radius. The slab waveguide mode is being transformed by this transition into a field configuration of the optical fiber. However, this field does not become a guided mode of the fiber but belongs to the leaky wave solutions. It is a wave beyond cutoff. The rays associated with this field impinge on the dielectric interface safely above the critical angle for total internal reflection. The loss of power is caused by the evanescent field outside the waveguide core. The wave traveling around the circumference of the cylinder drags its evanescent field along. At increasing distances from the fiber the field is forced to move faster. At a certain critical radius it would exceed the velocity of light in the medium outside the fiber core. At this point the field detaches itself and radiates away. This picture would suggest that any wave with $\nu \neq 0$ would suffer the same fate, since all waves of this type have a nonvanishing component of the propagation vector (2.5-6) in circumferential direction. In order to understand why some modes lose power by radiation while others are properly guided, we consider the field of the modes outside the fiber core. The Bessel function is now replaced by a Hankel function. Using the large argument approximation (2.2-62), we have the following expression for the field (2.2-25) outside the fiber core.

$$E_y = BH_\nu^{(1)}(i\gamma r)\cos\nu\phi\, e^{i(\omega t-\beta z)}$$

$$\approx \tfrac{1}{2}(2/\pi i\gamma r)^{1/2} B e^{i\psi} e^{-\gamma r}\left[e^{i(\nu\phi-\beta z)}+e^{-i(\nu\phi+\beta z)}\right]e^{i\omega t} \qquad (2.5\text{-}15)$$

The phase angle ψ assumes the constant value $\psi = (\nu+1/2)(\pi/2)$. For guided modes γ is real, so that there are now only two quasi-plane waves spiraling like right- and left-handed screws around the fiber. The propagation vector of one of these plane waves is given by

$$\mathbf{K} = (\nu/r)\,\mathbf{e}_\phi + \beta\mathbf{e}_z \qquad (2.5\text{-}16)$$

This vector does not have a radial component. Its circumferential component decreases with increasing distance from the fiber axis. At $r = \infty$ the quasi-plane waves move parallel to the waveguide axis. The mode field outside the fiber core has the tendency to lose its circumferential component of travel and to follow the axis of the fiber at large distances. Whether the field loses power by radiation or whether it can be a properly guided wave depends on the rate at which the circumferential velocity component decreases. We use the following expression for the velocity of the quasi-plane wave associated with the evanescent field outside the fiber:

$$v(r) = \omega/|\mathbf{K}| = kc/[(\nu/r)^2+\beta^2]^{1/2} \qquad (2.5\text{-}17)$$

Equation (2.5-17) represents the phase velocity (not the group velocity) of the quasi-plane wave. However, the explanation of the loss mechanism of bent dielectric waveguide fields also involves the phase velocity.

Since we have for guided waves

$$\beta > n_2 k \qquad (2.5\text{-}18)$$

we see that for all radii

$$v(r) < c/n_2 \tag{2.5-19}$$

This shows that guided waves manage to maintain the velocity of their fields outside the core below the velocity of light, c/n_2. If β becomes too small, the velocity of light is exceeded at some radius that depends on the value of ν. In particular, the wave (discussed above) with $\beta = 0$ travels in circumferential direction without an axial velocity component, so that from Eq. (2.5-17) we have

$$v(r) = rkc/\nu, \qquad \text{for} \quad \beta = 0 \tag{2.5-20}$$

At $r = \nu/(n_2 k)$ this wave exceeds the velocity of light and loses power by radiation. By maintaining inequality (2.5-18) the guided mode fields do not reach quasi-plane wave velocities in excess of the light velocity outside the waveguide core. However, Eq. (2.5-17) shows clearly that all guided modes reach the conditions

$$v(r) = c/n_2 \tag{2.5-21}$$

at $r = \infty$ at their cutoff points $\beta = n_2 k$. It is thus clear that cutoff means that the guided mode field is just beginning to move at the velocity of light at $r = \infty$. At cutoff the guided mode field detaches itself at infinite distances and radiates away. This discussion shows that even beyond cutoff the mode losses may not be very high in very thick optical fibers, since bent slab waveguides are known to have low radiation losses if their bending radius is large.

For modes with $\nu = 0$, cutoff can be explained in the same way as for slab waveguide modes. Expressions (2.5-6) and (2.5-16) for the propagation constants of the quasi-plane waves show that there are no circumferential components of wave propagation if $\nu = 0$. The field is thus not moving around the axis of the guide. The cutoff condition for $\nu = 0$ modes follows from the critical angle argument. However, we can explain the critical angle of total internal reflection also in terms of light velocity. A guided mode field in a slab (or in a fiber with $\nu = 0$) is accompanied by an evanescent wave outside the core region. As cutoff is approached the velocity of the evanescent field is increasing. The direction of the quasi-plane wave propagation vectors outside the core is parallel to the interface. No power is radiating away. However, as the critical angle (or cutoff) is reached, the evanescent field is just reaching the velocity of light in the outside medium. If the direction of the quasi-plane wave propagation vector would remain parallel to the surface, the phase velocity of the field would exceed the light velocity c/n_2. This "catastrophe" is avoided by a change of the direction of the propagation vector of the quasi-plane waves. They now acquire a component perpendicular to the interface so that power is radiated away from the surface as the mode goes beyond cutoff. In all cases it is the onset of power loss by radiation that determines cutoff. The cutoff phenomenon that is related to the critical angle of total internal reflection and the cutoff for fiber modes with $\nu \neq 0$ are distinguished by the location where the quasiplane waves would exceed the velocity of light if no radiation would occur. For slab waveguide modes and fiber modes with $\nu = 0$, there is no critical distance. Below and right at cutoff, the evanescent field is moving with uniform velocity everywhere. For modes with $\nu \neq 0$ the critical distance is located at $r = \infty$.

3

Coupled Mode Theory

3.1 Introduction

The guided modes that we studied in the preceding chapters are waveforms that can actually be excited. These waves propagate along the axis of the dielectric waveguide undisturbed provided that the waveguide structure is free of imperfections. If the dielectric materials of the guide are lossy, the propagation constant β becomes complex. However, for reasonable low loss the solutions presented in the last two chapters are still valid. The losses of the modes of weakly guiding fibers are equal to the plane wave loss in the dielectric material of the guide provided that core and cladding have the same losses.

Actual waveguides are never perfect. There are always index inhomogeneities or slight changes of the core width. These imperfections cause the modes of the waveguide to couple among each other. If we excite a pure mode at the beginning of the guide, some of its power is transferred to other guided modes and also to the radiation modes. Power transfer to other guided modes results in signal distortion since each guided mode travels at its own characteristic group velocity. Transfer of power to the radiation modes results in waveguide losses since the power is carried away from the core regions into the (infinite) cladding. Therefore, it is very important to know the amount of coupling that is caused by the different types of waveguide imperfections. Knowledge of the coupling coefficients makes it possible to determine the tolerance requirements if bounds on the allowed radiation losses or on guided mode coupling have been established. Mode coupling may even be a desirable effect. It is possible to design dielectric antennas by providing a dielectric waveguide with periodic irregularities that couple power from the guided mode to a narrow spectrum of radiation modes. This process can also be used to excite a guided mode by collecting radiated power [OCS]. For multimode operation of waveguides it is possible, at least in principle, to reduce the pulse dispersion resulting from the different group velocities of the modes by coupling all guided modes among each other [PK1]. The theory of mode coupling is thus of fundamental importance for the design and operation of dielectric waveguides [Me13].

The description of mode coupling is not unique. There are many different ways to obtain infinite coupled equation systems. Of the many possibilities we will discuss two in detail. The different possibilities arise from the different ways of describing the modes of the structure. Consider an imperfect slab waveguide with varying core width as shown in Fig. 3.1.1. The dashed lines indicate the ideal core boundary, while the actual core boundary is shown by the solid lines. We can now expand the electromagnetic field of the actual waveguide in terms of modes of the ideal waveguide corresponding to the dashed lines. The expansion coefficients are functions of z, and it is possible to formulate an infinite system of coupled differential equations for them.

There is an alternate way of obtaining an infinite coupled system of differential equations. Instead of the modes of the ideal waveguide, we now use so-called local normal modes [So1, Sr3]. Figure 3.1.1 shows two short solid lines that intersect the core boundary of the actual waveguide at a particular coordinate z. We can now express the field at point z in terms of modes of a hypothetical waveguide that is formed by the two short lines. The hypothetical waveguide, whose modes are now used for the field expansion, varies in width as a function of z. It coincides locally with the width of the actual waveguide.

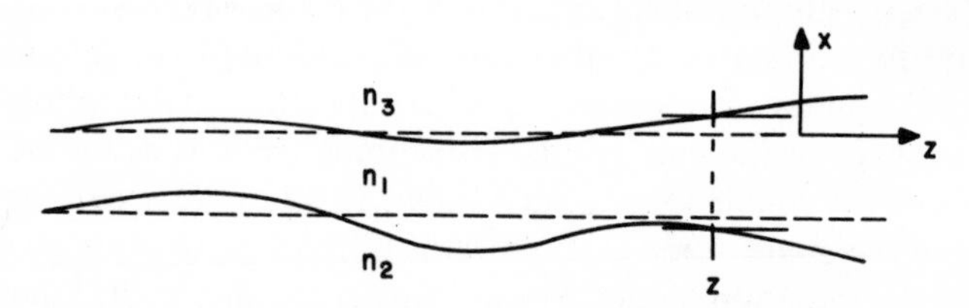

Fig. 3.1.1 *Dielectric waveguide with distorted core–cladding interface.*

The modes that are used for the field expansion are thus called local normal modes. These modes are not themselves solutions of Maxwell's equations since their parameters are functions of z. However, they can be superimposed to yield a solution of Maxwell's equations that represents the field of the actual waveguide that we want to describe. The actual procedure for deriving coupled wave equations with the help of ideal or local normal modes is the subject of this chapter.

Both types of mode expansions have their advantages for certain types of problems. The ideal mode expansion is particularly well suited for waveguides with perfect geometry but with refractive index inhomogeneities. The coupling coefficients for ideal modes have a simple general form. The coupling coefficients for local normal modes are somewhat more complicated, but this type of expansion is useful for the description of waveguides with imperfect geometry. Consider, for example, the waveguide taper shown in Fig. 3.1.2. If we want to describe the field of a guided wave, as it travels along the taper, in terms of ideal modes of the narrow guide, we find that the coupling coefficients are nonvanishing everywhere except in those regions where the actual guide and the ideal guide coincide. In particular, the ideal modes remain coupled on the wide portion of the straight waveguide that is connected to the tapered section. This description is correct, but it is certainly less advantageous than the local normal mode description which results in coupling coefficients that vanish everywhere except on the tapered portion connecting the two straight waveguide sections. Both descriptions are exact, but the local normal mode description is clearly preferable for tapers.

Local normal modes have an additional advantage for waveguides with changing core dimensions. We know that the tangential field components must remain continuous at the core–cladding boundary, while the normal electric field component is discontinuous at this surface. The local normal modes have field components that are either continuous or discontinuous at the same points in space as the total electromagnetic field. The only difficulty that arises consists in the fact that the tangential components of the local normal mode fields are not necessarily tangential to the actual core boundary. However, for core–cladding boundaries with gentle slopes this difference between the tangential direction of the actual core boundary and the coordinate directions (which are tangential to the ideal waveguide) is only slight. The expansion in terms of ideal waveguide modes encounters more serious

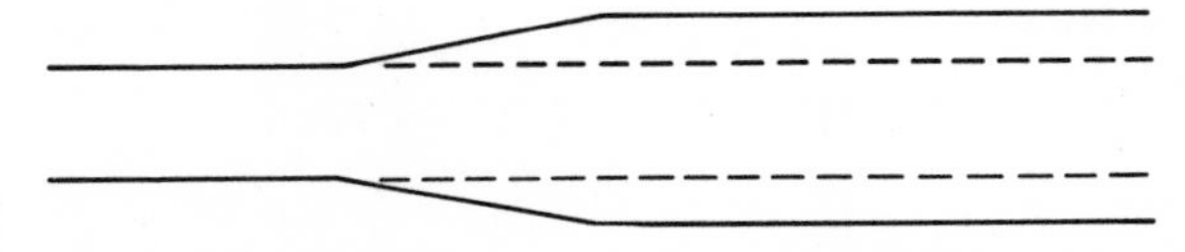

Fig. 3.1.2 *Dielectric waveguide with tapered core.*

difficulties. Since the core boundary of the actual guide does not usually coincide with the core boundary of the ideal waveguide, the terms of the series expansion are discontinuous at points where their superposition must result in a continuous field component, but they are continuous where their superposition must result in a discontinuous field component. This problem causes the coupling coefficients of the ideal mode expansion to be different from the coupling coefficients of the local normal mode expansion even for very slight deviations of the core boundary from its ideal shape. The difference between the two expansions and their coupling coefficients decreases for decreasing values of the index difference between core and cladding.

3.2 Expansion in Terms of Ideal Modes

The guided and radiation modes of dielectric waveguides form a complete orthogonal set. Orthogonality can be proven either by direct calculation or by the general method described in [Me1]. Completeness is harder to prove. We are assuming that our mode systems are complete without attempting to prove this assertion.

The task of this section consists in finding solutions of Maxwell's equations for an arbitrary refractive index distribution $n = n(x, y, z)$ [Sr2]. However, we assume that the magnetic properties of the material are those of vacuum:

$$\nabla \times \mathbf{H} = i\omega\varepsilon_0 n^2 \mathbf{E} \tag{3.2-1}$$

$$\nabla \times \mathbf{E} = -i\omega\mu_0 \mathbf{H} \tag{3.2-2}$$

It is possible to express the longitudinal components E_z and H_z in terms of the transverse field components. This is accomplished by separating the vectors into transverse and longitudinal parts designated by the subscripts t and z:

$$\mathbf{E} = \mathbf{E}_t + \mathbf{E}_z \tag{3.2-3}$$

$$\mathbf{H} = \mathbf{H}_t + \mathbf{H}_z \tag{3.2-4}$$

The operator ∇ can also be decomposed into its transverse and logitudinal components:

$$\nabla = \nabla_t + \mathbf{e}_z\, \partial/\partial z$$

The curl of the magnetic field vector can now be expressed by its transverse and logitudinal parts:

$$(\nabla \times \mathbf{H})_t = \nabla_t \times \mathbf{H}_z + \mathbf{e}_z \times \partial\mathbf{H}_t/\partial z \tag{3.2-5}$$

and

$$(\nabla \times \mathbf{H})_z = \nabla_t \times \mathbf{H}_t \tag{3.2-6}$$

Similar expressions hold, of course, also for the curl of the electric vector. We can use Eq. (3.2-6) and the corresponding equation for the electric field vector to write down the z components of Maxwell's equations. By regrouping the equations we find from Eqs. (3.2-1), (3.2-2), and (3.2-6)

$$\mathbf{E}_z = (1/i\omega\varepsilon_0 n^2)\nabla_t \times \mathbf{H}_t \tag{3.2-7}$$

and

$$\mathbf{H}_z = -(1/i\omega\mu_0)\nabla_t \times \mathbf{E}_t \tag{3.2-8}$$

These equations show that the longitudinal components E_z and H_z are obtained from the tangential field vectors by differentiation. We need thus only find solutions of $\mathbf{E}_t$ and $\mathbf{H}_t$ since the longitudinal components are known automatically once the transverse field components are known. The transverse parts of Maxwell's equations follow from Eqs. (3.2-1), (3.2-2), and (3.2-5) and a corresponding equation for the curl of $\mathbf{E}$. In order to eliminate all longitudinal field vectors from the equations, we replace $\mathbf{H}_z$ (and $\mathbf{E}_z$) in Eq. (3.2-5) with the help of Eqs. (3.2-7) and (3.2-8):

$$-(1/i\omega\mu_0)\nabla_t \times (\nabla_t \times \mathbf{E}_t) + (\mathbf{e}_z \times \partial\mathbf{H}_t/\partial z) = i\omega\varepsilon_0 n^2 \mathbf{E}_t \tag{3.2-9}$$

$$(1/i\omega\varepsilon_0)\nabla_t \times [(1/n^2)\nabla_t \times \mathbf{H}_t] + (\mathbf{e}_z \times \partial\mathbf{E}_t/\partial z) = -i\omega\mu_0 \mathbf{H}_t \tag{3.2-10}$$

The tangential Maxwell Eqs. (3.2-9) and (3.2-10) are general and hold for any nonmagnetic material of arbitrary refractive index distribution

$$n = n(x, y, z) \tag{3.2-11}$$

It is our purpose to express the general field solutions of Eqs. (3.2-9) and (3.2-10) in terms of modes of the ideal waveguide. Modes are characterized by the fact that their z dependence can be expressed as

$$e^{-i\beta z} \tag{3.2-12}$$

The ideal waveguide is defined by a refractive index distribution

$$n_0 = n_0(x, y) \tag{3.2-13}$$

that is independent of z. The modes must, of course, also satisfy Maxwell's equations with n replaced by n_0. We thus obtain from Eqs. (3.2-9) and (3.2-10) the following equations for the transverse electric and magnetic field vectors of the ideal normal modes:

$$-(1/i\omega\mu_0)\nabla_t \times (\nabla_t \times \mathscr{E}_{\nu t}) - i\beta_\nu(\mathbf{e}_z \times \mathscr{H}_{\nu t}) = i\omega\varepsilon_0 n_0^2 \mathscr{E}_{\nu t} \tag{3.2-14}$$

$$(1/i\omega\varepsilon_0)\nabla_t \times [(1/n_0^2)\nabla_t \times \mathscr{H}_{\nu t}] - i\beta_\nu(\mathbf{e}_z \times \mathscr{E}_{\nu t}) = -i\omega\mu_0 \mathscr{H}_{\nu t} \tag{3.2-15}$$

The longitudinal components follow from the transverse components by differentiation.

$$\mathscr{E}_{\nu z} = (1/i\omega\varepsilon_0 n_0^2)\nabla_t \times \mathscr{H}_{\nu t} \tag{3.2-16}$$

$$\mathscr{H}_{\nu z} = -(1/i\omega\mu_0)\nabla_t \times \mathscr{E}_{\nu t} \tag{3.2-17}$$

The solutions of these difficult looking equations have been found for several types of ideal waveguides in the preceding two chapters. The script letters indicate that the field vectors belong to normal modes of the waveguide. The subscript ν is used to label the modes. The single subscript is a shorthand notation for the total number of subscripts needed to label all the modes. This label also includes TE and TM modes of the slab waveguide or *EH* and *HE* modes of round optical fibers. Since we need to find solutions only for the transverse parts of the fields, we use series expansions of the transverse field vectors:

$$\mathbf{E}_t = \sum_{\nu=1}^{N} a_\nu \mathscr{E}_{\nu t} + \sum \int_0^\infty a_\rho \mathscr{E}_{\rho t}\, d\rho \tag{3.2-18}$$

$$\mathbf{H}_t = \sum_{\nu=1}^{N} b_\nu \mathscr{H}_{\nu t} + \sum \int_0^\infty b_\rho \mathscr{H}_{\rho t}\, d\rho \tag{3.2-19}$$

It should be noted that we have used a single amplitude coefficient for both the E_x and E_y components of the electric field and a different amplitude coefficient for both types of transverse magnetic field components. We shall see shortly that the amplitude coefficients for the electric and magnetic fields of the same mode are actually the same. However, the sign of backward traveling transverse magnetic field components is reversed. This makes it necessary to give the magnetic field a different coefficient from the outset. We could have used a field expansion that shows forward and backward traveling modes explicitly. In this case we could have used the same amplitude coefficient for the transverse electric and magnetic field component of each forward traveling mode and also for each backward traveling mode if we accommodated the sign change of the backward traveling transverse magnetic field component explicitly in the series expansion.

The reader may wonder why it is not necessary to give each field component its own amplitude coefficient. This seems particularly logical since we would have four different amplitude coefficients and also four equations—there are two scalar equations contained in Eq. (3.2-9) and two more in Eq. (3.2-10). The use of only one coefficient for all four transverse vector components of each forward and each backward traveling mode is justified by the fact that each mode must behave as an entity. All field components of each mode must change at the same rate since otherwise the mode would loose its identity.

The summation in Eqs. (3.2-18) and (3.2-19) extends over all guided modes, while the integral is extended over the entire range of the ρ parameter that is used to label radiation modes. The sum in front of the integral serves as a reminder that the contributions of the different types of radiation modes (TE and TM modes in the slab case) must be added. The fields and expansion coefficients of radiation modes actually need an additional label to distinguish between these different types of radiation modes. This extra label is omitted for simplicity of notation. In fact, we now simplify the notation even more by writing Eqs. (3.2-18) and (3.2-19) in the form

$$\mathbf{E}_t = \sum_\nu a_\nu \mathscr{E}_{\nu t} \tag{3.2-20}$$

$$\mathbf{H}_t = \sum_\nu b_\nu \mathscr{H}_{\nu t} \tag{3.2-21}$$

using the single summation symbol to indicate both the sum over guided modes and the sum and integral over radiation modes.

The expansion coefficients can be determined by substitution of the field expansions (3.2-20) and (3.2-21) into Eqs. (3.2-9) and (3.2-10). The question may be raised whether it is permissible to interchange summation (or integration) and differentiation since the series expansion expresses discontinuous field components in case of discontinuous index distributions. This difficulty can be avoided by assuming that all refractive index distributions are actually continuous even though the spatial variation may be quite rapid. The fields belonging to the "smoothed-out" and the actual discontinuous index distributions can be arbitrarily close to each other if the "smoothed-out" index distribution approaches the actual discontinuous distribution arbitrarily closely. We can even argue that discontinuous index distributions are idealizations that do not actually occur in nature.

We now substitute the field expansions (3.2-20) and (3.2-21) into the differential Eqs. (3.2-9) and (3.2-10). Use of mode Eqs. (3.2-14) and (3.2-15) allows the resulting equations to be expressed in this form:

$$\sum_\nu \{[(db_\nu/dz) + i\beta_\nu a_\nu](\mathbf{e}_z \times \mathscr{H}_{\nu t}) - i\omega\varepsilon_0(n^2 - n_0{}^2)\, a_\nu \mathscr{E}_{\nu t}\} = 0 \tag{3.2-22}$$

$$\sum_\nu \{[(da_\nu/dz) + i\beta_\nu b_\nu](\mathbf{e}_z \times \mathscr{E}_{\nu t}) + (1/i\omega\varepsilon_0)\, b_\nu \nabla_t \times [(n^{-2} - n_0^{-2})(\nabla_t \times \mathscr{H}_{\nu t})]\} = 0 \tag{3.2-23}$$

The ideal mode fields are solutions of Eqs. (3.2-14) and (3.2-15) and are independent of z. The propagation factor (3.2-12) is not included in the mode field expressions. We have followed the practice of omitting Eq. (3.2-12) from the mode equations of the preceding two chapters, but it was understood that this factor had to be added to the field expressions. The script fields used to denote the modes of the field expansion differ from this procedure in that the factor (3.2-12) is actually not included in the field expressions and not just omitted for convenience.

In order to proceed further with our derivation of a coupled equation system for the mode amplitudes, we need the orthogonality relation (1.4-17):

$$\int_{-\infty}^{\infty}\int_{-\infty}^{\infty} \mathbf{e}_z \cdot (\mathscr{E}_{\nu t} \times \mathscr{H}_{\mu t}^*)\, dx\, dy = 2s_\mu \beta_\mu{}^* P \delta_{\nu\mu}/|\beta_\mu| \qquad (3.2\text{-}24)$$

Note that the longitudinal parts of the vectors do not contribute to the left-hand side of this expression. The symbol $\delta_{\nu\mu}$ indicates Kronecker's delta expression for discrete indices ν and μ. It stands for the Dirac delta function if both ν and μ are labels of continuum modes and is zero if one of the labels indicates a guided mode while the other indicates a mode of the continuum. The factor s_μ is always $s_\mu = 1$ for guided as well as radiation modes with real propagation constants. In this case, the asterisk indicating complex conjugation has no effect. However, for evanescent continuum modes with imaginary values of the propagation constant β_μ, we may need $s_\mu = -1$ in order to keep P positive. However, for imaginary β_μ both sides of Eq. (3.2-24) become imaginary. The proper choice of sign for s_μ follows from expressions like Eqs. (1.4-36), (1.4-41), and (2.4-33). The coefficients s_μ must be made either $+1$ or -1 to keep the square of these amplitude coefficients positive. Since it is seldom necessary to consider coupling to the evanescent modes of the continuous spectrum, the need to choose a sign other than the positive one for $s_\mu = \pm 1$ rarely arises. The propagation constant β_μ is also inherently a positive quantity so that for real values of β_μ we have $\beta_\mu{}^*/|\beta_\mu| = +1$.

We take the scalar product of Eq. (3.2-22) with $\mathscr{E}_{\mu t}^*$ and Eq. (3.2-23) with $\mathscr{H}_{\mu t}^*$. After integration over the infinite cross section, we obtain with the help of the orthogonality relation Eq. (3.2-24)

$$(db_\mu/dz) + i\beta_\mu a_\mu = 2\sum_\nu \overline{K}_{\mu\nu} a_\nu \qquad (3.2\text{-}25)$$

$$(da_\mu/dz) + i\beta_\mu b_\mu = 2\sum_\nu \overline{k}_{\mu\nu} b_\nu \qquad (3.2\text{-}26)$$

with

$$\overline{K}_{\mu\nu} = \frac{\omega\varepsilon_0}{4is_\mu P}\frac{|\beta_\mu|}{\beta_\mu}\int_{-\infty}^{\infty}\int_{-\infty}^{\infty}(n^2 - n_0{}^2)\mathscr{E}_{\mu t}^* \cdot \mathscr{E}_{\nu t}\, dx\, dy \qquad (3.2\text{-}27)$$

$$\overline{k}_{\mu\nu} = \frac{-1}{4s_\mu i\omega\varepsilon_0 P}\frac{|\beta_\mu|}{\beta_\mu{}^*}\int_{-\infty}^{\infty}\int_{-\infty}^{\infty}\mathscr{H}_{\mu t}^* \cdot \nabla_t \times \left[\left(\frac{1}{n^2} - \frac{1}{n_0{}^2}\right)\nabla_t \times \mathscr{H}_{\nu t}\right] dx\, dy \qquad (3.2\text{-}28)$$

The field expansion method has thus yielded two sets of infinite integro-differential equations (the summation symbol indicates summation and integration) for the amplitude coefficients a_μ and b_μ. The coupling coefficient

(3.2-28) appears in an awkward form. In order to simplify it we substitute Eq. (3.2-16) into Eq. (3.2-28):

$$\bar{k}_{\mu\nu} = (1/4s_\mu P)(|\beta_\mu|/\beta_\mu{}^*)\int_{-\infty}^{\infty}\int_{-\infty}^{\infty} \mathscr{H}_{\mu t}^* \cdot \nabla_t \times [(1/n^2)(n^2 - n_0{}^2)\mathscr{E}_{\nu z}]\, dx\, dy \tag{3.2-29}$$

It is not hard to see that we can further transform this expression into the form

$$\bar{k}_{\mu\nu} = (1/4s_\mu P)(|\beta_\mu|/\beta_\mu{}^*)\int_{-\infty}^{\infty}\int_{-\infty}^{\infty} (1/n^2)(n^2 - n_0{}^2)\mathscr{E}_{\nu z} \cdot (\nabla_t \times \mathscr{H}_{\mu t}^*)\, dx\, dy \tag{3.2-30}$$

The validity of this transformation can be checked most easily by writing the vector expression out in x, y components and performing a partial integration on each term. The additional terms resulting from the partial integration vanish directly if one of the two modes is a guided mode. If both modes are radiation modes the remaining term vanishes in the sense of a delta function with non-zero argument. By using Eq. (3.2-16) once more we finally obtain

$$\bar{k}_{\mu\nu} = (\omega\varepsilon_0/4is_\mu P)(|\beta_\mu|/\beta_\mu{}^*)\int_{-\infty}^{\infty}\int_{-\infty}^{\infty} (n_0{}^2/n^2)(n^2 - n_0{}^2)\mathscr{E}_{\mu z}^* \cdot \mathscr{E}_{\nu z}\, dx\, dy \tag{3.2-31}$$

This expression is now of essentially the same form as Eq. (3.2-27) except that the longitudinal instead of the transverse parts of the electric vector are being used.

Before proceeding further it is useful to consider the coupled equation system, Eqs. (3.2-25) and (3.2-26), in the absence of coupling, $\bar{K}_{\mu\nu} = 0$ and $\bar{k}_{\mu\nu} = 0$. Taking the z derivative of Eq. (3.2-26) and substituting the result into Eq. (3.2-25) leads to the expression

$$(d^2 a_\mu/dz^2) + \beta_\mu{}^2 a_\mu = 0 \tag{3.2-32}$$

Similarly by differentiating Eq. (3.2-25) and substituting into Eq. (3.2-26), we obtain

$$(d^2 b_\mu/dz^2) + \beta_\mu{}^2 b_\mu = 0 \tag{3.2-33}$$

These equations have solutions of the form

$$a_\mu = a_\mu^{(+)} = c_\mu^{(+)} \exp(-i\beta_\mu z) \tag{3.2-34}$$

and

$$a_\mu = a_\mu^{(-)} = c_\mu^{(-)} \exp(i\beta_\mu z) \tag{3.2-35}$$

The first of these solutions represents a wave traveling in positive z direction (the time dependence is expressed by $e^{i\omega t}$), and the second is a wave traveling in negative z direction. The corresponding solution for b_μ follows from Eq. (3.2-26):

$$b_\mu = c_\mu^{(+)} \exp(-i\beta_\mu z) = a_\mu^{(+)} \tag{3.2-36}$$

and

$$b_\mu = -c_\mu^{(-)} \exp(i\beta_\mu z) = -a_\mu^{(-)} \tag{3.2-37}$$

These forms of the solutions—in the absence of coupling—suggest that we try to separate the variables in Eqs. (3.2-25) and (3.2-26) by introducing the transformation

$$a_\mu = a_\mu^{(+)} + a_\mu^{(-)} \tag{3.2-38}$$

and

$$b_\mu = a_\mu^{(+)} - a_\mu^{(-)} \tag{3.2-39}$$

Substitution of these two expressions into Eqs. (3.2-25) and (3.2-26) results in

$$(da_\mu^{(+)}/dz) + i\beta_\mu a_\mu^{(+)} - [(da_\mu^{(-)}/dz) - i\beta_\mu a_\mu^{(-)}] = 2\sum_\nu \overline{K}_{\mu\nu}(a_\nu^{(+)} + a_\nu^{(-)}) \tag{3.2-40}$$

$$(da_\mu^{(+)}/dz) + i\beta_\mu a_\mu^{(+)} + [(da_\mu^{(-)}/dz) - i\beta_\mu a_\mu^{(-)}] = 2\sum_\nu \overline{k}_{\mu\nu}(a_\nu^{(+)} - a_\nu^{(-)}) \tag{3.2-41}$$

Addition and subtraction of these two equations finally yields the following system of coupled wave equations [Mr2]:

$$da_\mu^{(+)}/dz = -i\beta_\mu a_\mu^{(+)} + \sum_\nu (K_{\mu\nu}^{(+,+)} a_\nu^{(+)} + K_{\mu\nu}^{(+,-)} a_\nu^{(-)}) \tag{3.2-42}$$

$$da_\mu^{(-)}/dz = i\beta_\mu a_\mu^{(-)} + \sum_\nu (K_{\mu\nu}^{(-,+)} a_\nu^{(+)} + K_{\mu\nu}^{(-,-)} a_\nu^{(-)}) \tag{3.2-43}$$

with the coupling coefficients

$$\begin{aligned} K_{\mu\nu}^{(p,q)} &= p\overline{K}_{\mu\nu} + q\overline{k}_{\mu\nu} \\ &= (\omega\varepsilon_0/4is_\mu P)\int_{-\infty}^{\infty}\int_{-\infty}^{\infty}(n^2 - n_0{}^2) \\ &\quad \times [(|\beta_\mu|/\beta_\mu^{(p)})\mathscr{E}_{\mu t}^{(p)*} \cdot \mathscr{E}_{\nu t}^{(q)} + (n_0{}^2/n^2)(|\beta_\mu|/\beta_\mu^{(p)*})\mathscr{E}_{\mu z}^{(p)*} \cdot \mathscr{E}_{\nu z}^{(q)}]\, dx\, dy \end{aligned} \tag{3.2-44}$$

The sign of $s_\mu = \pm 1$ is discussed in connection with Eq. (3.2-24). p and q are $(+)$ or $(-)$ if they appear as superscripts or $+1$ and -1 as factors. We use the definitions

$$\beta_\mu^{(+)} = \beta_\mu \tag{3.2-45}$$

$$\beta_\mu^{(-)} = -\beta_\mu \tag{3.2-46}$$

$$\mathscr{E}_{\mu t}^{(-)} = \mathscr{E}_{\mu t}^{(+)} = \mathscr{E}_{\mu t} \tag{3.2-47}$$

and

$$\mathscr{E}_{\mu z}^{(-)} = -\mathscr{E}_{\mu z}^{(+)} = -\mathscr{E}_{\mu z} \tag{3.2-48}$$

The amplitude coefficients $a_\mu^{(+)}$ and $a_\mu^{(-)}$ are rapidly varying functions of z. This can be seen from the solutions (3.2-34) and (3.2-35) in the absence of coupling. For perturbation solutions of the coupled wave equations it is more convenient to introduce slowly varying mode amplitudes. In the absence of coupling, the amplitude coefficients $c_\mu^{(+)}$ and $c_\mu^{(-)}$ in Eqs. (3.2-34) and (3.2-35) are actually constant. For the general case of coupled modes we let $c_\mu^{(+)}$ and $c_\mu^{(-)}$ be functions of z and obtain by substitution of Eqs. (3.2-34) and (3.2-35) into Eqs. (3.2-42) and (3.2-43) the system of coupled equations

$$dc_\mu^{(+)}/dz = \sum_\nu \{K_{\mu\nu}^{(+,+)} c_\nu^{(+)} \exp[i(\beta_\mu - \beta_\nu)z] + K_{\mu\nu}^{(+,-)} c_\nu^{(-)} \exp[i(\beta_\mu + \beta_\nu)z]\} \tag{3.2-49}$$

$$dc_\mu^{(-)}/dz = \sum_\nu \{K_{\mu\nu}^{(-,+)} c_\nu^{(+)} \exp[-i(\beta_\mu + \beta_\nu)z] + K_{\mu\nu}^{(-,-)} c_\nu^{(-)} \exp[-i(\beta_\mu - \beta_\nu)z]\} \tag{3.2-50}$$

We conclude this section with two remarks. The reader may have wondered why it was not necessary to use boundary conditions to obtain the general solutions (3.2-18) and (3.2-19) of Maxwell's equations. The explicit use of boundary conditions in the derivation of the mode field expressions in the preceding two chapters was necessary only since we did not have any other way to obtain solutions of Maxwell's equations for the discontinuous refractive index distribution $n_0 = n_0(x, y)$ which defines the dielectric waveguide. The normal mode expansion technique generates solutions of Maxwell's equations for arbitrary distributions $n = n(x, y, z)$ of the refractive index. Since the boundary conditions are themselves derived from Maxwell's equations, solutions of these equations must automatically satisfy the boundary conditions at dielectric discontinuities. Our construction of general solutions was based on Maxwell's equations and holds for any distribution of refractive index. It is thus not necessary to invoke boundary conditions separately.

The second remark concerns the appearance of backward traveling waves in our derivation. It is, of course, natural to expect backward traveling waves to appear in any general solution of a wave problem. However, there are many applications where backward traveling waves can be neglected. Furthermore we have seen that the longitudinal electric field components are often very much smaller than the transverse components, so that the right-hand side of Eq. (3.2-41) is very much smaller than the right-hand side of Eq. (3.2-40). Neglecting the longitudinal field components for the moment allows us to write Eqs. (3.2-40) and (3.2-41) in terms of the "slowly varying" field amplitudes as follows:

$$(dc_\mu^{(+)}/dz)\exp(-i\beta_\mu z) - (dc_\mu^{(-)}/dz)\exp(i\beta_\mu z) = 2\sum_\nu K_{\mu\nu}[c_\nu^{(+)}\exp(-i\beta_\nu z) + c_\nu^{(-)}\exp(i\beta_\nu z)] \tag{3.2-51}$$

and

$$(dc_\mu^{(+)}/dz)\exp(-i\beta_\mu z) + (dc_\mu^{(-)}/dz)\exp(i\beta_\mu z) = 0 \qquad (3.2\text{-}52)$$

If we now apply this equation system to a waveguide that provides coupling between the forward modes but does not cause appreciable power to flow in backward modes, we may be tempted to set $c_\mu^{(-)} = 0$. This, however, leads immediately to a contradiction, since the equation system now states that the z derivative of $c_\mu^{(+)}$ must be equal to the nonvanishing right-hand side in Eq. (3.2-51) but must vanish in Eq. (3.2-52). It thus appears as though our solution of the problem is not internally consistent. The fallacy in this argument consists in neglecting the derivative of the field amplitude $c_\mu^{(-)}$. Even if $c_\mu^{(-)}$ itself is actually negligible, its derivative is not. We see this immediately if we express $c_\mu^{(-)}$ in terms of $c_\mu^{(+)}$, with the help of Eq. (3.2-52),

$$c_\mu^{(-)}(z) = -\int_{z_1}^{z} (dc_\mu^{(+)}/du)\exp(-2i\beta_\mu u)\,du \qquad (3.2\text{-}53)$$

Equation (3.2-52) requires the derivative of $c_\mu^{(-)}$ to be of the same order of magnitude as the derivative of $c_\mu^{(+)}$. But according to Eq. (3.2-53) the field amplitude $c_\mu^{(-)}$ cannot build up to appreciable values if the derivative $c_\mu^{(+)}$ is a slowly varying function of z. The factor $\exp(-2i\beta_\mu z)$ causes cancellations by its rapid oscillation, so that the contribution of the integral does not result in a growing backward wave amplitude. However, it is not permissible to neglect the derivatives of $c_\mu^{(-)}$ even though the amplitude coefficient itself does not grow to appreciable values. A physical argument helps to understand the situation. Coupling at a particular point z along the waveguide feeds power at the same rate into the forward and backward waves. Whether this power builds up or remains insignificant depends on the phase relationship of the incremental amounts of power fed into the waves at other points. A wave builds up only if the incremental contributions at all points along the waveguide add up in phase.

3.3 Expansion in Terms of Local Normal Modes

It was explained in the introduction of this chapter that we can express an arbitrary field in the waveguide also in terms of modes belonging to a fictitious waveguide that coincides in width locally at the point z at which the field expansion is being considered. This means that instead of n_0 we now use

$$n_0 = n(x, y, z) \qquad (3.3\text{-}1)$$

in Eqs. (3.2-14)–(3.2-17). Since there is no z derivative in these mode equations, z appears simply as a parameter in the solutions. The mathematical form of the local normal mode field solutions is thus identical in form to the ideal normal mode expressions and can again be obtained from Chaps. 1 and 2. However, the waveguide dimensions appearing implicitly in these field expressions are now functions of z. The local normal modes are orthogonal among each other at each cross section z along the waveguide, even though they depend on z as a parameter. They also form a complete set of modes, but they are not themselves solutions of Maxwell's equations.

We use again the field expansions (3.2-18) and (3.2-19) or their short-hand versions Eqs. (3.2-20) and (3.2-21). When we substitute these field expansions into Eqs. (3.2-9) and (3.2-10), we must remember that local normal modes are functions of z contrary to ideal modes. Keeping this in mind and using the mode field expressions (3.2-14) and (3.2-15) with Eq. (3.3-1), we obtain

$$\sum_\nu \{[(db_\nu/dz)+i\beta_\nu a_\nu](\mathbf{e}_z \times \mathscr{H}_{\nu t}) + b_\nu(\mathbf{e}_z \times \partial\mathscr{H}_{\nu t}/\partial z)\} = 0 \qquad (3.3\text{-}2)$$

$$\sum_\nu \{[(da_\nu/dz)+i\beta_\nu b_\nu](\mathbf{e}_z \times \mathscr{E}_{\nu t}) + a_\nu(\mathbf{e}_z \times \partial\mathscr{E}_{\nu t}/\partial z)\} = 0 \qquad (3.3\text{-}3)$$

Scalar multiplication of these equations with the complex conjugate transverse field components and integration over the infinite cross section in the same manner as in the preceding section yields

$$(db_\mu/dz) + i\beta_\mu a_\mu = 2\sum_\nu \bar{R}_{\mu\nu} b_\nu \qquad (3.3\text{-}4)$$

$$(da_\mu/dz) + i\beta_\mu b_\mu = 2\sum_\nu \bar{S}_{\mu\nu} a_\nu \qquad (3.3\text{-}5)$$

with

$$\bar{R}_{\mu\nu} = -(1/4s_\mu P)(|\beta_\mu|/\beta_\mu)\int_{-\infty}^{\infty}\int_{-\infty}^{\infty} \mathbf{e}_z \cdot (\mathscr{E}_\mu^* \times \partial\mathscr{H}_\nu/\partial z)\,dx\,dy \qquad (3.3\text{-}6)$$

and

$$\bar{S}_{\mu\nu} = -(1/4s_\mu P)(|\beta_\mu|/\beta_\mu^*)\int_{-\infty}^{\infty}\int_{-\infty}^{\infty} \mathbf{e}_z \cdot [(\partial\mathscr{E}_\nu/\partial z) \times \mathscr{H}_\mu^*]\,dx\,dy \qquad (3.3\text{-}7)$$

The subscript t is not necessary to label the transverse field components under the integral signs since the longitudinal components do not contribute in any case. The transformation to slowly varying mode amplitudes follows the same procedure that was used in the last section. We thus obtain the following set of integrodifferential equations (the sum indicates summation as well as integration) for the amplitudes of local modes:

$$dc_\mu^{(+)}/dz = \sum_\nu \left\{R_{\mu\nu}^{(+,+)} c_\nu^{(+)} \exp\left[i\int_0^z (\beta_\mu - \beta_\nu)\,dz\right] + R_{\mu\nu}^{(+,-)} c_\nu^{(-)} \exp\left[i\int_0^z (\beta_\mu + \beta_\nu)\,dz\right]\right\} \qquad (3.3\text{-}8)$$

$$dc_\mu^{(-)}/dz = \sum_\nu \left\{ R_{\mu\nu}^{(-,+)} c_\nu^{(+)} \exp\left[-i \int_0^z (\beta_\mu + \beta_\nu)\, dz \right] \right.$$

$$\left. + R_{\mu\nu}^{(-,-)} c_\nu^{(-)} \exp\left[-i \int_0^z (\beta_\mu - \beta_\nu)\, dz \right] \right\} \tag{3.3-9}$$

with

$$R_{\mu\nu}^{(p,q)} = pq\bar{R}_{\mu\nu} + \bar{S}_{\mu\nu}$$

$$= -(1/4s_\mu P) \int_{-\infty}^{\infty} \int_{-\infty}^{\infty} \mathbf{e}_z \cdot \{(|\beta_\mu|/\beta_\mu^{(p)}) (\mathscr{E}_\mu^{(p)*} \times \partial \mathscr{H}_\nu^{(q)}/\partial z)$$

$$+ (|\beta_\mu|/\beta_\mu^{(p)*}) [(\partial \mathscr{E}_\nu^{(q)}/\partial z) \times \mathscr{H}_\mu^{(p)*}]\}\, dx\, dy \tag{3.3-10}$$

The integrals appearing in the exponents of the exponential functions are necessary since β_ν is a function of z. It was explained in connection with Eq. (3.2-24) that $s_\mu = \pm 1$. The superscripts and factors p and q have the same meaning as in the preceding section. Since only the transverse electric field components contribute to the mixed products under the integral sign in Eq. (3.3-10) the superscripts p and q could actually be omitted from the electric field vectors, since the sign of the forward and backward traveling transverse electric field is the same. The transverse magnetic field changes its sign on changing direction of propagation:

$$\mathscr{H}_{\mu t}^{(-)} = -\mathscr{H}_{\mu t}^{(+)} = -\mathscr{H}_{\mu t} \tag{3.3-11}$$

$$\mathscr{H}_{\mu z}^{(-)} = \mathscr{H}_{\mu z}^{(+)} = \mathscr{H}_{\mu z} \tag{3.3-12}$$

The sign of $\beta_\mu^{(p)}$ follows from Eqs. (3.2-45) and (3.2-46).

It is necessary to keep in mind that the summation symbol in Eqs. (3.3-8) and (3.3-9) indicates summation over guided and integration over continuum modes. The equation system (3.3-8) and (3.3-9) has the same form as the corresponding set of Eqs. (3.2-49) and (3.2-50) for the ideal modes. The transformations

$$c_\mu^{(+)} = a_\mu^{(+)} \exp\left[i \int_0^z \beta_\mu\, dz \right] \tag{3.3-13}$$

and

$$c_\mu^{(-)} = a_\mu^{(-)} \exp\left[-i \int_0^z \beta_\mu\, dz \right] \tag{3.3-14}$$

if substituted into Eqs. (3.3-8) and (3.3-9) result in the form (3.2-42) and (3.2-43) of the coupled set of integrodifferential equations for the rapidly varying mode amplitudes.

We have used the mode equations in the form (3.2-14) and (3.2-15), with n_0 given by Eq. (3.3-1) for local normal modes. This form of the equations is awkward for some applications. We thus introduce the longitudinal components again and obtain from Eqs. (3.2-14)–(3.2-17) with the use of Eq. (3.3-1) the following simpler form of the mode equations from which the local normal mode solutions are obtained:

$$\nabla_t \times \mathcal{H}_\mu - i\beta_\mu(\mathbf{e}_z \times \mathcal{H}_\mu) = i\omega n^2 \varepsilon_0 \mathcal{E}_\mu \tag{3.3-15}$$

and

$$\nabla_t \times \mathcal{E}_\mu - i\beta_\mu(\mathbf{e}_z \times \mathcal{E}_\mu) = -i\omega\mu_0 \mathcal{H}_\mu \tag{3.3-16}$$

The coupling coefficients (3.3-10) for the local normal mode field expansion are much more complicated in form than the coupling coefficients (3.2-44) for the ideal modes. However, it is possible to remove the derivatives of the field components from Eq. (3.3-10), at least for propagating modes with real values of the propagation constants. Since evanescent radiation modes with imaginary propagation constants are of little practical importance, this transformation is well worthwhile.

In order to transform expression (3.3-10) to a more useful form, we take the scalar product of the complex conjugate of Eq. (3.3-15) with

$$-(|\beta_\mu|/\beta_\mu^{(p)*})\, \partial\mathcal{E}_\nu^{(q)}/\partial z$$

and of the complex conjugate of Eq. (3.3-16) with

$$(|\beta_\mu|/\beta_\mu^{(p)})\, \partial\mathcal{H}_\nu^{(q)}/\partial z$$

and add the resulting equations. After regrouping of terms we obtain

$$\begin{aligned} i\beta_\mu^{(p)*}\{&(|\beta_\mu|/\beta_\mu^{(p)})(\mathcal{E}_\mu^{(p)*} \times \partial\mathcal{H}_\nu^{(q)}/\partial z) + (|\beta_\mu|/\beta_\mu^{(p)*})[(\partial\mathcal{E}_\nu^{(q)}/\partial z) \times \mathcal{H}_\mu^{(p)*}]\} \cdot \mathbf{e}_z \\ &= \nabla_t \cdot \{(|\beta_\mu|/\beta_\mu^{(p)*})(\overset{\downarrow}{\mathcal{H}}{}_\mu^{(p)*} \times \partial\mathcal{E}_\nu^{(q)}/\partial z) + (|\beta_\mu|/\beta_\mu^{(p)})[(\partial\mathcal{H}_\nu^{(q)}/\partial z) \times \overset{\downarrow}{\mathcal{E}}{}_\mu^{(p)*}]\} \\ &\quad + [in^2\omega\varepsilon_0(|\beta_\mu|/\beta_\mu^{(p)*})\mathcal{E}_\mu^{(p)*} \cdot \partial\mathcal{E}_\nu^{(q)}/\partial z] \\ &\quad + [i\omega\mu_0(|\beta_\mu|/\beta_\mu^{(p)})\mathcal{H}_\mu^{(p)*} \cdot \partial\mathcal{H}_\nu^{(q)}/\partial z] \end{aligned} \tag{3.3-17}$$

The vertical arrows pointing down on the field vectors indicate that these vectors are being differentiated by the ∇_t operator. The terms without arrows are understood to be unaffected by this operator. The left-hand side of Eq. (3.3-17) is of the desired form. On the right-hand side we would like to obtain a complete divergence expression with ∇_t operating on every term to its right. To achieve this type of expression we differentiate Eqs. (3.3-15) and (3.3-16) with respect to z and multiply Eq. (3.3-15) with

$$(|\beta_\mu|/\beta_\mu^{(p)})\,\mathcal{E}_\mu^{(p)*}$$

and Eq. (3.3-16) with

$$-(|\beta_\mu|/\beta_\mu^{(p)*})\,\mathcal{H}_\mu^{(p)*}$$

Addition of the resulting equations results in

$$-i\beta_\nu^{(q)}\{(|\beta_\mu|/\beta_\mu^{(p)})(\mathscr{E}_\mu^{(p)*}\times\partial\mathscr{H}_\nu^{(q)}/\partial z)+(|\beta_\mu|/\beta_\mu^{(p)*})[(\partial\mathscr{E}_\nu^{(q)}/\partial z)\times\mathscr{H}_\mu^{(p)*}]\}\cdot\mathbf{e}_z$$

$$=\nabla_t\cdot\{(|\beta_\mu|/\beta_\mu^{(p)*})(\mathscr{H}_\mu^{(p)*}\times\partial\overset{\downarrow}{\mathscr{E}}{}_\nu^{(q)}/\partial z)+(|\beta_\mu|/\beta_\mu^{(p)})[(\partial\overset{\downarrow}{\mathscr{H}}{}_\nu^{(q)}/\partial z)\times\mathscr{E}_\mu^{(p)*}]\}$$
$$+i(\partial\beta_\nu^{(q)}/\partial z)\{(|\beta_\mu|/\beta_\mu^{(p)})(\mathscr{E}_\mu^{(p)*}\times\mathscr{H}_\nu^{(q)})+(|\beta_\mu|/\beta_\mu^{(p)*})(\mathscr{E}_\nu^{(q)}\times\mathscr{H}_\mu^{(p)*})\}\cdot\mathbf{e}_z$$
$$-[i\omega\varepsilon_0 n^2(|\beta_\mu|/\beta_\mu^{(p)})\mathscr{E}_\mu^{(p)*}\cdot\partial\mathscr{E}_\nu^{(q)}/\partial z]$$
$$-[i\omega\mu_0(|\beta_\mu|/\beta_\mu^{(p)*})\mathscr{H}_\mu^{(p)*}\cdot\partial\mathscr{H}_\nu^{(q)}/\partial z]-i\omega\varepsilon_0(\partial n^2/\partial z)(|\beta_\mu|/\beta_\mu^{(p)})\mathscr{E}_\mu^{(p)*}\cdot\mathscr{E}_\nu^{(q)} \tag{3.3-18}$$

The left-hand side of this expression is again of the form of the integrand of Eq. (3.3-10). If we add Eqs. (3.3-17) and (3.3-18) we obtain a complete divergence expression, since the terms that are operated on by the ∇_t operator complement each other in each of the two expressions. After integration over the infinite xy plane the divergence can be converted to an integral over the circumference of the infinite circle that surrounds the area of integration in the xy plane. This line integral vanishes if at least one of the two modes μ or ν is a guided mode. If both modes are radiation modes the integral vanishes in the sense of a delta function with nonzero argument. Thus the divergence does not contribute to the coupling coefficient. The two terms next to the divergence terms in Eqs. (3.3-17) and (3.3-18) yield, after addition of the two expressions,

$$i\omega\left(\frac{|\beta_\mu|}{\beta_\mu^{(p)*}}-\frac{|\beta_\mu|}{\beta_\mu^{(p)}}\right)\left(n^2\varepsilon_0\mathscr{E}_\mu^{(p)*}\cdot\frac{\partial\mathscr{E}_\nu^{(q)}}{\partial z}-\mu_0\mathscr{H}_\mu^{(p)*}\cdot\frac{\partial\mathscr{H}_\nu^{(q)}}{\partial z}\right) \tag{3.3-19}$$

This term is zero for real values of the propagation constant but does not vanish if $\beta_\mu^{(p)}$ is imaginary. This is the reason why our present derivation fails for evanescent modes of the continuous spectrum.

The factor of the $\partial\beta_\nu^{(q)}/\partial z$ term vanishes after integration over the infinite cross section because of the orthogonality relation Eq. (3.2-24). Addition of Eqs. (3.3-17) and (3.3-18) and integration over the infinite xy plane thus allows us to convert the coupling coefficient Eq. (3.3-10) to the more useful form

$$R_{\mu\nu}^{(p,q)}=\frac{p\omega\varepsilon_0}{4P(\beta_\mu^{(p)}-\beta_\nu^{(q)})}\int_{-\infty}^{\infty}\int_{-\infty}^{\infty}\frac{\partial n^2}{\partial z}\mathscr{E}_\mu^{(p)*}\cdot\mathscr{E}_\nu^{(q)}\,dx\,dy \tag{3.3-20}$$

The factor p is $+1$ or -1 as the superscript p is $(+)$ or $(-)$. This form of the coupling coefficient holds only for modes with real values of β. We could thus set $s_\mu=1$ and omit the complex conjugation sign from one of the $\beta_\mu^{(p)}$. Equation (3.3-20) is further restricted to off-diagonal terms $\mu\neq\nu$ or $\mu=\nu$ and $p\neq q$. The diagonal elements with $\mu=\nu$ and $p=q$ can be combined with β_μ in Eqs. (3.2-42) and (3.2-43) so that they simply lead to a slight change of the propagation constant.

In its form (3.3-20), the coupling coefficient of the local normal modes shows much more resemblance to the coupling coefficient Eq. (3.2-44) of the ideal modes. Both coupling coefficients contain the scalar product of the electric fields of the two coupled modes in combination with the refractive index under an integral over the infinite cross section.

It appears as though the difference of the propagation constants in the denominator of Eq. (3.3-20) is a profound departure from the form Eq. (3.2-44) of the coupling coefficient of the ideal modes. However, we shall see in the next section that this feature is actually common to both coupling coefficients even though this is not immediately obvious. For small departures from the ideal waveguide and for slight index differences, both coupling coefficients Eqs. (3.2-44) and (3.3-20) become nearly identical.

The power factor P in the denominator of Eqs. (3.2-44) and (3.3-20) vanishes from the theory, since the mode amplitude coefficients [see, for example, Eq. (2.2-42)] are expressed in terms of this same factor.

3.4 Perturbation Solution of the Coupled Amplitude Equations

We consider the coupled amplitude equations in the form (3.2-49) for slowly varying amplitudes. The form (3.2-42) can be reduced to Eq. (3.2-49) by means of the transformation Eq. (3.2-34).

The coupled amplitude equations describe mode coupling and radiation loss phenomena of imperfect dielectric waveguides. In a typical application we may assume that two sections of perfect waveguide are connected to each other with a piece of imperfect waveguide as shown in Fig. 3.4.1.

The two perfect guide sections need not have the same core diameters shown in this figure. If two waveguides with different core diameters are connected to each other, we obtain the taper shown in Fig. 3.1.2. In a typical case we may be interested in the fate of a guided mode that enters the section of imperfect waveguide from the left. At $z = 0$, the amplitude coefficient $c_i^{(+)}$ is different from zero. The other amplitude coefficients with positive superscripts vanish, since no other modes are assumed to be initially excited. However, there may be power leaving the imperfect waveguide at $z = 0$ traveling

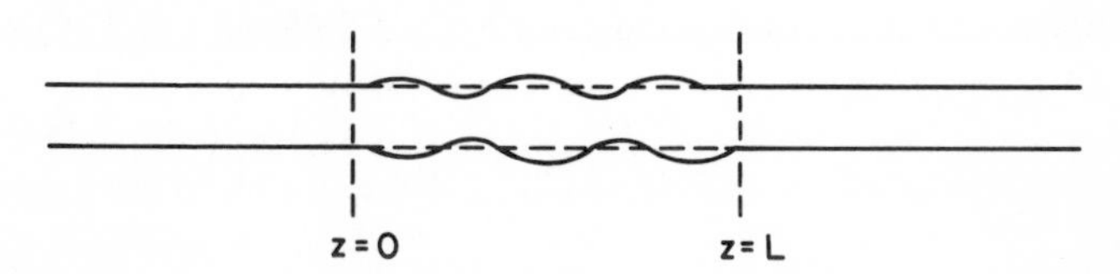

Fig. 3.4.1 *Dielectric waveguide with core–cladding interface distortions connected to two pieces of ideal guide.*

to the left, so that at least some of the coefficients $c_\nu^{(-)}$ may be nonvanishing. If the imperfect waveguide section is long, all the wave amplitudes that are excited by the departure from perfect geometry or index distribution may be large, and the entire coupled set of equations would have to be solved. This is an intractable problem unless some special conditions prevail. If only a purely sinusoidal deviation from perfect geometry or index distribution exists it is sufficient to consider only two of the infinitely many wave amplitudes and solve two instead of infinitely many simultaneous equations.

Another case that can be solved, even if the imperfect waveguide section is long, consists of a single-mode guide that loses power by radiation. In this case it is safe to assume that the radiated power is lost and no longer interacts with the guided mode. A perturbation solution can then be used to calculate the attenuation coefficient of the guided mode and to obtain a solution that holds for arbitrarily long waveguides.

Finally, we can treat the statistical case of a dielectric waveguide with random imperfections. This is actually the case of most practical interest. We show in Chap. 5 how the coupled amplitude equations can be used to derive approximate coupled equations for the average power carried by the modes. This theory is also applicable to arbitrarily long waveguides provided the coupling is reasonably weak.

This enumeration of the cases for which solutions can be obtained shows that most problems of practical interest can be treated. The situation is thus far more hopeful than one might have suspected by looking at the complicated set of integrodifferential equations. The only restriction that we must almost always impose is the requirement that the departure of the waveguide from perfect geometry is either slight or at least slow, so that the change of the slowly varying field amplitudes, which is caused by the imperfections, is very slow compared to the wavelength or, in the statistical case, compared to the correlation length.

The assumption that allows us to consider a perturbation solution of the coupled equation system consists in assuming that the coefficient $c_i^{(+)}$ of the incident mode is large compared to all other amplitudes:

$$|c_i^{(+)}| \gg |c_\nu^{(\pm)}| \tag{3.4-1}$$

with $\nu \neq i$. For weak coupling, this assumption allows us to neglect all amplitude coefficients on the right-hand side of Eqs. (3.2-49) and (3.2-50) and consider only the term with $\nu = i$. Inequality (3.4-1) implies that the amplitude $c_i^{(+)}$ does not vary very much, so that we treat it as approximately constant. We can then integrate Eqs. (3.2-49) and (3.2-50) and obtain the approximate solutions

$$c_\mu^{(+)}(z) = c_i^{(+)}(0) \int_0^z K_{\mu i}^{(+,+)}(u) \exp[i(\beta_\mu - \beta_i)u]\, du \tag{3.4-2}$$

and

$$c_\mu^{(-)}(z) = c_i^{(+)}(L) \int_L^z K_{\mu i}^{(-,+)}(u) \exp[-i(\beta_\mu + \beta_i)u] \, du \tag{3.4-3}$$

(The use of i as an index for the incident wave and also as $\sqrt{-1}$ should cause no confusion.) These perturbation solutions contain the initial conditions

$$c_\mu^{(+)} = 0, \qquad \text{at} \quad z = 0, \quad \text{for} \quad \mu \neq i \tag{3.4-4}$$

and

$$c_\mu^{(-)} = 0, \qquad \text{at} \quad z = L \tag{3.4-5}$$

According to our assumption that the amplitude of the incident wave does not change appreciably in the imperfect waveguide section of length L, we must insist that

$$c_i^{(+)}(0) \approx c_i^{(+)}(L) \tag{3.4-6}$$

When we evaluate the coupling coefficients (3.2-44) in the next two sections, it will become apparent that we always have

$$K_{\mu\nu}^{(p,q)}(z) = \hat{K}_{\mu\nu} f(z) \tag{3.4-7}$$

with constant $\hat{K}_{\mu\nu}$. The function $f(z)$ describes the actual geometrical shape of the deformed core boundary. We can thus write Eqs. (3.4-2) and (3.4-3) in the form

$$c_\mu^{(+)}(L) = \sqrt{L}\, c_i^{(+)}(0)\, F(\beta_i - \beta_\mu)\, \hat{K}_{\mu i} \tag{3.4-8}$$

and

$$c_\mu^{(-)}(0) = -\sqrt{L}\, c_i^{(+)}(0)\, F(\beta_i + \beta_\mu)\, \hat{K}_{\mu i} \tag{3.4-9}$$

with the Fourier transform of the deformation function $f(z)$

$$F(\beta_i - \beta_\mu) = \sqrt{L}^{-1} \int_0^L f(z) \exp[-i(\beta_i - \beta_\mu)z] \, dz \tag{3.4-10}$$

This form of the perturbation solution brings out an important point. It shows that the mode amplitudes $c_\mu^{(\pm)}$ are proportional to the Fourier transform of the deformation function $f(z)$ taken at the difference in the propagation constants of the incident and the spurious modes. If the Fourier transform at this spatial frequency vanishes, the corresponding spurious mode amplitude remains unexcited.

This important result allows us immediately to draw several conclusions. If only one Fourier component exists, that is, if the deformation function is purely sinusoidal, than only two modes are coupled, provided that the differences $\beta_i \pm \beta_\mu$ are all different from each other. In this case it is sufficient to reduce the coupled system to only two equations and ignore all the other modes. This case will be studied in Sect. 4.2.

We can now show that the presence of the difference of the two propagation constants in Eq. (3.3-20) does not mean that this form of the coupling coefficient is substantially different from Eq. (3.2-44). It will become apparent in the next section that the z derivative of n^2 appearing in Eq. (3.3-20) leads to the form

$$R_{\mu\nu}^{(p,q)} = \hat{R}_{\mu\nu}(df/dz)/i(\beta_\nu^{(q)} - \beta_\mu^{(p)}) \tag{3.4-11}$$

of the coupling coefficient. The theory of coupled local normal modes thus yields expressions for the wave amplitudes that are proportional to (we assume that the waveguide is only slightly perturbed so that β_ν is nearly constant)

$$G = [i\sqrt{L}(\beta_i - \beta_\mu)]^{-1} \int_0^L (df/dz) \exp[-i(\beta_i - \beta_\mu)z]\, dz \tag{3.4-12}$$

A partial integration results in

$$G = [i\sqrt{L}(\beta_i - \beta_\mu)]^{-1} \Big\{ f(L) \exp[-i(\beta_i - \beta_\mu)L] - f(0) + i(\beta_i - \beta_\mu) \int_0^L f(z) \exp[-i(\beta_i - \beta_\mu)z]\, dz \Big\} \tag{3.4-13}$$

If both waveguides, which are connected by the imperfect guide section, have the same core diameter, we can assume that $f(0) = f(L) = 0$. We thus find that $G = F(\beta_i - \beta_\mu)$ in complete agreement with the coupling formula obtained from the ideal mode theory. There may be a residual difference between the two theories, since $\hat{K}_{\mu\nu}$ may not be identical to $\hat{R}_{\mu\nu}$. But the appearance of the propagation constants in the denominator of Eq. (3.3-20) does not indicate a profoundly different coupling process for local normal modes.

Finally, we derive a useful expression for the attenuation constant of a guided mode losing power by coupling to radiation modes. We calculate the power carried by the field:

$$P_t = \tfrac{1}{2} \operatorname{Re} \int_{-\infty}^{\infty} \int_{-\infty}^{\infty} (\mathbf{E}_t \times \mathbf{H}_t^*) \cdot \mathbf{e}_z \, dx\, dy \tag{3.4-14}$$

We now substitute the series expansions Eqs. (3.2-18) and (3.2-19) and express the mode amplitudes in terms of forward and backward traveling waves, Eqs. (3.2-38) and (3.2-39). The rapidly varying field amplitudes are finally expressed in terms of slowly varying amplitudes using Eqs. (3.2-34) and (3.2-35). The following expression for the power flow through a plane perpendicular to the waveguide axis is thus obtained:

$$P_T = P\left[\sum_\nu (|c_\nu^{(+)}|^2 - |c_\nu^{(-)}|^2) + \sum \int_0^{n_2 k} (|c_\rho^{(+)}|^2 - |c_\rho^{(-)}|^2)\, d\rho\right] \tag{3.4-15}$$

The upper limit on the integral indicates that the evanescent modes do not carry power. The integral is extended only over propagating radiation modes. Power flowing in positive z direction appears positive, while power flowing in negative z direction is counted negative. For the moment we are interested only in power loss caused by coupling to radiation modes $\mu = \rho$. We must count all power as positive that is lost from the imperfect waveguide section of length L. We thus use the expression

$$\Delta P = P \sum \int_0^{n_2 k} (|c_\rho^{(+)}(L)|^2 + |c_\rho^{(-)}(0)|^2)\, d\rho \tag{3.4-16}$$

The mode amplitudes with positive superscripts account for power outflow at $z = L$, and the mode amplitudes with negative superscript carry power away from the imperfect waveguide section through the plane at $z = 0$. With the help of Eq. (2.4-3) we can transform the integral to the integration variable β,

$$\int_0^{n_2 k} d\rho = \int_0^{n_2 k} |\beta|\, d\beta/\rho \tag{3.4-17}$$

According to Eqs. (3.4-8) and (3.4-9), the two terms under the integral sign in Eq. (3.4-16) differ only in the sign of $\beta_\rho = \beta$. The z independent term $\hat{K}_{\rho i}$ is also a function of β_i and β. This functional dependence is understood but is not indicated explicitly to simplify the notation. When $\beta_\mu = \beta$ changes sign in going from Eq. (3.4-8) to Eq. (3.4-9), the same sign change must also be made in the implicit dependence of $\hat{K}_{\rho i}$ on β. It is therefore convenient to write Eq. (3.4-16) in the form

$$\Delta P = P \sum \int_{-n_2 k}^{n_2 k} |c_\rho^{(+)}(L)|^2 \, |\beta|\, d\beta/\rho \tag{3.4-18}$$

The quantity $|c_i^{(+)}(0)|^2 P$ represents the power carried by the incident mode. We thus obtain the power loss coefficient $2\alpha_i$ from Eqs. (3.4-8) and (3.4-18):

$$2\alpha_i = \Delta P/|c_i^{(+)}|^2 PL = \sum \int_{-n_2 k}^{n_2 k} |\hat{K}_{\rho i}|^2 \, |F(\beta_i - \beta)|^2 \, |\beta|\, d\beta/\rho \tag{3.4-19}$$

The factor 2 on the left-hand side is necessary, since we interpret α_i as the amplitude loss coefficient. This expression allows us to calculate the radiation loss suffered by mode i as a result of coupling to the continuous spectrum of radiation modes. The integral extends over all propagating radiation modes, and the summation sign reminds us to sum the contributions from the various types of radiation modes. The power in mode i as a function of z can then be calculated with the help of the formula

$$P_i(z) = P_i(0) \exp(-2\alpha_i z) \tag{3.4-20}$$

Perturbation theory thus allows us to calculate the power loss coefficient, which in turn can be used to calculate the actual power loss even for long dielectric waveguides. The justification for this extension of perturbation theory to cases of substantial power loss is based on the observation that the incident guided mode is not altered in its shape by the coupling to the radiation field. The power that is transferred to the radiation modes radiates away into the space outside the waveguide core. We can calculate the small amount of power loss for a short piece of waveguide using perturbation theory and obtain Eq. (3.4-19). However, we now apply the same procedure to the next short section of waveguide and once more obtain Eq. (3.4-19), provided we interpret $|c_i^{(+)}|^2$ as the power incident on this next waveguide section. Proceeding in this manner with small length increments allows us to calculate any amount of loss that actually occurs. The result of summing all these loss contributions is Eq. (3.4-20).

3.5 Coupling Coefficients for the Asymmetric Slab Waveguide

In this and the next section we present a collection of coupling formulas for slab waveguides and for round optical fibers with core boundary imperfections. These formulas are derived by substitution of the mode field expressions into the coupling formula for the local normal mode theory.

There are very many different problems of imperfect optical waveguides which can be solved with the coupled mode theory presented in Sects. 3.1–3.4. We limit the derivation of specific coupling expressions to slight core boundary irregularities and impose condition (1.3-4). Large deviations from the geometry of the perfect waveguide are very hard to evaluate since perturbation theory is not applicable. However, slight irregularities are of great practical importance to evaluate tolerance requirements. In case of tapers the core dimensions are allowed to change appreciably, provided these changes are gradual. We do not discuss refractive index inhomogeneities. These waveguide imperfections can easily be treated with the help of the general theory presented in Sects. 3.2 and 3.3.

Before we begin to apply our coupling formulas, we discuss the relationship between the coupling coefficient Eq. (3.2-44) for ideal modes and Eq. (3.3-20) for local normal modes.

The index distributions $n_0(x,y)$ and $n(x,y,z)$ define the ideal and actual dielectric waveguide. For a waveguide that is deformed as indicated in Fig. 3.1.1, the two index distributions coincide everywhere except in the vicinity of the core boundary. When the core boundary moves outward, the core index n_1 extends further than the ideal core, so that we have in the region between the dashed and solid line

$$n^2 - {n_0}^2 = {n_1}^2 - n_{2,3}^2 \tag{3.5-1}$$

In regions where the core boundary moves inward we have similarly

$$n^2 - n_0{}^2 = -(n_1{}^2 - n_{2,3}^2) \tag{3.5-2}$$

in the region between the dashed and solid lines. Everywhere else the equation $n^2 - n_0{}^2 = 0$ holds. The assumption of only very slight deviations from the perfect core geometry allows us to consider the field components to be approximately constant over the region between the dashed and solid lines in Fig. 3.1.1. The departure of the core boundary is given by the functions $f(z)$ for the upper core–cladding interface and $h(z)$ for the lower core–cladding interface. The actual core boundaries are thus

$$x = f(z) \tag{3.5-3}$$

and

$$x = -d + h(z) \tag{3.5-4}$$

The tangential field components are continuous at the core boundary. We thus take the tangential components out of the integral sign [the integral extends only over the very narrow region of width $f(z)$ or $h(z)$ of core displacement] and replaces them by their value at the ideal core boundaries $x = 0$ and $x = -d$. The normal electric field component is discontinuous at the core boundary. Assuming that the slopes of $f(z)$ and $h(z)$ are very gentle, we approximate the normal field component by replacing it with $\mathscr{E}_x$. The normal component just outside the core is related to the normal component just inside by the equation

$$(\mathscr{E}_x)_{\text{outside}} = (n_1{}^2/n_{2,3}^2)(\mathscr{E}_x)_{\text{inside}} \tag{3.5-5}$$

which expresses the continuity of the normal component of the displacement vector $\mathbf{D} = n^2\varepsilon_0\mathbf{E}$. When we refer the field component $\mathscr{E}_x$ to the value that it assumes just inside the ideal waveguide core, we must use $\mathscr{E}_x$ when the core boundary moves inward, but $(n_1{}^2/n_{2,3}^2)\mathscr{E}_x$ when the core boundary moves outward. The values of $n_0{}^2/n^2$ that are associated with the $\mathscr{E}_z$ components are also different for inward or outward moving core boundaries. We thus obtain from Eq. (3.2-44) with Eq. (1.3-4) for slight core boundary deviations and real values of β_μ

$$\begin{aligned} K_{\mu\nu}^{(p,q)} &= \frac{p\omega\varepsilon_0}{4iP} \\ &\quad \times \{(n_1{}^2 - n_3{}^2)f(z)[\eta_f(n_1{}^2/n_3{}^2)\mathscr{E}_{\mu x}^*\mathscr{E}_{\nu x} + \mathscr{E}_{\mu y}^*\mathscr{E}_{\nu y} + (1/\eta_f)\mathscr{E}_{\mu z}^{(p)*}\mathscr{E}_{\nu z}^{(q)}]_{x=0} \\ &\quad - (n_1{}^2 - n_2{}^2)h(z)[\eta_h(n_1{}^2/n_2{}^2)\mathscr{E}_{\mu x}^*\mathscr{E}_{\nu x} + \mathscr{E}_{\mu y}^*\mathscr{E}_{\nu y} + (1/\eta_h)\mathscr{E}_{\mu z}^{(p)*}\mathscr{E}_{\nu z}^{(q)}]_{x=-d}\} \end{aligned} \tag{3.5-6}$$

In y direction the integration was extended only over the unit length instead of $-\infty$ to ∞. The functions η_f and η_h are defined as follows:

$$\eta_f = \begin{cases} n_3^2/n_1^2, & \text{if } f(z) < 0 \\ n_1^2/n_3^2, & \text{if } f(z) > 0 \end{cases} \tag{3.5-7}$$

$$\eta_h = \begin{cases} n_2^2/n_1^2, & \text{if } h(z) > 0 \\ n_1^2/n_2^2, & \text{if } h(z) < 0 \end{cases} \tag{3.5-8}$$

These discontinuous functions make the coupling formula rather awkward. For slight index differences between core and cladding, η_f and η_h are nearly unity. A reasonable approximation for slight index differences consists in replacing η_f and η_h by their geometric mean values

$$\eta_f = [(n_3^2/n_1^2)(n_1^2/n_3^2)]^{1/2} = 1 \tag{3.5-9}$$

and

$$\eta_h = [(n_2^2/n_1^2)(n_1^2/n_2^2)]^{1/2} = 1 \tag{3.5-10}$$

The η functions are now treated as constants, which simplifies the evaluation of Eq. (3.5-6) considerably. For large index difference the approximations (3.5-9) and (3.5-10) are inaccurate, and the more precise formulas must be used. It is important to remember that, whereas the coupled mode theory with the coupling coefficient Eq. (3.2-44) is precise, Eq. (3.5-6) is only an approximation that holds for very slight core boundary deflections. The approximation becomes poorer as Eq. (3.5-6) is applied to higher-order modes.

The coupled mode theory for local normal modes is accurate only if evanescent continuum modes are not needed when the coupling coefficient Eq. (3.3-20) is used. However, for many practical applications it is convenient to work out an approximation of this expression that holds for small values of the derivatives of the deformation functions. We thus assume

$$|df/dz| \ll 1, \qquad \text{and} \qquad |dh/dz| \ll 1 \tag{3.5-11}$$

The local normal mode theory has the advantage that the conditions $|f(z)| \ll 1$ and $|h(z)| \ll 1$—which had to be imposed for Eq. (3.5-6) to be applicable—are now replaced by the more liberal requirements (3.5-11). This makes even the simplified form of the coupling coefficient (3.3-20), which we are about to derive, applicable to waveguide tapers.

The derivation of the simplified formula is based on Fig. 3.5.1. The coordinate system x, z corresponds to the coordinates shown in Fig. 3.1.1. The coordinate z' is tangential to the core boundary at the origin of the coordinate systems. The problem in the evaluation of Eq. (3.3-20) is caused by the fact that the refractive index distribution of the slab waveguide is discontinuous at

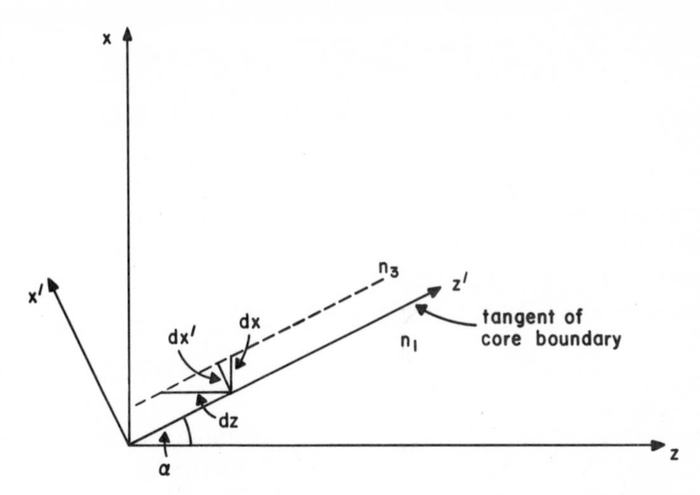

Fig. 3.5.1 *Graphical aid in the evaluation of the integral in Eq. (3.3-20).*

the core boundary, so that the derivative of the index distribution does not exist. It is thus necessary to replace the abrupt index change by a smoothed-out but arbitrary distribution. Figure 3.5.1 shows the upper core boundary of the slab waveguide. Using restrictions (3.5-11) we assume that the x components of the field are approximately normal to the interface, while the y and z components are (approximately) tangential. The angle α can be expressed by the derivative of the function $f(z)$,

$$\tan\alpha = df/dz \tag{3.5-12}$$

Because of the discontinuity of the normal electric field components at the interface, we must treat the tangential and normal parts of the scalar product [appearing in Eq. (3.3-20)] separately. The y integration cannot be extended over the infinite range shown in Eq. (3.3-20) since this would lead to an infinite result. We regard the coupling coefficient as a quantity that is defined per unit length (in y direction) and omit the y integral. For the tangential part of the scalar product, we now obtain

$$I_1 = \int_{-\infty}^{\infty} (\partial n^2/\partial z)\, \mathscr{E}_{\mu t}^{(p)*} \mathscr{E}_{\nu t}^{(q)}\, dx \tag{3.5-13}$$

where the subscript t indicates the y and z components of the field. We consider the refractive index distribution as a function of x' (locally at the origin of the coordinate system) but not of z':

$$n = n(x') \tag{3.5-14}$$

We may thus write

$$\partial n^2/\partial z = (\partial n^2/\partial x')(\partial x'/\partial z) = -(\partial n^2/\partial x')\sin\alpha \tag{3.5-15}$$

The integration variable is changed from x to x' with the help of

$$dx = (dx/dx')\,dx' = dx'/\cos\alpha \tag{3.5-16}$$

Since the tangential electric field component is constant across the core boundary, we can take the field components out of the integral over the infinitesimal range over which the derivative of the refractive index is different from zero and obtain, from Eqs. (3.5-13)–(3.5-16),

$$\int_{-\infty}^{\infty} (\partial n^2/\partial z)\,\mathscr{E}_{\mu t}^{(p)*}\mathscr{E}_{\nu t}^{(q)}\,dx = (n_1{}^2 - n_3{}^2)\,\mathscr{E}_{\mu t}^{(p)*}\mathscr{E}_{\nu t}^{(q)}\tan\alpha \tag{3.5-17}$$

The actual shape of the smoothed-out index distribution does not appear in the final result. We are thus allowed to let the refractive index change become as abrupt as desired.

The part of the scalar product containing the normal electric field components can also be evaluated, if we use the fact that the field on the cladding side of the interface is related to the field on the core side by the relation

$$\mathscr{E}_x(x') = n_1{}^2\mathscr{E}_x(0)/n^2(x') \tag{3.5-18}$$

The field component is a function of x'. The point $x' = 0$ is assumed to be located just inside the core at a point where $n(x') = n_1$. The refractive index distribution $n(x')$ is supposed to vary very rapidly but continuously, so that a very short distance from $x' = 0$ we have $n(x') = n_3$. Relation (3.5-18) follows from the fact that the normal component of the electric displacement vector $\mathbf{D} = \varepsilon_0 n^2 \mathbf{E}$ remains continuous at the core boundary. The normal part of the scalar product in Eq. (3.3-20) thus contributes the integral

$$I_2 = \int_{-\infty}^{\infty} (\partial n^2/\partial z)\,\mathscr{E}_{\mu x}^{(p)*}\mathscr{E}_{\nu x}^{(q)}\,dx$$

$$= -n_1{}^4\mathscr{E}_{\mu x}^{(p)*}(0)\,\mathscr{E}_{\nu x}^{(q)}(0)\tan\alpha \int_{-\infty}^{\infty} [1/n^4(x')](\partial n^2/\partial x')\,dx' \tag{3.5-19}$$

The integral on the right-hand side of Eq. (3.5-19) can easily be solved with the result

$$\int_{-\infty}^{\infty} (\partial n^2/\partial z)\,\mathscr{E}_{\mu x}^{(p)*}\mathscr{E}_{\nu x}^{(q)}\,dx = (n_1{}^2/n_3{}^2)(n_1{}^2 - n_3{}^2)\,\mathscr{E}_{\mu x}^{(p)*}\mathscr{E}_{\nu x}^{(q)}\tan\alpha \tag{3.5-20}$$

Using Eq. (3.5-12) and a similar procedure for the lower core boundary finally results in the desired approximation of the coupling coefficient for local normal modes:

$$R_{\mu\nu}^{(p,q)} = \frac{-p\omega\varepsilon_0}{4P(\beta_\nu^{(q)} - \beta_\mu^{(p)})} \left\{ (n_1^{\,2} - n_3^{\,2}) \frac{df}{dz} \left[\frac{n_1^{\,2}}{n_3^{\,2}} \mathscr{E}_{\mu x}^* \mathscr{E}_{\nu x} + \mathscr{E}_{\mu y}^* \mathscr{E}_{\nu y} + \mathscr{E}_{\mu z}^{(p)*} \mathscr{E}_{\nu z}^{(q)} \right]_{x = f(z)} \right.$$
$$\left. - (n_1^{\,2} - n_2^{\,2}) \frac{dh}{dz} \left[\frac{n_1^{\,2}}{n_2^{\,2}} \mathscr{E}_{\mu x}^* \mathscr{E}_{\nu x} + \mathscr{E}_{\mu y}^* \mathscr{E}_{\nu y} + \mathscr{E}_{\mu z}^{(p)*} \mathscr{E}_{\nu z}^{(q)} \right]_{x = -d + h(z)} \right\} \tag{3.5-21}$$

The field components in this formula are evaluated right at the upper or lower core boundary inside the core.

The most conspicuous difference between the two coupling coefficients Eqs. (3.5-6) and (3.5-21) is the fact that Eq. (3.5-21) contains the derivatives of $f(z)$ and $h(z)$ and has the difference of the propagation constants of the two coupled modes in its denominator. We have discussed in the previous section that these two forms are equivalent for slight core boundary deviations. By using the equivalence

$$\frac{1}{\beta_\nu^{(q)} - \beta_\mu^{(p)}} \frac{df}{dz} \to if(z) \tag{3.5-22}$$

it is apparent that the two coupling coefficients become identical in the limit of slight core–cladding index differences and slight core boundary deviations from the geometry of the perfect waveguide. This result is gratifying and should have been expected since local and ideal modes become identical for vanishing departures of the core boundary from the ideal shape. It is actually more surprising that this agreement of the two coupling coefficients is achieved only if we consider waveguides with very slight core–cladding index differences where approximations (3.5-9) and (3.5-10) are valid. Since ideal and local normal modes become identical for very slight core boundary deflections, one should expect that the agreement of the two coupling coefficients is independent of the index difference. However, this naive assumption is justified only if the two series expansions for ideal and local normal modes become identical term for term in the limit of vanishing core boundary distortion. An order of magnitude estimate of these series shows that those terms, which we would like to neglect in the perturbation treatment of the infinite equation system, make a contribution that can be of the same magnitude as the term containing the c_i coefficient of the incident wave.

Let us ignore the backward traveling waves for simplicity and substitute the approximate solution Eq. (3.4-8) into Eq. (3.2-49). We obtain, with $h(z) = 0$,

$$dc_\mu/dz = \hat{K}_{\mu i} f(z) c_i(z) \exp[i(\beta_\mu - \beta_i)z]$$
$$+ c_i(0) \sum_{\nu \neq i} \left\{ \hat{K}_{\mu\nu} \hat{K}_{\nu i} f(z) \exp[i(\beta_\mu - \beta_\nu)z] \int_0^z f(u) \exp[i(\beta_\nu - \beta_i)u]\, du \right\} \tag{3.5-23}$$

The term under the integral sign is actually more complicated than is shown here; it should contain all the z-dependent terms in Eq. (3.5-6). The function $f(z)$ is considered to be a small quantity of first order. The sum term in Eq. (3.5-23) is thus a second-order term, provided that only a small number of terms of the summation contribute. However, this may not be the case. Convergence of the sum can be caused by the fact that the Fourier transform of the function $f(u)$ appearing under the summation sign decreases in magnitude as the mode number ν increases. However, if the Fourier spectrum of the function $f(z)$ is very wide, convergence of the sum is caused by another mechanism. We have assumed that the field components do not vary appreciably over the range $0 \leqslant x \leqslant f(z)$, so that we replaced the electric fields under the integral sign in Eq. (3.2-44) with their constant values taken at the position of the core boundary of the ideal waveguide. For very high-order modes, the field variation over the range $0 \leqslant x \leqslant f(z)$ becomes important and causes cancellation of the integral. For sufficiently large values of ν, the coupling coefficient decreases much faster than is apparent from its approximation (3.5-6). The value at which the high-order modes begin to contribute to convergence of the sum are obtained from

$$\rho f(z) \approx \pi \tag{3.5-24}$$

because the parameter ρ determines the transverse field variation of the radiation modes, as is apparent from Eq. (1.4-6). When the product of ρ times the maximum value of x, which occurs within the integration range, reaches π the field begins to reverse its sign within the integration range, thus forcing rapid decrease of the magnitude of the integral. The sum over radiation modes is actually an integral over ρ, so that we have

$$\sum_{\nu} = \int_0^{\infty} d\rho \tag{3.5-25}$$

The range $\Delta\rho = \rho$ over which the integral extends undiminished follows from Eq. (3.5-24):

$$\Delta\rho = \pi / f(z) \tag{3.5-26}$$

It is thus apparent that the summation contributes a factor $1/f(z)$ to the right-hand side of Eq. (3.5-23), so that the two terms are of equal order of magnitude, contrary to our initial assumption. This conclusion is valid provided that convergence is caused not by the Fourier transform term but by the $\hat{K}_{\mu\nu}$ factors.

We see that it can happen that the neglected terms make a considerable contribution and that it is not sufficient to compare the two series expansions for the ideal and local normal mode theories term by term since there are other hidden factors that influence the results. This explains why the coupling coefficients do not become identical, as would be expected from the observation that the local and ideal modes approach each other arbitrarily closely when $f(z)$ becomes vanishingly small. The two series have different convergence behavior and do not become identical in the limit $f(z) \rightarrow 0$.

The preceding discussion may cause the reader to wonder about the value of the perturbation solutions of the coupled mode theory. Actually the local normal mode theory gives excellent results for functions $f(z)$ and $h(z)$ with reasonably narrow Fourier spectra. However the Fourier spectrum that follows from Eq. (3.5-6) cannot be arbitrarily narrow since it contains the discontinuous functions η_f and η_h. Only if we are allowed to approximate these functions by Eqs. (3.5-9) and (3.5-10) is the convergence of the series in Eq. (3.5-23) determined by the width of the Fourier spectrum of $f(z)$. This explains why the two types of coupling coefficients become identical only in the limit $n_1/n_{2,3} \rightarrow 1$.

Since the approximations (3.5-9) and (3.5-10) cause the coupling coefficients (3.5-6) of the ideal mode theory to agree with the coupling coefficients (3.5-21) of the local normal modes, we shall always use this approximation. Whether we use form (3.5-6) [with Eqs. (3.5-9) and (3.5-10)] or form (3.5-21) depends on the problem at hand. For slight random imperfections of the core boundary, form (3.5-6) is more convenient to use since it contains the functions $f(z)$ and $h(z)$ and not their derivatives. For tapers we must use Eq. (3.5-21) since Eq. (3.5-6) is not applicable to this case.

We conclude this section by listing the coupling coefficients for guided modes and guided and radiation modes caused by core boundary imperfections of the slab waveguide. The actual derivations of these equations from Eq. (3.5-21) and the mode equations of Chap. 1 are straightforward and will be omitted.

Coupling between two guided TE modes μ and ν is governed by the coupling coefficient

$$R_{\mu\nu}^{(p,q)} = \frac{-p\kappa_\mu \kappa_\nu [(df/dz) - (\sin\kappa_\mu d \sin\kappa_\nu d/|\sin\kappa_\mu d \sin\kappa_\nu d|)(dh/dz)]}{(\beta_\nu^{(q)} - \beta_\mu^{(p)})\{|\beta_\mu \beta_\nu| [d + (1/\gamma_\mu) + (1/\delta_\mu)][d + (1/\gamma_\nu) + (1/\delta_\nu)]\}^{1/2}} \tag{3.5-27}$$

Equations (1.3-27) and (1.3-28) were used to achieve some simplification. The parameters κ, γ, and δ are defined by Eqs. (1.2-13)–(1.2-15). The propagation constants β_μ and β_ν are solutions of the eigenvalue Eq. (1.3-26). If we use Eq. (3.5-27) to describe a taper, the core width d is a function of z, so that all the parameters are also z dependent. Coupling of guided modes to radiation modes must be described by two different coupling coefficients because there are two types of radiation modes. Coupling of a guided TE mode ν to a TE radiation mode in the range

$$0 \leqslant \rho \leqslant (n_2{}^2 - n_3{}^2)^{1/2} k \tag{3.5-28}$$

is described by

$$R_{\rho\nu}^{(p,q)} = \frac{-p(n_1{}^2 - n_3{}^2)^{1/2} k\kappa_\nu \rho}{(\beta_\nu^{(q)} - \beta_\rho^{(p)})\{\pi|\beta_\nu \beta_\rho|[d + (1/\gamma_\nu) + (1/\delta_\nu)]\}^{1/2}}$$
$$\cdot (\sigma\, df/dz) - (\sin\kappa_\nu d/|\sin\kappa_\nu d|)[(n_1{}^2 - n_2{}^2)/(n_1{}^2 - n_3{}^2)]^{1/2}$$
$$\cdot \frac{[\sigma\cos\sigma d + (\Delta/i)\sin\sigma d](dh/dz)}{\{\rho^2[\sigma\cos\sigma d + (\Delta/i)\sin\sigma d]^2 + \sigma^2[\sigma\sin\sigma d - (\Delta/i)\cos\sigma d]^2\}^{1/2}} \tag{3.5-29}$$

The parameters σ, ρ, and Δ are defined by Eqs. (1.4-7)–(1.4-9). A check of the dimensions reveals that the coupling coefficient (3.5-27) has the dimension of inverse length, while Eq. (3.5-29) has the dimension of the square root of an inverse length. This result is indeed correct. Equation (3.3-8) shows clearly that the coupling coefficients between guided modes must have the dimension of an inverse length. However, coupling coefficients from guided to radiation modes or *vice versa* appear under an integral sign. The dimension of the coupling coefficients as well as the amplitude coefficient for radiation modes is different from those of guided modes. Equation (3.4-18) shows that the dimension of the amplitude coefficients of radiation modes is the square root of length, since ρ and β have the dimension of inverse length. The integral

$$\int_0^\infty R_{\mu\rho}\, c_\rho\, d\rho \tag{3.5-30}$$

that replaces the sum in Eq. (3.3-8) must have the dimension of inverse length. This is only possible if the coupling coefficient between a guided and a radiation mode has the dimension of the square root of an inverse length in agreement with the dimension of Eq. (3.5-29).

The coupling coefficient from a guided TE mode ν to a TE radiation mode in the range

$$(n_2{}^2 - n_3{}^2)^{1/2} k \leqslant \rho < n_2 k \tag{3.5-31}$$

is

$$R_{\rho\nu}^{(p,q)} = \frac{-p(n_1{}^2 - n_3{}^2)^{1/2} k\kappa_\nu}{(\beta_\nu^{(q)} - \beta_\rho^{(p)})\{\pi|\beta_\nu \beta_\rho|[d + (1/\gamma_\nu) + (1/\delta_\nu)]\}^{1/2}}$$
$$\cdot \{(df/dz) - (\sin\kappa_\nu d/|\sin\kappa_\nu d|)[(n_1{}^2 - n_2{}^2)/(n_1{}^2 - n_3{}^2)]^{1/2}$$
$$\cdot (\cos\sigma d - F_i \sin\sigma d)(dh/dz)\}$$
$$\cdot \{(\cos\sigma d - F_i\sin\sigma d)^2 + (\sigma^2/\rho^2)(\sin\sigma d + F_i\cos\sigma d)^2$$
$$+ [1 + (\sigma^2/\Delta^2) F_i^2](\Delta/\rho)\}^{-1/2} \tag{3.5-32}$$

The coefficient F_i is obtained from Eq. (1.4-24) with $i = 1$ and 2 indicating two different types of radiation modes.

In isotropic media there is no coupling between TE and TM modes of the slab waveguide. Since we restricted the discussion in this book to waveguides made of isotropic media, all coupling coefficients between TE and TM modes vanish.

The coupling coefficients that mediate coupling between the different types of TM modes are more complicated in appearance than the corresponding TE mode coupling coefficients. For this reason we refrain from substituting the amplitude coefficients of the mode fields explicitly into the coupling coefficients.

Coupling between two guided TM modes is governed by the coupling coefficient

$$R_{\mu\nu}^{(p,q)} = \frac{-pC_\mu C_\nu}{4P\omega\varepsilon_0(\beta_\nu^{(q)}-\beta_\mu^{(p)})\,n_1{}^2 n_3{}^4}$$
$$\cdot\left\{(n_1{}^2-n_3{}^2)\,\frac{df}{dz}(n_3{}^2|\beta_\mu\beta_\nu|+pqn_1{}^2\delta_\mu\delta_\nu)-(n_1{}^2-n_2{}^2)\,\frac{dh}{dz}\right.$$
$$\cdot\frac{\sin\kappa_\mu d\sin\kappa_\nu d}{|\sin\kappa_\mu d\sin\kappa_\nu d|}\left[\frac{(n_3{}^4\kappa_\mu{}^2+n_1{}^4\delta_\mu{}^2)(n_3{}^4\kappa_\nu{}^2+n_1{}^4\delta_\nu{}^2)}{(n_2{}^4\kappa_\mu{}^2+n_1{}^4\gamma_\mu{}^2)(n_2{}^4\kappa_\nu{}^2+n_1{}^4\gamma_\nu{}^2)}\right]^{1/2}$$
$$\left.\cdot(n_2{}^2|\beta_\mu\beta_\nu|+pqn_1{}^2\gamma_\mu\gamma_\nu)\right\} \qquad (3.5\text{-}33)$$

The factors p and q are $+1$ or -1 depending on the sign of the superscripts p and q. The mode amplitude coefficients C_μ and C_ν are given by Eq. (1.3-73). This coupling coefficient was again obtained by substitution of the field expressions for the guided TM modes of Sect. 1.3 into Eq. (3.5-21). Expressions (1.3-64) and (1.3-65) were used to achieve some simplification.

Coupling between a guided TM mode and a TM radiation mode in the range

$$0 \leqslant \rho \leqslant (n_2{}^2-n_3{}^2)^{1/2}k \qquad (3.5\text{-}34)$$

is governed by

$$R_{\rho\nu}^{(p,q)} = \frac{-pC_\nu D_{r\rho}}{4P\omega\varepsilon_0(\beta_\nu^{(q)}-\beta_\rho^{(p)})\,n_1{}^2 n_3{}^2}$$
$$\cdot\left\{(n_1{}^2-n_3{}^2)\,\frac{df}{dz}\left(|\beta_\nu\beta_\rho|+pq\frac{n_1{}^2}{n_3{}^2}\delta_\nu\frac{\Delta}{i}\right)\right.$$
$$-(n_1{}^2-n_2{}^2)\,\frac{dh}{dz}\,\frac{\sin\kappa_\nu d}{|\sin\kappa_\nu d|}\left(\frac{n_3{}^4\kappa_\nu{}^2+n_1{}^4\delta_\nu{}^2}{n_2{}^4\kappa_\nu{}^2+n_1{}^4\gamma_\nu{}^2}\right)^{1/2}$$
$$\left.\cdot\left[|\beta_\nu\beta_\rho|\left(\cos\sigma d+\frac{n_1{}^2}{n_3{}^2}\frac{\Delta}{i\sigma}\sin\sigma d\right)+pq\gamma_\nu\sigma\left(\sin\sigma d-\frac{n_1{}^2}{n_3{}^2}\frac{\Delta}{i\sigma}\cos\sigma d\right)\right]\right\}$$
$$(3.5\text{-}35)$$

The parameters κ_ν, γ_ν, and δ_ν are defined by Eqs. (1.2-13)–(1.2-15). The parameter σ is obtained from Eq. (1.4-8). The parameter Δ/i is real and positive for ρ values in the range (3.5-34) and is defined by Eq. (1.4-7). The amplitude coefficients are obtained from Eqs. (1.3-73) and (1.4-36).

The coupling coefficient for a guided mode and a TM radiation mode in the range

$$(n_2^2-n_3^2)^{1/2}k \leqslant \rho \leqslant n_2 k \tag{3.5-36}$$

is, finally,

$$\begin{aligned} R_{\rho\nu}^{(p,q)} = & \frac{-pC_\nu S_{r\rho}}{4P\omega\varepsilon_0(\beta_\nu^{(q)}-\beta_\rho^{(p)})n_1^2 n_3^2} \\ & \cdot \left\{(n_1^2-n_3^2)\frac{df}{dz}(|\beta_\nu\beta_\rho|-pq\delta_\nu\sigma R_i)\right. \\ & -(n_1^2-n_2^2)\frac{dh}{dz}\frac{\sin\kappa_\nu d}{|\sin\kappa_\nu d|}\left(\frac{n_3^4\kappa_\nu^2+n_1^4\delta_\nu^2}{n_2^4\kappa_\nu^2+n_1^4\gamma_\nu^2}\right)^{1/2} \\ & \left.\cdot\,[|\beta_\nu\beta_\rho|(\cos\sigma d-R_i\sin\sigma d)+pq\gamma_\nu\sigma(\sin\sigma d+R_i\cos\sigma d)]\right\} \end{aligned} \tag{3.5-37}$$

Many of the parameters in this equation are the same as those of the coupling coefficients listed above. The amplitude coefficient $S_{r\rho}$ is given by Eq. (1.4-41). R_i assumes two values that correspond to the two choices of the sign of the square root in Eq. (1.4-40). The two signs determine two kinds of radiation modes as explained in Sect. 1.4.

We have expressed all coupling coefficients in terms of the derivatives of the deformation functions $f(z)$ and $h(z)$. These derivatives can be replaced by the functions themselves with the help of the substitution (3.5-22), which is valid for waveguides with only slight core boundary perturbations. For gentle tapers the coupling coefficients must be used in the form given in this section.

3.6 Coupling Coefficients for the Optical Fiber

Our aim in this section is to collect formulas for coupling coefficients between guided modes and guided and radiation modes of the weakly guiding optical fiber for slight core boundary imperfections. We assume that the core boundary is determined by the function

$$r(x,y,z) = a + f(z)\cos(m\phi+\psi) \tag{3.6-1}$$

The most general core boundary deformation would be obtained if we attach the subscript m to $f(z)$ and ψ and sum over m from 0 to ∞. However, for the sake of simplicity we refrain from using a Fourier series expansion of the most general core boundary distortion but consider the effect of each term of the form (3.6-1) separately.

A similar argument as the one leading from Eq. (3.2-44) to Eq. (3.5-6) allows us to write

$$K_{\mu\nu}^{(p,q)} = \hat{K}_{\mu\nu}^{(p,q)} f(z) \tag{3.6-2}$$

with

$$\hat{K}_{\mu\nu}^{(p,q)} = (p\omega\varepsilon_0 a/4iP)(n_1{}^2 - n_2{}^2) \int_0^{2\pi} \mathscr{E}_{\mu t}^* \cdot \mathscr{E}_{\nu t} \cos(m\phi + \psi)\, d\phi \tag{3.6-3}$$

The integral in Eq. (3.2-44) extends only over the region of the core displacement (with $dx\, dy = a\, dr\, d\phi$). In the spirit of the weakly guiding fiber approximation, we have used $n_1/n_2 \approx 1$ wherever it would have appeared as a factor in Eq. (3.6-3). The longitudinal electric field components have been omitted from the scalar product appearing under the integral sign in Eq. (3.6-3). It is apparent from a comparison of the $\mathscr{E}_z$ component (2.2-28) and the E_x component (2.2-30) and using (2.2-65) that to order of magnitude

$$\mathscr{E}_z/\mathscr{E}_t \approx \gamma/\beta = (\beta^2 - n_2{}^2k^2)^{1/2}/\beta = (n_e{}^2 - n_2{}^2)^{1/2}/n_e \ll 1 \tag{3.6-4}$$

with the effective refractive index n_e defined by

$$\beta = n_e k \tag{3.6-5}$$

The effective refractive index must satisfy the relation

$$n_2 < n_e < n_1 \tag{3.6-6}$$

Since the difference $n_1 - n_2$ is assumed to be very small the inequality on the right-hand side of Eq. (3.6-4) is justified. For coupling between guided modes the contribution of the $\mathscr{E}_z$ component to the scalar product of the electric vectors can safely be neglected.

For coupling between a guided and a radiation mode the situation is less favorable, since the $\mathscr{E}_z$ components of radiation modes need not be small. The contribution of the $\mathscr{E}_z$ components to the scalar product of the electric fields becomes more important as the angle of propagation, α, of the radiation modes increases. By "angle of propagation," we mean the angle that the direction of propagation of the outward traveling component of the radial standing wave makes with the axis of the waveguide. In order to estimate the contribution of the $\mathscr{E}_z$ term we consider the expression

$$(\mathscr{E}_z/\mathscr{E}_t)_g(\mathscr{E}_z/\mathscr{E}_t)_r \approx (\gamma/\beta_g)(\rho/\beta_r) = (\gamma/\beta_g)\tan\alpha \tag{3.6-7}$$

The subscripts g and r indicate guided and radiation modes. For the lowest-order guided mode (HE_{11} mode)—close to the cutoff of the next guided mode—we obtain from Eq. (2.2-68) with $V = 2.4$ the values $\kappa a = 1.58$ and $\gamma a = 1.8$.

With the help of Eqs. (1.3-92) and (2.2-48) we can express the propagation constant in terms of κa and V as

$$\beta a = \{V^2[1-(n_2{}^2/n_1{}^2)]^{-1} - (\kappa a)^2\}^{1/2} \tag{3.6-8}$$

For a typical weakly guiding fiber, we may have $n_1/n_2 = 1.005$. We thus obtain $\beta a = 24.04$. The ratio of Eq. (3.6-4) is thus $\gamma/\beta = 0.075$. It requires an angle of $\alpha = 53°$ to make the ratio of Eq. (3.6-7) equal to 0.1. This ratio becomes 0.5 for $\alpha = 81.5°$. Depending on the desired accuracy of the result, we must restrict the application of our coupling formula that neglects $\mathscr{E}_z$ to radiation that is mainly forward directed. However, our estimate shows that we obtain at least the correct order of magnitude for coupling to radiation that approaches the 90° angle quite closely. This estimate was based on the assumption of single-mode operation of a weakly guiding fiber. For low-order guided modes in multimode waveguides, the situation is much less favorable. The γa value of the lowest-order mode approaches V for large values of V. We thus obtain, for $V = 20$, the ratio $\gamma/\beta = 0.1$. The angle α may now approach $\alpha = 45°$ until Eq. (3.6-7) becomes 0.1, or $\alpha = 79°$ until Eq. (3.6-7) becomes 0.5.

Neglecting the $\mathscr{E}_z$ component in Eq. (3.6-3), as indicated by limiting the scalar product of the electric fields to their transverse parts, provides an excellent approximation for coupling between weakly guided modes. It is still a good approximation for coupling between a weakly guided and a radiation mode that is forward directed. As the radiation mode approaches 90°, the approximation becomes poorer. However an estimate of the correct order of magnitude is obtained at all angles except for a very narrow region around 90°. The coupling formulas simplify substantially if we neglect the longitudinal components. We are thus restricting ourselves to this approximation. But it is important to keep in mind that the following formulas in this section are not applicable to radiation problems in which the radiation, which is excited by a guided mode, escapes at a 90° angle with respect to the waveguide axis.

We continue our practice of labeling the modes with the single index μ or ν. This index coincides with the integral factor appearing in the argument of the sine or cosine functions in Eqs. (2.2-28)–(2.2-35) and similar mode expressions. However, in addition to the integer that indicates the ϕ symmetry of the modes, we need another label that indicates their radial variation, a third label to indicate the sine or cosine dependence of the mode, and finally a label indicating its polarization. Attaching all these indices to the modes would overburden the notation. We thus follow the practice of omitting most of the cumbersome mode labels, retaining only the label indicating the ϕ symmetry.

The ϕ integration in Eq. (3.6-3) restricts coupling to modes whose ϕ symmetries are related by

$$\mu \pm \nu = \pm m \tag{3.6-9}$$

Substitution of the mode field expressions for guided modes into Eq. (3.6-3) leads to the following coupling coefficient between guided modes ($n \approx n_1 \approx n_2$):

$$\hat{K}_{\mu\nu}^{(p,q)} = \frac{e_{\mu\nu m}}{(e_\mu e_\nu)^{1/2}} \frac{p\gamma_\nu \gamma_\mu J_\nu(\kappa_\nu a) J_\mu(\kappa_\mu a)}{2iank[|J_{\nu-1}(\kappa_\nu a) J_{\nu+1}(\kappa_\nu a) J_{\mu-1}(\kappa_\mu a) J_{\mu+1}(\kappa_\mu a)|]^{1/2}} \tag{3.6-10}$$

The parameters appearing in this and the following coupling coefficients are defined in Chap. 2. In particular, we find the definition of κ_ν in Eq. (2.2-10), of γ_ν in Eq. (2.2-27), and of e_ν in Eq. (2.2-42a). The factor p becomes $+1$ or -1 depending on the sign of the superscript on the left-hand side of the equation. The factor $e_{\nu\mu m}$ determines which modes couple to each other and is tabulated in Table 3.6.1. It can assume only the values 0, 1, 2, and 4. The last column in Table 3.6.1 indicates the form of the core boundary distortion. The sine or cosine functions result by using either $\psi = \pi/2$ or $\psi = 0$ in Eq. (3.6-3).

Coupling of a guided mode to the forward (or backward) directed radiation modes Eqs. (2.4-5)–(2.4-22) is governed by the coupling coefficient

$$\hat{K}_{\mu\nu}^{(p,q)} = \frac{e_{\mu\nu m}}{(e_\nu e_\mu)^{1/2}} \cdot \frac{p[(n_1/n_2)-1]^{1/2}\gamma_\nu \rho^{1/2} J_\nu(\kappa_\nu a) J_\mu(\sigma a)}{i\pi a|J_{\nu-1}(\kappa_\nu a) J_{\nu+1}(\kappa_\nu a)|^{1/2}|\sigma J_{\mu-1}(\sigma a) H_\mu^{(1)}(\rho a) - \rho J_\mu(\sigma a) H_{\mu-1}^{(1)}(\rho a)|} \tag{3.6-11}$$

Coupling between a guided mode and the "free space" radiation modes Eqs. (2.4-23)–(2.4-32) is determined by

$$\hat{K}_{\mu\nu}^{(p,q)} = \frac{e_{\mu\nu m}}{(e_\nu e_\mu)^{1/2}} \frac{p\{n[(n_1/n_2)-1]k\rho\beta\}^{1/2}\gamma_\nu J_\nu(\kappa_\nu a) J_\mu(\rho a)}{i[2(\beta^2+n^2k^2)|J_{\nu-1}(\kappa_\nu a) J_{\nu+1}(\kappa_\nu a)|]^{1/2}} \tag{3.6-12}$$

Coupling to continuum modes with imaginary propagation constants is, of course, excluded by the approximations involved in arriving at Eq. (3.6-12).

To appreciate the simplicity of the coupling coefficients, Eqs. (3.6-10)–(3.6-12), the reader should compare the present theory with the theory presented in [Me2], which was based on the exact form of the mode field expressions. However, for many applications the coupling coefficients can be simplified even further. For multimode applications most of the guided modes are far from their cutoff point, so that we can utilize far from cutoff approximations. We write Eq. (2.2-65) in the form

$$\gamma J_\nu(\kappa a) = -\kappa J_{\nu-1}(\kappa a) \tag{3.6-13}$$

TABLE 3.6.1

$e_{\mu\nu m}$[a, b]

Incident mode	Spurious mode	Distortion
$\cos \nu\phi$	$\cos \mu\phi$ $e_{\mu\nu m} = \begin{cases} 4, & \nu = \mu = m = 0 \\ 2, & \begin{cases} \nu = 0, & \mu = m \\ \mu = 0, & \nu = m \end{cases} \\ 1, & \begin{cases} 0 < \mu = \nu \pm m \\ 0 < \mu = m - \nu \end{cases} \end{cases}$	$\cos m\phi$
$\sin \nu\phi$	$\sin \mu\phi$ $e_{\mu\nu m} = \begin{cases} 0, & \nu \text{ or } \mu = 0 \\ 1, & 0 < \mu = \nu \pm m \\ -1, & 0 < \mu = m - \nu \end{cases}$	$\cos m\phi$
$\cos \nu\phi$	$\sin \mu\phi$ $e_{\mu\nu m} = \begin{cases} 0, & \mu = 0 \\ 2, & \nu = 0, \ \mu = m \\ 1, & \mu = \nu + m \\ -1, & 0 < \mu = \nu - m \\ 1, & 0 < \mu = m - \nu \end{cases}$	$\sin m\phi$
$\cos \nu\phi$	$\sin \mu\phi$ $e_{\mu\nu m} = 0$	$\cos m\phi$
$\cos \nu\phi$	$\cos \mu\phi$ $e_{\mu\nu m} = 0$	$\sin m\phi$
$\sin \nu\phi$	$\sin \mu\phi$ $e_{\mu\nu m} = 0$	$\sin m\phi$

[a] The parameter $e_{\mu\nu m}$ determines which modes can couple to each other. The symmetries of the modes and of the core deformation are indicated by the sine and cosine functions listed in the table ($e_{\mu\nu m} = 0$, unless specified otherwise).

[b] *From* D. Marcuse, "Coupled mode theory of round optical fibers," *Bell. Syst. Tech. J.* **52**, 817–842 (1973). Copyright 1973, The American Telephone and Telegraph Co., reprinted by permission.

This equation was derived from Eq. (2.2-39) with the help of Eq. (2.2-62). Proceeding in the same manner from Eq. (2.2-38) results in

$$\gamma J_\nu(\kappa a) = \kappa J_{\nu+1}(\kappa a) \tag{3.6-14}$$

The square root of the product of these two equations is

$$\gamma J_\nu(\kappa a) = \kappa |J_{\nu-1}(\kappa a) J_{\nu+1}(\kappa a)|^{1/2} \tag{3.6-15}$$

Substitution of Eq. (3.6-15) into Eq. (3.6-10) yields the far from cutoff approximation of the coupling coefficient between two guided modes:

$$\hat{K}_{\mu\nu}^{(p,q)} = p \frac{e_{\mu\nu m}}{(e_\nu e_\mu)^{1/2}} \frac{\kappa_\mu \kappa_\nu}{2iank} \tag{3.6-16}$$

The parameter κ_ν is obtained either from Eq. (2.2-67), for $\nu \neq 0$, or from Eq. (2.2-68), for $\nu = 0$.

The coupling coefficients, Eqs. (3.6-11) and (3.6-12), can also be simplified with the help of Eq. (3.6-15). We obtain the far from cutoff approximation of Eq. (3.6-11) in the form

$$\hat{K}_{\mu\nu}^{(p,q)} = \frac{p e_{\mu\nu m}}{(e_\nu e_\mu)^{1/2}} \frac{[(n_1/n_2)-1]^{1/2}(\rho)^{1/2}\kappa_\nu J_\mu(\sigma a)}{i\pi a|\sigma J_{\mu-1}(\sigma a) H_\mu^{(1)}(\rho a) - \rho J_\mu(\sigma a) H_{\mu-1}^{(1)}(\rho a)|} \tag{3.6-17}$$

The far from cutoff approximation of Eq. (3.6-12) is

$$\hat{K}_{\mu\nu}^{(p,q)} = \frac{p e_{\mu\nu m}}{(e_\nu e_\mu)^{1/2}} \frac{\{n[(n_1/n_2)-1]k\rho\beta\}^{1/2}\kappa_\nu J_\mu(\rho a)}{i[2(\beta^2+n^2k^2)]^{1/2}} \tag{3.6-18}$$

These far from cutoff approximations of the coupling coefficients for weakly guiding round fibers are even simpler than the coupling coefficients for the TE modes of the asymmetric slab waveguide.

4

Applications of the Coupled Mode Theory

4.1 Introduction

The coupled mode theory of Chap. 3 is capable of describing, at least in principle, any type of waveguide imperfection. The general coupling coefficients Eq. (3.2-44) for ideal modes and Eq. (3.3-10) for local normal modes can be used to determine the transfer of power from one mode to any other mode. Equation (3.3-20) is already restricted to coupling between modes with real propagation constants excluding the evanescent modes of the continuum of radiation modes. Equations (3.5-6) and (3.5-21) apply only to core boundary deflections with small values of the deformation functions $f(z)$ and $h(z)$ or their derivatives. The examples that we consider in this chapter are likewise restricted to slight core boundary deviations from the geometry of the perfect dielectric waveguide. A fairly detailed discussion of core boundary imperfections for symmetric slab waveguides has already been given in [Me1]. We are limiting the discussion in this chapter to core boundary imperfections of asymmetric slab waveguides and of weakly guiding optical fibers. These examples are important for the determination of scattering losses of integrated optics waveguides and fibers and can also be used for the design of grating couplers.

The coupled mode theory considers coupling between modes that are members of a complete orthogonal set. The waves that propagate in hollow dielectric waveguides are not part of an orthogonal set of normalizable modes. The coupling coefficients derived in Chap. 3 are thus not directly applicable to hollow dielectric waveguides. However, we show in Sect. 4.3 how coupling between modes and radiation losses of leaky waves in hollow dielectric waveguides can be described by means of a plane wave analysis.

4.2 Slab Waveguide with Sinusoidal Deformation

As a first application of the general coupled mode theory, we consider an asymmetric slab waveguide with sinusoidal deformation of one core boundary

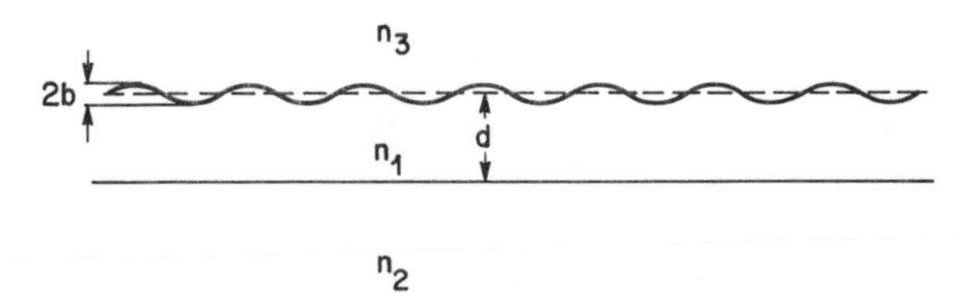

Fig. 4.2.1 *Slab waveguide with sinusoidal deformation of one core–cladding interface.*

as shown in Fig. 4.2.1. The function $f(z)$ is now given as

$$f(z) = b \sin \theta z \tag{4.2-1}$$

while we assume $h(z) = 0$. From Eq. (3.5-27) we obtain the coupling coefficient for two guided modes:

$$R_{\mu\nu}^{(p,q)} = \pm 2p\hat{R}_{\mu\nu} \cos \theta z \tag{4.2-2}$$

with

$$\hat{R}_{\mu\nu} = \frac{-\kappa_\mu \kappa_\nu b}{2\{|\beta_\mu \beta_\nu|[d+(1/\gamma_\mu)+(1/\delta_\mu)][d+(1/\gamma_\nu)+(1/\delta_\nu)]\}^{1/2}} \tag{4.2-3}$$

It was shown in Sect. 3.4, Eq. (3.4-8), that the amplitude of a spurious mode can grow to appreciable values only if there is a Fourier coefficient of the deformation function at the spatial frequency (we ignore the slight z dependence of the β_ν of local normal modes):

$$\beta_\nu^{(q)} - \beta_\mu^{(p)} = \pm \theta \tag{4.2-4}$$

The plus or minus sign in Eq. (4.2-2) corresponds to the plus or minus sign in Eq. (4.2-4). This relation is assumed to hold between the two guided modes and the spatial frequency θ of the deformation function. Modes whose propagation constants violate relation (4.2-4) are not coupled effectively. It is thus possible to neglect all the modes except the two that obey relation (4.2-4) and limit the infinite equation system (3.3-8) and (3.3-9) to only two equations. If more than two modes satisfy relation (4.2-4) simultaneously, all these modes would have to be considered. However, for a waveguide supporting only a few guided modes, we can assume that relation (4.2-4) is satisfied for only two of them. The validity of this assumption can be verified by considering the separation $\Delta\beta$ in β space of adjacent guided modes. From Eqs. (1.3-89) and (1.3-92) we obtain

$$\Delta\beta d \approx -(\nu+1)\pi^2/\beta d \tag{4.2-5}$$

The separation between adjacent modes is thus not constant but increases with increasing mode number. It is unlikely that Eq. (4.2-4) will hold for more than two modes simultaneously. Labeling the two modes by the subscripts 1 and 2, we obtain

$$dc_1^{(+)}/dz = 2\hat{R}_{11}c_1^{(+)}\cos\theta z + \hat{R}_{12}(1+e^{-2i\theta z})\,c_2^{(p)} \tag{4.2-6}$$

and

$$dc_2^{(p)}/dz = -p\hat{R}_{12}(1+e^{2pi\theta z})\,c_1^{(+)} + 2\hat{R}_{22}c_2^{(p)}\cos\theta z \tag{4.2-7}$$

Mode 1 is supposed to travel in positive z direction, while mode 2 travels either in positive (p positive) or negative (p negative) direction. The mode amplitudes c_1 and c_2 are slowly varying functions of z. We may take a spatial average over a distance in the order of $1/\theta$. The mode amplitudes and their derivatives do not change much over this distance, so that we consider their averages equal to their values at the point z. The spatial averaging eliminates the sinusoidal functions from the equation, so that we obtain

$$dc_1^{(+)}/dz = \hat{R}_{12}c_2^{(p)} \tag{4.2-8}$$

and

$$dc_2^{(p)}/dz = -p\hat{R}_{12}c_1^{(+)} \tag{4.2-9}$$

The solutions of these equations depend on whether p is positive or negative. Considering first the case that p is positive, which means that both modes travel in positive z direction, we obtain the solutions

$$c_1^{(+)}(z) = A\cos\hat{R}_{12}z + B\sin\hat{R}_{12}z \tag{4.2-10}$$

and

$$c_2^{(+)}(z) = -A\sin\hat{R}_{12}z + B\cos\hat{R}_{12}z \tag{4.2-11}$$

These solutions show that the two modes exchange power periodically as they travel along the waveguide. The amplitudes A and B depend on the initial conditions at $z = 0$. For $B = 0$ we have $c_1^{(+)}(0) = A$ and $c_2^{(+)}(0) = 0$. After the distance L given by

$$L = \pi/2\hat{R}_{12} \tag{4.2-12}$$

we find $c_1^{(+)}(L) = 0$ and $c_2^{(+)}(L) = -A$. The power has been exchanged and mode 1 is completely depleted, while all the power resides in mode 2.

To get a feeling for the strength of the coupling process, we consider an example. We use $n_1 = 1.6$, $n_2 = 1.5$, and $n_3 = 1$. The V parameter is chosen so that two guided TE modes can be supported by the waveguide. We thus use $V = 7$, according to Eq. (1.3-42). The cutoff value of the third mode with $v = 2$ is $V_c = 7.392$, and the second mode cuts off at $V_c = 4.25$. $V = 7$ thus guarantees that only two TE modes can exist, but that both modes are well guided. From the eigenvalue Eq. (1.3-26) and relations (1.2-13)–(1.2-15), we obtain the solutions: $\kappa_1 d = 2.562$, $\gamma_1 d = 6.514$, $\delta_1 d = 12.31$, $\beta_1 d = 19.95$, and $\kappa_2 d = 5.061$, $\gamma_2 d = 4.836$, $\delta_2 d = 11.51$, $\beta_2 d = 19.47$. With these parameters we obtain from Eqs. (4.2-3) and (4.2-12)

$$L/d = 6.03 d/b \tag{4.2-13}$$

If we assume a vacuum wavelength of $\lambda = 1\ \mu\text{m}$, we have with the above values $d = 2\ \mu\text{m}$. A total conversion length of $L = 1$ cm is obtained with an amplitude of $b = 0.0024\ \mu\text{m}$ of the sinusoidally deformed top surface of the asymmetric slab waveguide. This example shows how very effectively a sinusoidal deformation of the core boundary couples two guided modes to each other.

The situation is different if mode 2 travels in the negative z direction. For $p = -1$, both Eqs. (4.2-8) and (4.2-9) have the same sign on the right-hand side. The solution of the equation system is now

$$c_1^{(+)}(z) = A \cosh \hat{R}_{12} z + B \sinh \hat{R}_{12} z \tag{4.2-14}$$

$$c_2^{(-)}(z) = A \sinh \hat{R}_{12} z + B \cosh \hat{R}_{12} z \tag{4.2-15}$$

The circular functions of the solution Eqs. (4.2-10) and (4.2-11) have now been replaced with hyperbolic functions. If we consider mode 1 as the incident wave and mode 2 as the reflected wave, we impose the following boundary conditions:

$$c_1^{(+)}(0) = A \tag{4.2-16}$$

and

$$c_2^{(-)}(\infty) = 0 \tag{4.2-17}$$

From Eqs. (4.2-15) and (4.2-17), we find $B = -A$, so that Eqs. (4.2-14) and (4.2-15) assume the form

$$c_1^{(+)}(z) = A \exp(-\hat{R}_{12} z) \tag{4.2-18}$$

and

$$c_2^{(-)}(z) = -A \exp(-\hat{R}_{12} z) \tag{4.2-19}$$

Both waves decay in positive z direction. However, whereas mode 1 travels in positive z direction so that it decreases in amplitude in its direction of propagation, mode 2 grows as it travels in negative z direction. Mode 2 builds up at the expense of mode 1. At the beginning of the periodically deformed waveguide at $z = 0$, both waves have the same amplitude. The waveguide is a periodic structure with a stop band. Condition (4.2-4) ensures that the incident wave enters the periodic structure at the peak of the stop band so that it is totally reflected. However, if the periodic perturbation is only slight, the incident wave is able to penetrate a considerable distance into the waveguide. At the distance L given by Eq. (4.2-12), the amplitude of mode 1 has decayed to $0.21A$, so that the power of the wave has decreased to 4.3% of its initial value. If we use the same numbers that applied to the first example, we find that the incident wave decays to 4.3% of its initial power in 1 cm if the amplitude of the core boundary ripple is 0.0024 μm.

If condition (4.2-4) applies to two guided modes, it means that the mode with the larger propagation constant can not couple to the radiation modes. However, the mode with the smaller propagation constant, mode 2, couples to radiation modes if

$$\beta_2 - n_2 k \leqslant \theta \tag{4.2-20}$$

However, both modes may be sufficiently far removed from the continuous spectrum of radiation modes that neither one suffers radiation loss. In the two-mode numerical example given above, we had $\theta d = (\beta_1 - \beta_2)d = 0.48$, and $(\beta_2 - n_2 k)d = 0.61$, so that even mode 2 is too far from the continuous spectrum of modes to be able to couple to it.

Coupling of a guided mode to the continuum of radiation modes results in loss. We again use the sinusoidal deformation function (4.2-1) but assume that θ is large enough, so that condition (4.2-4) holds for the guided mode v and the radiation mode $\mu = \rho$. We limit the discussion to the case that ρ lies in the interval (3.5-28), so that the guided mode couples to radiation modes that decay exponentially in x direction into region 3 of the asymmetric slab waveguide. Radiation losses caused by sinusoidal core boundary perturbations have been studied for symmetric slab waveguides in [Me1]. However, symmetric slab waveguides do not have exponentially decaying radiation modes since the range (3.5-28) shrinks to zero for $n_2 = n_3$. The radiation caused by a sinusoidal core boundary ripple emerges on both sides of the symmetric slab. For radiation modes of the asymmetric slab in the range (3.5-28), radiation is directed only into the substrate region 2. Since the process of radiation can be reversed to excite a guided mode with an incident radiation field, asymmetric slab waveguides offer the advantage of higher coupling efficiency, since radiation with only one radiation lobe is possible. One need only reverse the process of radiation from a guided mode to see that, in principle, all the irradiated power can be trapped into a guided mode if only one radiation lobe exists. With two radiation lobes, two properly phased radiation fields would be necessary to achieve 100% conversion efficiency. This condition is hard to realize for optical fields.

From Eqs. (3.5-29), (4.2-1), and (4.2-2) we obtain, with the help of Eq. (4.2-4),

$$\hat{R}_{\rho\nu} = -((n_1{}^2 - n_3{}^2)^{1/2} k\kappa_\nu \rho b / 2\{\pi |\beta_\nu \beta_\rho| [d + (1/\gamma_\nu) + (1/\delta_\nu)]\}^{1/2}) \cdot \sigma / \{\rho^2 [\sigma \cos \sigma d + (\Delta/i) \sin \sigma d]^2 + \sigma^2 [\sigma \sin \sigma d - (\Delta/i) \cos \sigma d]^2\}^{1/2} \tag{4.2-21}$$

The propagation constant β_ρ of the radiation modes is not a constant in this case but a continuous variable. Condition (4.2-4) can therefore be satisfied only for one exact point of the continuum. However, it will soon become apparent that only a very narrow range of radiation modes contributes to radiation losses, so that we make actually no appreciable error by using Eq. (4.2-4) to remove the difference of the propagation constants appearing in Eq. (3.5-29).

We obtain the radiation loss coefficient $2\alpha_\nu$ from Eq. (3.4-19) with $\hat{K}_{\rho\nu}$ replaced by $2\hat{R}_{\rho\nu}$. [The factor 2 stems from definition (4.2-2).] The Fourier coefficient follows from Eqs. (3.4-10) and (4.2-1):

$$\begin{aligned} F(\beta_\nu - \beta_\rho) &= \sqrt{L^{-1}} \int_0^L \cos \theta z \exp[-i(\beta_\nu - \beta_\rho) z]\, dz \\ &= \sqrt{L^{-1}} \{\exp\{i[\theta - (\beta_\nu - \beta_\rho)] L/2\} \sin[\theta - (\beta_\nu - \beta_\rho)](L/2)/[\theta - (\beta_\nu - \beta_\rho)] \\ &\quad + \exp\{-i[\theta + (\beta_\nu - \beta_\rho)] L/2\} \sin[\theta + (\beta_\nu - \beta_\rho)](L/2)/[\theta + (\beta_\nu - \beta_\rho)]\} \end{aligned} \tag{4.2-22}$$

In most cases of practical interest, L is much larger than the vacuum wavelength of light. The expression on the right-hand side of Eq. (4.2-22) is thus very small unless the argument of one of the functions of the form $(\sin x)/x$ vanishes. $(\sin x)/x$ has its maximum value at $x = 0$. Since $\theta > 0$ and $\beta_\nu - \beta_\rho > 0$, only the argument of the first $(\sin x)/x$ function can vanish. The second function makes only a very small contribution and can be neglected. We thus use the approximation

$$F(\beta_v - \beta_\rho) = \sqrt{L}^{-1}\{\sin[\theta - (\beta_v - \beta_\rho)](L/2)/[\theta - (\beta_v - \beta_\rho)]\} \cdot \exp\{i[\theta - (\beta_v - \beta_\rho)]L/2\} \quad (4.2\text{-}23)$$

For large values of L, the peak of the $(\sin x)/x$ function in Eq. (4.2-23) is very narrow so that the other terms, which appear under the integral sign in Eq. (3.4-19), are constant over the region that contributes to the integral. The remaining integral over the square of the Fourier coefficient is

$$\int_{-n_2k}^{n_2k} |F(\beta_v - \beta_\rho)|^2 \, d\beta \approx \tfrac{1}{2} \int_{-\infty}^{\infty} [(\sin x)/x]^2 \, dx = \pi/2 \quad (4.2\text{-}24)$$

The radiation loss coefficient of a TE mode for a sinusoidal core boundary deformation follows now directly from Eqs. (3.4-19), (4.2-21), and (4.2-24):

$$2\alpha = \{(n_1{}^2 - n_3{}^2)k^2\kappa_v{}^2 b^2/\beta_v[d + (1/\gamma_v) + (1/\delta_v)]\} \cdot \rho\sigma^2/2\{\rho^2[\sigma\cos\sigma d + (\Delta/i)\sin\sigma d]^2 + \sigma^2[\sigma\sin\sigma d - (\Delta/i)\cos\sigma d]^2\} \quad (4.2\text{-}25)$$

The parameters Δ, σ, and ρ are obtained from Eqs. (1.4-7)–(1.4-9), with

$$\beta = \beta_\rho = \beta_v - \theta \quad (4.2\text{-}25\text{a})$$

Similar expressions for the radiation loss coefficients for TE modes in the range (3.5-31) can be obtained from Eq. (3.5-32). However, in this case we must add the contributions from the two types of radiation modes corresponding to the two choices for the sign of the square root in Eq. (1.4-24)—as indicated by the summation symbol in Eq. (3.4-19). Radiation losses for TM modes can likewise be calculated from Eqs. (3.5-35) and (3.5-37).

If we regard θ as a continuous variable we can vary β according to Eq. (4.2-25a) and let it range throughout the interval

$$n_3 k < \beta < n_2 k \quad (4.2\text{-}26)$$

Figure 4.2.2 shows the normalized power loss coefficient $2\alpha d^3/b^2$ as a function of ρd in the interval (3.5-28) that corresponds to the interval (4.2-26). The curve was drawn for the second guided mode used in the two-mode example earlier in this section. The guided mode has the propagation constant $\beta_2 d = 19.47$, and the waveguide is operated at the value $V = 7$. The oscillations in the curve are caused by the fact that the core boundaries reflect the radiation originating at the sinusoidally deformed upper core boundary. Depending on the direction at which the radiation is injected into the core region, the reflected and the direct beams interfere constructively or destructively.

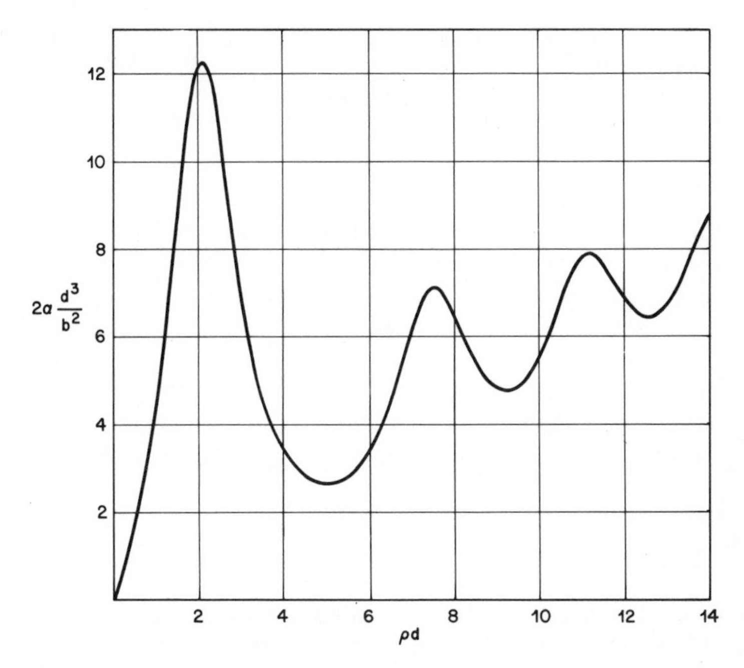

Fig. 4.2.2 *Power loss coefficient 2α of a slab waveguide with sinusoidal core–cladding interface distortion. The parameter ρ is related to the spatial frequency θ of the sinusoidal distortion, $\rho = [n_2{}^2k^2 - (\beta_v - \theta)^2]^{1/2}$. The loss peaks are leaky wave resonances.*

We can look at the interference phenomenon in a different way. We have seen in Sect. 1.5 that waves which travel in the core region with angles exceeding the total internal reflection angle (measured with respect to the core boundary) can be regarded as leaky waves. From Eq. (1.5-2) we find the following solutions of the transcendental equation, with $n_1 = 1.6$, $n_2 = 1.5$, $n_3 = 1$, and $V = 7$:

$$\begin{aligned} N &= 2, \quad \kappa d = 7.366, \quad \rho d = 2.293 \\ N &= 3, \quad \kappa d = 10.282, \quad \rho d = 7.531 \\ N &= 4, \quad \kappa d = 13.145, \quad \rho d = 11.126 \end{aligned} \tag{4.2-27}$$

The parameter ρd listed in the last column of Eq. (4.2-27) corresponds to the abscissa in Fig. 4.2.2 and is related to κ by the expression

$$\rho d = [(\kappa d)^2 - V^2]^{1/2} \tag{4.2-28}$$

The values of ρd listed in Eq. (4.2-27) coincide with the maxima of the loss curve in Fig. 4.2.2. The radiation loss peaks can thus be interpreted as "resonant" power transfer from the guided wave to leaky waves. The coupling between the guided and leaky waves is provided by the sinusoidal core boundary ripple. The amount of radiation loss that the guided mode suffers depends on the angle at which the radiation is emitted. Using an average value of

$$2\alpha d^3/b^2 = 6 \tag{4.2-29}$$

and the parameters of the last example, $d = 2\ \mu\text{m}$ and $b = 0.0024\ \mu\text{m}$, we obtain a radiation loss of $2\alpha = 0.043\ \text{cm}^{-1} = 0.19\ \text{db cm}^{-1}$. When we consider that the same ripple amplitude was sufficient to cause complete power transfer between two guided modes in 1 cm of length, it appears that sinusoidal core boundary deformations couple guided modes to radiation modes less effectively. A loss of $2\alpha = 1\ \text{cm}^{-1}$ is caused by a ripple amplitude of $b = 0.012\ \mu\text{m}$. Since the vacuum wavelength in our example was chosen to be $\lambda = 1\ \mu\text{m}$, these ripple amplitudes are safely within the limits of applicability of the approximation (4.2-21) for the coupling coefficient.

Our theory allowed us to calculate the radiation losses caused by a sinusoidal core boundary ripple without ever considering the process of radiation in detail. The radiation losses could be calculated, since we have a perturbation solution for the amplitude of the radiation modes that are coupled to the guided mode. We were thus able to account for the power loss of the guided mode simply by calculating the power that is carried by the radiation modes. We conclude our discussion of asymmetric slab waveguides with sinusoidal core boundary ripples by calculating the far field of the radiation carrying the power that is coupled out of the guided mode.

We have already derived all the necessary equations and only need to collect our results. Using the traveling wave representation Eq. (3.2-34), we obtain the radiation field from the integral expression (3.2-18):

$$E_y = \int_0^\infty c_\rho \mathscr{E}_{\rho y} e^{-i\beta z}\, d\rho \tag{4.2-30}$$

The expression for the radiation mode field is given in Eq. (1.4-6). Since the field in region 3 decays exponentially, it does not radiate, so that we can restrict the discussion to region 2. After substitution of the amplitude coefficients, we obtain

$$\mathscr{E}_{\rho y} = [2(\omega\mu_0 P)^{1/2}/(\pi|\beta|)^{1/2}] \cos[\rho(x+d) - \Psi] \tag{4.2-31}$$

with

$$\tan\Psi = [\sigma \sin\sigma d - (\Delta/i)\cos\sigma d]/[\sigma\cos\sigma d + (\Delta/i)\sin\sigma d] \tag{4.2-32}$$

The expansion coefficient in Eq. (4.2-30) follows from Eqs. (3.4-8) (with

$\hat{K}_{\rho\nu}$ replaced by $2\hat{R}_{\rho\nu}$), (4.2-21), and (4.2-23):

$$c_\rho = c_\nu(0)\left(\frac{2}{\pi}\frac{\rho}{|\beta|}2\alpha\right)^{1/2}\frac{\sin[\theta-(\beta_\nu-\beta)]L/2}{\theta-(\beta_\nu-\beta)}\exp\{i[\theta-(\beta_\nu-\beta)]L/2\} \tag{4.2-33}$$

Formula (4.2-25) was used to abbreviate the notation.

The amplitude coefficient c_ρ is actually a function of z. However, we have tacitly assumed that the position z at which the field is to be evaluated is located somewhat outside the range $0 < z < L$. This means that we are assuming that the sinusoidal ripple persists only over a length L and is followed by an infinite length of perfect waveguide. The integral in Eq. (3.4-2), which extends from 0 to the point z where the field is evaluated, contributes only over the range from 0 to L, since for $z > L$ we have assumed $f(z) = 0$. We can also state our procedure by saying that we want to consider only the radiation that is contributed by the section of waveguide of length L. If the imperfect waveguide is actually longer than L, we ignore the radiation that is contributed by any other part of it.

The integral (4.2-30) is very complicated, since β is a function of ρ, and α depends on ρ in a complex way. Therefore, we can not hope to solve the integral exactly. However, far from the source of the radiation the problem simplifies, since for very large values of x and z there are rapidly oscillating functions under the integral sign that cause almost complete cancellation. The only contribution to the integral stems from a very narrow region of the integration range where the rapid oscillations cease. This point is called a stationary point, and the method of utilizing the rapid oscillations of the integrand and evaluating only the region around the stationary point is called "the method of stationary phase" [MW1].

In order to be able to use the method of stationary phase, we move the field point far from the source of the radiation and assume that $z \gg L$ and $x \gg L$. Compared to the rapid variation of the oscillatory functions that contain x and z, all other terms in the integrand of Eq. (4.2-30) change only slowly. This statement does not contradict the assumption made earlier that L is much larger than the wavelength. We concentrate now only on the rapidly varying terms

$$\cos[\rho(x+d)-\Psi]e^{-i\beta z} = \tfrac{1}{2}e^{i(\rho d-\Psi)}e^{i(\rho x-\beta z)} + \tfrac{1}{2}e^{-i(\rho d-\Psi)}e^{-i(\rho x+\beta z)} \tag{4.2-34}$$

Because of the very large values of x and z, the two exponential functions containing these variables vary so rapidly that their oscillations cancel out any contribution to the integral. However, there are points on the ρ axis where the rapid oscillations cease locally. These stationary points occur where the derivative of the exponent of the exponential function vanishes. Using Eq. (1.4-9) we take the derivative of the exponent and set it equal to zero:

$$d(\rho x - \beta z)/d\rho = x + (\rho/\beta)z = 0 \tag{4.2-35}$$

Equation (4.2-35) defines the stationary point $\rho = \rho_s$. In order to evaluate the integral, we expand the exponent in a Taylor series around the stationary point. The Taylor expansion is carried to the second-order term. We thus need also the second derivative of the exponent

$$d^2(\rho x - \beta x)/d\rho^2 = (\beta^2 + \rho^2)z/\beta^3 = n_2{}^2k^2z/\beta^3 \tag{4.2-36}$$

Equation (1.4-9) has again been used. In the vicinity of the stationary point we have approximately

$$\rho x - \beta z = \rho_s x - \beta_s z + (n_2{}^2k^2/2\beta_s{}^3)z(\rho - \rho_s)^2 + \cdots \tag{4.2-37}$$

The subscript s indicates that the quantities are evaluated at the stationary point $\rho = \rho_s$. The linear term is absent from this series expansion because of Eq. (4.2-35).

Equation (4.2-35) has a solution for real positive values of ρ if x is negative. Since we are looking for a radiation field in region 2 where we have $x < 0$ and $z > 0$, we are assured that the exponent has a stationary point. By the same argument we see that the other exponent occurring in Eq. (4.2-34), $\rho x + \beta z$, does not have a stationary point. To the approximation considered here, this exponent and the terms containing it do not contribute to the electric field (4.2-30). This is a very interesting observation. The field of the radiation mode (4.2-31) is a standing wave in x direction. We are now demonstrating that this standing wave field contributes traveling waves if the radiation modes are superimposed on each other as in Eq. (4.2-30). The standing wave is decomposed into the two traveling wave components Eq. (4.2-34). As we have just seen, only one of these traveling wave components has a stationary point in the region $z > 0$ and $x < 0$, so that it can contribute to the radiation field. The presence of the stationary point means that it is here where all the radiation modes superimpose each other in phase. At all other points we have destructive interference. The method of stationary phase makes it clear that only the outward traveling parts of the standing wave superimpose each other constructively, while the inward traveling parts are destroyed by destructive interference. In this way the radiation modes provide a traveling wave radiation field even though each of them individually consists of a standing wave.

The region over which the stationary part of the oscillatory function contributes is so narrow that all other factors appearing in the integrand of Eq. (4.2-30) can be regarded as constant and can be taken out of the integration sign. The remaining integration is of the form

$$\int_{-\infty}^{\infty} \exp(iau^2)\, du = (\pi/a)^{1/2} \exp(i\pi/4) \tag{4.2-38}$$

The finite limits on the integral were extended to infinity since the rapidly varying parts of the integrand do not contribute in any case. Collecting our results we obtain the following expression for the radiation far field:

$$E_y = c_v(0) \frac{2(\omega\mu_0 P)^{1/2}}{\pi^{1/2}} (2\alpha \sin\phi)^{1/2} \exp(i\pi/4) \exp[i(n_2 kd \cos\phi - \Psi)]$$
$$\cdot \exp[i(\theta - \beta_v + n_2 k \cos\phi) L/2] \frac{\sin(\theta - \beta_v + n_2 k \cos\phi) L/2}{\theta - \beta_v + n_2 k \cos\phi} \frac{\exp(-in_2 kr)}{r^{1/2}} \tag{4.2-39}$$

In addition to the equations listed above, the following relations were used:

$$x = -r \sin\phi \tag{4.2-40}$$

$$z = r \cos\phi \tag{4.2-41}$$

From these two equations and Eqs. (1.4-9) and (4.2-35), we obtain

$$\rho_s = n_2 k \sin\phi \tag{4.2-42}$$

and

$$\beta_s = n_2 k \cos\phi \tag{4.2-43}$$

The radiation field is proportional to the amplitude $c_v(0)$ of the guided wave that excited it *via* the sinusoidal core boundary ripple. It is also proportional to $(\sin\phi)^{1/2}$, where ϕ is the angle at which the field point appears as seen from the source of the radiation. The various phase factors are of no particular importance. However, the factor $[\exp(-in_2 kr)]/r^{1/2}$ identifies the radiation field as a cylindrical wave. That we obtain a cylindrical instead of a spherical wave is caused by the shape of the source. Our problem is two-dimensional in nature, since we have assumed that the slab waveguide is infinitely extended in y direction. The ripply core boundary is thus a line source as seen from an infinite distance. The factor of the form $(\sin x)/x$ in Eq. (4.2-39) modulates the amplitude of the radiation field and produces a very narrow radiation lobe for large values of L.

The sinusoidal core boundary ripple is thus seen to scatter power out of the incident guided mode into a direction that is obtained by setting the argument of the $(\sin x)/x$ function equal to zero. The peak of the radiation lobe appears at

$$\cos\phi = (\beta_v - \theta)/n_2 k \tag{4.2-44}$$

According to this derivation, 2α of Eq. (4.2-25) appears in Eq. (4.2-39) simply as an amplitude factor. In order to identify its meaning in the context of the radiation field study, we determine the power that is carried away by the radiation field,

$$\Delta P = (n_2 k/2\omega\mu_0)\int_0^{\pi} |E_y|^2 r \, d\phi \tag{4.2-45}$$

We introduce the new integration variable

$$n_2 k \cos\phi = u \tag{4.2-46}$$

and obtain from Eqs. (4.2-39) and (4.2-45), with the help of Eq. (4.2-24),

$$\Delta P/|c_\nu(0)|^2 PL = 2\alpha \tag{4.2-47}$$

Since $|c_\nu(0)|^2 P$ is the power carried by the guided mode and $\Delta P/L$ is the power radiated per unit length of sinusoidally deformed waveguide, we see that Eq. (4.2-47) is indeed the correct expression for the radiation loss. The calculation of the radiation field thus leads to an independent determination of the radiation loss formula (4.2-25), since 2α was used in the present derivation only as a convenient abbreviation.

Radiation losses caused by more general core boundary deformations can be described by superimposing sinusoidal core deformations on each other. However, we shall not pursue this procedure since a loss formula for general core boundary deformations has already been derived in Sect. 3.4, Eq. (3.4-19). Radiation losses for random core boundary perturbations of symmetric slab waveguides have been discussed in some detail in Sect. 9.4 of [Me1]. In this book we restrict the discussion of radiation effects caused by random core boundary deviations to weakly guiding round optical fibers (see Sect. 4.6).

We have shown in this section that two guided modes exchange their power completely if they are coupled by means of a sinusoidal core boundary perturbation. From the fact that only one Fourier component of a general boundary distortion function is instrumental in coupling two guided modes, the reader might conclude that two modes would still exchange their power completely provided that a suitable Fourier component of a more general coupling function existed. That complete power transfer does not occur if the distortion function is random can be seen by the following argument: A narrow-band random function can be described by a sinusoidal function with randomly varying amplitude and phase. However, the phase relationship between the deformation function and the two guided modes is crucial for the power transfer process. Whether the power flows from mode 1 to mode 2 or *vice versa* depends on the relative phases of the three functions—the two modes and the sinusoidal coupling function. If we assume that the phase relationship is such that at a given point on the z axis power is being transferred from mode 1 to mode 2, this phase relationship would have to be preserved throughout the entire exchange length if the power transfer is to be complete. A random function with its randomly changing phase does not provide the proper phase relationship for complete power transfer. Instead the direction of power transfer fluctuates randomly with the random phase of the coupling function. Complete power transfer occurs only with low probability as a result of random fluctuations. On the average the two randomly coupled modes will carry equal power. Power exchange caused by random coupling will be discussed in Chap. 5.

4.3 Hollow Dielectric Waveguide with Sinusoidal Deformation

Hollow dielectric waveguides were discussed in Sect. 1.6, where we saw that hollow dielectric waveguides do not support proper guided modes. Their complete set of orthogonal modes consists entirely of radiation modes. However, wave guidance is still possible in these structures. The guided waves of hollow dielectric waveguides are leaky waves that cannot exist without losing power by radiation. The leaky wave solutions of the eigenvalue equation are not normalizable and do not belong to the complete set of orthogonal modes. The derivation of the coupling coefficients in Chap. 3 thus does not apply to leaky waves.

Hollow dielectric waveguides are important for laser cavities. Coupling between forward and backward traveling modes of hollow dielectric waveguides can be used as a feedback mechanism for laser oscillators [KS1, KS2, SBK, SSS]. It is thus worthwhile to derive expressions for the coupling coefficient between the leaky waves of hollow dielectric waveguides. Our derivation is based on an approximate plane wave analysis [Me3].

We assume that only one of the core boundaries of the hollow slab waveguide is sinusoidally deformed and begin the study of coupling between two leaky waves by considering the problem of diffraction, reflection, and scattering of a plane wave at the sinusoidally distorted interface between two dielectric media shown in Fig. 4.3.1. The arrows shown in the figure indicate the propagation

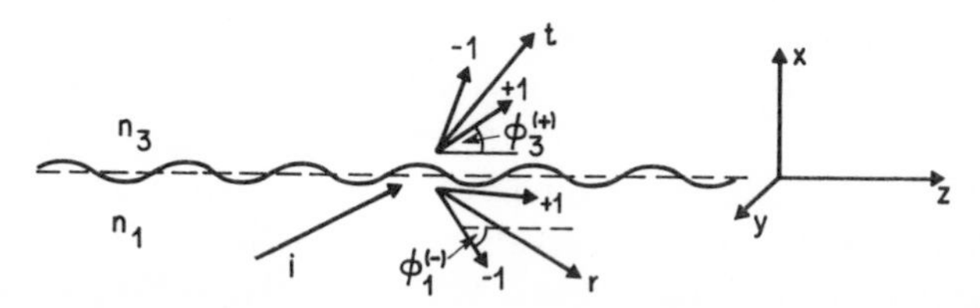

Fig. 4.3.1 *The sinusoidally distorted core–cladding interface acts like a phase grating. The incident, reflected, transmitted, and diffracted rays are shown.*

vectors of the incident, reflected, and transmitted plane waves labeled i, r, and t. The shorter arrows labeled 1 and -1 indicate scattered plane waves. We limit the discussion to TE waves because they have lower losses than TM waves and are thus more important for distributed feedback lasers. TE waves have only one electric field component, E_y, if their propagation vectors lie in the xz plane. We consider the problem to first order of perturbation theory. The fields of zero order are the incident, reflected, and transmitted plane waves. The scattered waves are assumed to be small of first order. The zero-order field is obtained by considering a plane interface,

$$E_y = A_i \exp[-i(\sigma_0 x + \beta_0 z)] + A_r \exp[i(\sigma_0 x - \beta_0 z)], \qquad \text{for} \quad x < 0 \tag{4.3-1}$$

$$E_y = A_t \exp[-i(\Delta_0 x + \beta_0 z), \qquad \text{for} \quad x > 0 \tag{4.3-2}$$

The parameters Δ_0 and σ_0 are defined by Eqs. (1.4-7) and (1.4-8). The magnetic field components are obtained from Eqs. (1.3-13) and (1.3-14). The requirement of continuity of E_y and H_z at the dielectric interface at $x = 0$ yields the following equations:

$$A_r = \{(\sigma_0 - \Delta_0)/(\Delta_0 + \sigma_0)\} A_i \tag{4.3-3}$$

and

$$A_t = \{2\sigma_0/(\sigma_0 + \Delta_0)\} A_i \tag{4.3-4}$$

The fields (4.3-1) and (4.3-2) do not satisfy the boundary condition if the dielectric interface is not plane but is described by the function

$$x = b \sin \theta z \tag{4.3-5}$$

We must add an additional scattered field so that the superposition of Eqs. (4.3-1) and (4.3-2) with the additional field satisfies the boundary conditions approximately. The scattered field is assumed to be small and is given by the equations

$$e_y = \int_{-\infty}^{\infty} B_1(\beta) \exp[i(\sigma x - \beta z)]\, d\beta, \qquad \text{for} \quad x < 0 \tag{4.3-6}$$

$$e_y = \int_{-\infty}^{\infty} B_3(\beta) \exp[-i(\Delta x + \beta z)]\, d\beta, \qquad \text{for} \quad x > 0 \tag{4.3-7}$$

Since we do not know *a priori* what type of scattered wave will result from the interface distortion, we must express the additional field in the most general way as a superposition of plane waves. The field in region 1, at $x < 0$, travels downward, while the field in region 3, at $x > 0$, travels upward. The field expressions are thus chosen to result in scattered waves traveling away from the dielectric interface. The magnetic field components are again obtained from Eqs. (1.3-13) and (1.3-14). The superposition of the zero-order and first-order fields must now be required to satisfy the boundary condition at the surface, Eq. (4.3-5), at least approximately. The field amplitudes $B_i(\beta)$ and the amplitude b of the sinusoidal ripple, Eq. (4.3-5), are both assumed to be quantities that are small of first order. Products of first-order quantities are neglected. When we substitute Eq. (4.3-5) into Eqs. (4.3-6) and (4.3-7) we can expand the exponential function $\exp(i\sigma b \sin \theta z)$ in a series expansion. This results in products of $B(\beta)$ with powers of b. All these products are of second and higher order so that they are all negligibly small in the context of first-order perturbation theory. We thus use $x = 0$ as the boundary of the first-order field, Eqs. (4.3-6) and (4.3-7). Continuity of the E_y and H_z components of the superposition field is achieved if the following equations are satisfied:

$$A_i(1 - i\sigma_0 b \sin \theta z) + A_r(1 + i\sigma_0 b \sin \theta z) + \int_{-\infty}^{\infty} B_1(\beta) \exp[-i(\beta - \beta_0)z]\, d\beta$$
$$= A_t(1 - i\Delta_0 b \sin \theta z) + \int_{-\infty}^{\infty} B_3(\beta) \exp[-i(\beta - \beta_0)z]\, d\beta \qquad (4.3\text{-}8)$$

and

$$-\sigma_0 A_i(1 - i\sigma_0 b \sin \theta z) + \sigma_0 A_r(1 + i\sigma_0 b \sin \theta z)$$
$$+ \int_{-\infty}^{\infty} \sigma B_1(\beta) \exp[-i(\beta - \beta_0)z]\, d\beta$$
$$= -\Delta_0 A_t(1 - i\Delta_0 b \sin \theta z) - \int_{-\infty}^{\infty} \Delta B_3(\beta) \exp[-i(\beta - \beta_0)z]\, d\beta \qquad (4.3\text{-}9)$$

Equation (4.3-8) expresses the continuity of the electric field components, and Eq. (4.3-9) results from the requirement that the z component of the magnetic field remain continuous at the boundary (4.3-5). The exponential functions with b in their arguments have been expanded up to first-order terms. The zero-order terms cancel in Eqs. (4.3-8) and (4.3-9) on account of relations (4.3-3) and (4.3-4). In addition, we find that the first-order terms in Eq. (4.3-8) containing A_i, A_r, and A_t also cancel each other. Equation (4.3-8) thus reduces to

$$\int_{-\infty}^{\infty} B_1(\beta) \exp[-i(\beta - \beta_0)z]\, d\beta = \int_{-\infty}^{\infty} B_3(\beta) \exp[-i(\beta - \beta_0)z]\, d\beta \qquad (4.3\text{-}10)$$

so that we obtain

$$B_1(\beta) = B_3(\beta) \qquad (4.3\text{-}11)$$

With the help of Eqs. (4.3-3) and (4.3-4), Eq. (4.3-9) simplifies to the form

$$\int_{-\infty}^{\infty} (\sigma+\Delta)\, B_1(\beta)\, e^{-i\beta z}\, d\beta = b\sigma_0(\Delta_0-\sigma_0)\, A_i (e^{i\theta z} - e^{-i\theta z}) \exp(-i\beta_0 z) \tag{4.3-12}$$

Equation (4.3-12) can be solved for B_1 by multiplication with $\exp(i\beta' z)$ and integration over z. Using the definition of the delta function

$$\delta(\beta'-\beta) = (1/2\pi) \int_{-\infty}^{\infty} \exp[i(\beta'-\beta)z]\, dz \tag{4.3-13}$$

we obtain

$$\int_{-\infty}^{\infty} (\sigma+\Delta)\, B_1(\beta)\, \delta(\beta'-\beta)\, d\beta = b\sigma_0(\Delta_0-\sigma_0)\, A_i [\delta(\beta'-\beta_0+\theta) - \delta(\beta'-\beta_0-\theta)] \tag{4.3-14}$$

With the help of the relation

$$\int F(\beta)\, \delta(\beta'-\beta)\, d\beta = F(\beta') \tag{4.3-15}$$

$B_1(\beta)$ is found to be

$$B_1(\beta) = bA_i [\sigma_0(\Delta_0-\sigma_0)/(\sigma+\Delta)][\delta(\beta-\beta_0+\theta) - \delta(\beta-\beta_0-\theta)] \tag{4.3-16}$$

The scattered field is finally obtained by substitution of Eq. (4.3-16) into Eq. (4.3-6):

$$e_y = bA_i \sigma_0(\Delta_0-\sigma_0) \left[\frac{\exp[i(\sigma^{(-)}x - \beta^{(-)}z)]}{\sigma^{(-)}+\Delta^{(-)}} - \frac{\exp[i(\sigma^{(+)}x - \beta^{(+)}z)]}{\sigma^{(+)}+\Delta^{(+)}} \right], \quad \text{for} \quad x > 0 \tag{4.3-17}$$

and from Eqs. (4.3-7) and (4.3-11)

$$e_y = bA_i \sigma_0(\Delta_0-\sigma_0) \left[\frac{\exp[-i(\Delta^{(-)}x + \beta^{(-)}z)]}{\sigma^{(-)}+\Delta^{(-)}} - \frac{\exp[-i(\Delta^{(+)}x + \beta^{(+)}z)]}{\sigma^{(+)}+\Delta^{(+)}} \right], \quad \text{for} \quad x > 0 \tag{4.3-18}$$

with the definitions

$$\beta^{(\pm)} = \beta_0 \pm \theta \tag{4.3-19}$$

$$\sigma^{(\pm)} = (n_1{}^2 k^2 - \beta^{(\pm)2})^{1/2} \tag{4.3-20}$$

$$\Delta^{(\pm)} = (n_3{}^2 k^2 - \beta^{(\pm)2})^{1/2} \tag{4.3-21}$$

$$\sigma_0 = (n_1{}^2 k^2 - \beta_0{}^2)^{1/2} \tag{4.3-22}$$

$$\Delta_0 = (n_3{}^2 k^2 - \beta_0{}^2)^{1/2} \tag{4.3-23}$$

The first-order theory of plane wave scattering from a sinusoidally deformed dielectric interface is interesting in its own right. It shows that a sinusoidally deformed dielectric interface scatters the incident radiation. Two scattered waves emerge on either side of the interface. One set of scattered waves accompanies the reflected wave in medium 1 and another set accompanies the transmitted wave in medium 3. The angle ϕ of the direction of each wave indicated in Fig. 4.3.1 is obtained from the equation

$$\cos\phi = \beta/n_i k \tag{4.3-24}$$

The propagation constant β indicates either β_0, $\beta^{(+)}$, or $\beta^{(-)}$, and n_i stands for n_1 or n_3. The angle is not always real. It can happen that $\beta > n_i k$. In this case we do not obtain a propagating scattered wave: Either σ or Δ becomes imaginary, and the corresponding scattered wave decays exponentially from the dielectric interface in x direction. If the period of the spatial variation of Eq. (4.3-5) is too short, that is, if

$$|\theta \pm \beta_0| > n_i k \tag{4.3-25}$$

for $i = 1$ and $i = 3$, no propagating scattered wave exists. Far from the surface only the zero-order reflected and transmitted waves can be observed, so that the interface acts as if it were perfectly smooth. Only in the immediate vicinity of the distorted surface is there any field distortion.

We are now ready to apply our results to the theory of wave coupling and radiation losses. The present derivation does not only apply to leaky waves of hollow dielectric waveguides but to TE modes of any type of slab waveguide.

The amplitude coefficients that enter the coupled wave Eqs. (4.2-8) and (4.2-9) are normalized so that they are dimensionless. Their absolute square values represent the power carried by the wave relative to an arbitrary power unit P. The average power flow per unit area in z direction is given by

$$S_z = |\beta_0||E_y|^2/2\omega\mu_0 = |\beta_0||A_\mathrm{i}|^2/2\omega\mu_0 \tag{4.3-26}$$

for the incident plane wave. The total power inside a slab waveguide of width d follows from Eq. (1.5-11):

$$P_\mathrm{i} = |\beta_0| d |A_\mathrm{i}|^2/2\omega\mu_0 = |c_\mathrm{i}|^2 P \tag{4.3-27}$$

The normalized amplitude is thus

$$c_\mathrm{i} = (|\beta_0| d/2\omega\mu_0 P)^{1/2} A_\mathrm{i} \tag{4.3-28}$$

An analogous calculation leads to the normalized amplitude for the scattered waves Eq. (4.3-17):

$$c_\mathrm{s}^{(\mathrm{p})} = (|\beta^{(\mathrm{p})}| d/2\omega\mu_0 P)^{1/2} b A_\mathrm{i} \sigma_0 |\Delta_0 - \sigma_0| / (|\sigma^{(\mathrm{p})} + \Delta^{(\mathrm{p})}|) \tag{4.3-29}$$

with p indicating the plus or minus superscript.

We have now found the expressions for the normalized amplitudes of the incident and scattered waves. However, in a coupled wave theory the amplitudes change with z. This feature is still missing in our theory. We introduce it by a physical argument.

The element that is lacking in our discussion is the fact that the guided waves in a slab waveguide are composed of plane waves that are reflected back and forth between the two core boundaries. The scattered wave thus does not radiate away into space but keeps on interacting with the incident wave. In a first-order perturbation theory, we need not worry about the depletion of energy from the incident wave, but we must consider interference effects. Figure 4.3.2 aids in understanding the interaction between incident and

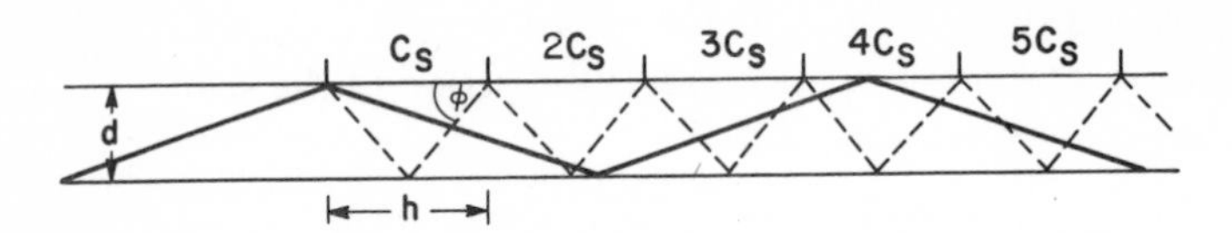

Fig. 4.3.2 *This figure indicates coupling between two guided waves. The solid line represents the incident wave. The dashed line indicates the second guided wave whose amplitude builds up as indicated by the numbers shown in the drawing.*

scattered waves. The solid line indicates the zigzag trajectory of the incident wave inside the waveguide core. The dashed line represents the trajectory of the scattered wave. The scattered wave starts at the first encounter of the incident wave with the sinusoidally distorted upper core boundary with an amplitude c_s. However, after one round trip the scattered wave arrives again at the upper core boundary and is reinforced by an additional scattered wave that is generated by the incident wave. If the relationship between the propagation constants of the two waves and the spatial frequency of the sinusoidal boundary distortion obeys Eq. (4.3-19), the newly created scattered wave adds in phase to the reflected scattered wave, and the amplitude of the combined scattered waves becomes $2c_s$. The scattered wave thus grows in amplitude as indicated by the amplitude values marked on top of the waveguide core in Fig. 4.3.2. The distance h required for one round trip between the two core boundaries is obtained from Fig. 4.3.2:

$$h = 2d/\tan\phi = 2d\beta^{(p)}/\sigma^{(p)} \tag{4.3-30}$$

The integer numbers associated with the growth of the scattered wave amplitude Nc_s are thus given by

$$N = z/h \tag{4.3-31}$$

For actual waves N is, of course, not an integer but a continuous variable, so that relation (4.3-31) can be used for any value of z. The scattered wave amplitude can now be written as a function of z:

$$c_s^{(p)}(z) = Nc_s^{(p)} = bc_i\sigma^{(p)}\sigma_0|\Delta_0 - \sigma_0|z/(2|\beta_0\beta^{(p)}|^{1/2}|\sigma^{(p)} + \Delta^{(p)}|d) \tag{4.3-32}$$

Equations (4.3-28)–(4.3-31) were used to obtain the scattered wave amplitude in this form. The coupling coefficient follows from Eqs. (4.2-9) and (4.3-32):

$$|(1/c_i)(dc_s^{(p)}/dz)| = \hat{R}_{is} = \sigma_0\sigma^{(p)}|\Delta_0 - \sigma_0|b/(2|\beta_0\beta^{(p)}|^{1/2}|\sigma^{(p)} + \Delta^{(p)}|d) \tag{4.3-33}$$

This coupling coefficient between two guided waves applies to any type of slab waveguide. For regular waveguides with $n_1 > n_2 \geqslant n_3$, the parameters Δ_0 and $\Delta^{(p)}$ are imaginary. We thus obtain, with the help of Eqs. (4.3-20)–(4.3-23):

$$|\Delta_0 - \sigma_0| = |\Delta^{(p)} + \sigma^{(p)}| = (n_1^2 - n_3^2)^{1/2}k \tag{4.3-34}$$

so that Eq. (4.3-33) specializes to

$$\hat{R}_{is} = \sigma_0\sigma^{(p)}b/2|\beta_0\beta^{(p)}|^{1/2}d \tag{4.3-35}$$

A comparison with Eq. (4.2-3) shows that we have rederived the coupling coefficient between two guided modes of the ordinary slab waveguide. The only difference between the correct expression (4.2-3) and the approximation (4.3-35) consists in the fact that the effective width of the core of the slab waveguide is not d, as we have assumed, but is modified by the penetration of the evanescent field outside the core region. The effective width of the guide is different for each mode. Our approximation (4.3-35) thus applies to tightly guided modes with very large values of $\gamma_\nu d$ and $\delta_\nu d$.

Since our derivation of Eq. (4.3-32) was not based on any assumption about mode orthogonality and completeness, Eq. (4.3-33) also holds for leaky waves of hollow dielectric waveguides. However, it is important to keep in mind that the derivation of the coupling formula was based on the assumption that the guided wave in the core region can be approximated by two plane waves. For very lossy leaky waves this assumption breaks down, since the parameters σ and β are complex quantities. The restriction to low-loss leaky waves is no series limitation for most applications, since only low-loss waves are of practical interest.

Specializing the following discussion to low-loss leaky waves, we assume that

$$n_1 < n_3 \leqslant n_2 \tag{4.3-36}$$

Low losses can only be achieved if the plane waves, whose superposition forms the guided wave, strike the core boundary at grazing angles. We thus have for the propagation constants of both guided waves

$$|\beta| \approx n_1 k \tag{4.3-37}$$

The parameters Δ_0 and $\Delta^{(p)}$ are both real, and Eq. (4.3-37) implies for both guided waves, according to Eqs. (4.3-21) and (4.3-23), that

$$\Delta \approx (n_3^2 - n_1^2)^{1/2} k \gg \sigma \tag{4.3-38}$$

The coupling formula, Eq. (4.3-33), can thus be approximated as follows:

$$\hat{R}_{is} = \sigma_0 \sigma^{(p)} b / 2 n_1 k d \tag{4.3-39}$$

That the same result can be obtained from Eqs. (4.3-37) and (4.3-35) and consequently also from the coupled mode theory Eq. (4.2-3) is perhaps not surprising, but it could not have been predicted with certainty. We have thus proved that low-loss leaky waves couple among each other in the same way as tightly guided modes of ordinary slab waveguides. The important case of coupling between a forward and backward traveling wave of the same kind occurs when we use

$$\theta = 2\beta_0 \tag{4.3-40}$$

so that Eq. (4.3-19) yields

$$\beta^{(-)} = -\beta_0 \tag{4.3-41}$$

and Eqs. (4.3-20) and (4.3-22) show that the relation holds,

$$\sigma^{(p)} = \sigma_0 \tag{4.3-42}$$

In ordinary dielectric waveguides coupling between a forward and backward wave of the same kind does not result in radiation losses. In hollow dielectric waveguides the coupling process is accompanied by radiation losses that must be added to the inevitable leaky wave losses discussed in Sect. 1.6. With Eq. (4.3-41) one of the two radiation lobes of Eq. (4.3-18) becomes an evanescent wave. The other lobe radiates into region 3 carrying away power. This additional scattering loss is easily calculated. The power loss coefficient $2\alpha_r$ is the ratio of power s_x, carried away per unit length, divided by the power $2S_z d$ carried inside the waveguide core. The factor 2 is necessary if S_z indicates the power carried by one of the plane wave components. We thus have

$$2\alpha_r = s_x / 2 S_z d \tag{4.3-43}$$

From Eqs. (1.3-14), (4.3-38), and (4.3-18), we obtain

$$s_x = \tfrac{1}{2} \operatorname{Re}(e_y h_z^*) = (\Delta^{(-)}/2\omega\mu_0)|e_y|^2 = b^2 |A_i|^2 \sigma_0^2 \Delta^{(-)}/2\omega\mu_0 \tag{4.3-44}$$

Substitution of Eqs. (4.3-26) and (4.3-44) into Eq. (4.3-43) results in the following formula for the radiation loss (coefficient) that accompanies coupling between a forward and backward traveling mode of a hollow dielectric waveguide:

$$2\alpha_r = b^2 \sigma_0^2 \Delta^{(-)} / 2 |\beta_0| d \tag{4.3-45}$$

With Eqs. (1.6-5) (note that we now use the symbol σ_0 instead of κ), (4.3-37), and (4.3-38), we can write the radiation loss formula in the form

$$2\alpha_r = (N\pi b)^2 (n_3{}^2 - n_1{}^2)^{1/2}/2n_1 d^3 \tag{4.3-46}$$

The loss coefficient of the leaky wave is, according to Eq. (1.6-4),

$$2\alpha = (2\sigma_0{}^2/|\beta_0| d)[(1/\rho_0)+(1/\Delta_0)] \tag{4.3-47}$$

We assume for simplicity $n_2 = n_3$, so that we have $\rho_0 = \Delta_0$ and form the ratio of scattering loss Eq. (4.3-45) to leaky wave loss Eq. (4.3-47):

$$2\alpha_r/2\alpha = \tfrac{1}{8}b^2\Delta_0\Delta^{(-)} \approx \tfrac{1}{8}(n_3{}^2 - n_1{}^2)k^2b^2 \tag{4.3-48}$$

Perturbation theory is only applicable if $kb \ll 1$. Equation (4.3-48) tells us that the scattering loss is negligible compared to the inevitable leaky wave loss for all cases to which our theory applies.

4.4 Fiber with Sinusoidal Diameter Changes

The behavior of round optical fibers with sinusoidal core radius changes is very similar to that of slab waveguides. Sinusoidal core radius changes couple two modes to each other so that all the power of one mode can be coupled to the other mode, provided that relation (4.2-4) holds for only two of the guided modes. In order to compare the effectiveness of core radius changes of optical fibers and slab waveguides, we assume that $f(z)$ in Eq. (3.6-1) is given by Eq. (4.2-1) and obtain from Eq. (3.6-10) in analogy with Eq. (4.2-2), for $v = 0$ modes,

$$K_{00}^{(pq)} = 2p\hat{R}_{12}\sin\theta z \tag{4.4-1}$$

with

$$\hat{R}_{12} = b\gamma_1\gamma_2 J_0(\kappa_1 a) J_0(\kappa_2 a)/2nakJ_1(\kappa_1 a) J_1(\kappa_2 a) \tag{4.4-2}$$

Since we have chosen $v = 0$ for both guided modes, we obtain coupling only if we set $m = 0$ in Eq. (3.6-1). From Table 3.6.1 we find $e_{000} = 4$, and from Eq. (2.2-42a) we obtain $e_0 = 2$. A factor 2 has been separated from $\hat{R}_{12}$ by definition (4.4-1). The indices 1 and 2 attached to κ and γ serve as a reminder that these parameters belong to mode 1 and 2; they are not related to the ϕ symmetry of the modes in this case.

The reader may have noticed that the $\pm$ sign which entered Eq. (4.2-2) is absent from Eq. (4.4-1). However, we do obtain the same form (4.2-8) and (4.2-9) also in this case. The difference in appearance is associated with the form (3.5-6) and (3.5-21) of the coupling coefficients. In Eq. (4.2-2) we obtained the two choices for the sign from the fact that either $\beta_1 - \beta_2$ or $\beta_2 - \beta_1$ appears in the denominator of Eq. (3.5-21). However, since the cosine function appears in Eq. (4.2-2), there is no sign change associated with the selection of either $e^{i\theta z}$ or $e^{-i\theta z}$ as the contributing factor to the final form, Eqs. (4.2-8) and (4.2-9), of the coupled wave equations. In the present case the sign of the coupling coefficient does not change, but the change in sign in Eqs. (4.2-8) and (4.2-9) comes about since we now have the sine function in Eq. (4.4-1), so that the sign changes as either $e^{i\theta z}$ or $e^{-i\theta z}$ contributes to the coupled equation system.

Far from cutoff, Eq. (3.6-15) can be used to simplify expression (4.4-2):

$$\hat{R}_{12} = \kappa_1 \kappa_2 b/2nak \tag{4.4-3}$$

In order to show the magnitude of the coupling strength and to compare the approximation Eq. (4.4-3) with expression (4.4-2), we have plotted in Fig. 4.4.1

$$(L/a)(b/a) = (\pi/2\hat{R}_{12}a)(b/a) \tag{4.4-4}$$

as a function of

$$V = (n_1{}^2 - n_2{}^2)^{1/2} ka \tag{4.4-5}$$

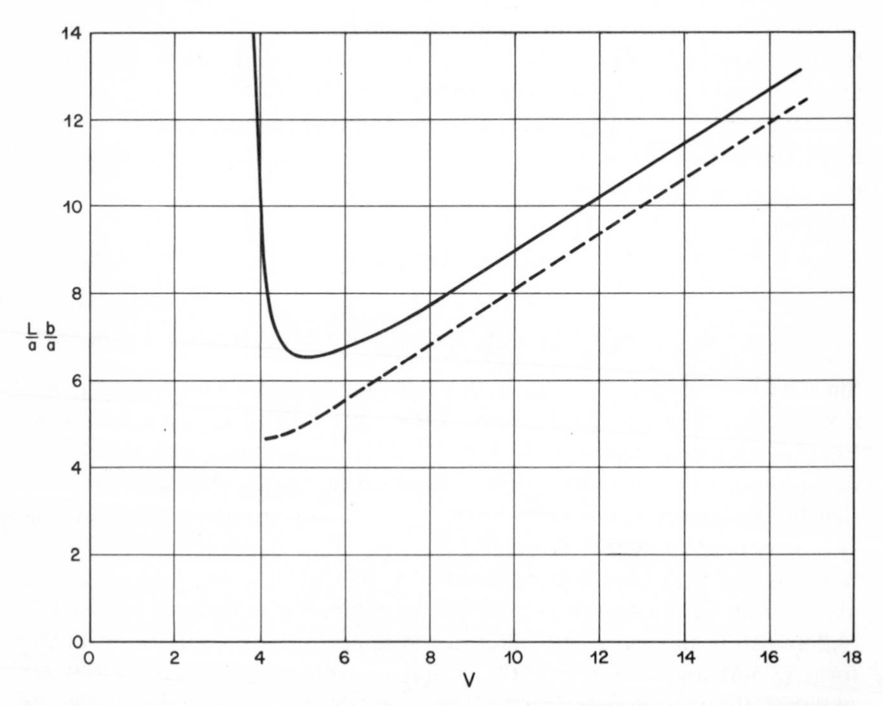

Fig. 4.4.1 *Complete energy exchange length L as a function of the frequency parameter V. The solid line represents Eq. (4.4-2), and the dashed line represents the approximate Eq. (4.4-3).*

The length L necessary for one complete exchange of power between the two modes was defined in Eq. (4.2-12). In order to compare the fiber to the slab waveguide of Sect. 4.2, we use $n_1 = 1.6$ and $n_2 = 1.5$. The solid line in the figure represents Eq. (4.4-2), and the dashed line was obtained from Eq. (4.4-3). The parameters κ and γ were calculated from the eigenvalue Eq. (2.2-39). The approximation, Eq. (4.4-3), improves for increasing values of V. For $V = 16$, the error is only 6%. For a free space wavelength of $\lambda = 1$ μm and $V = 5$, we obtain for this example a core radius of $a = 1.43$ μm. An exchange length of $L = 1$ cm requires a ripple amplitude of $b = 0.0013$ μm, according to the value $Lb/a^2 = 6.5$ obtained from Fig. 4.4.1. This amplitude of the sinusoidal core radius change is approximately half the value that was obtained for the corresponding example in Sect. 4.2. A direct comparison is hard to make because of the different geometries of the two waveguides. However, it might be expected that the coupling is more effective for the round fiber since its entire core boundary is deformed, while only one side of the core of the slab waveguide was assumed to have a sinusoidal deformation.

We discuss the radiation losses caused by sinusoidal deformations of the core radius for the case of the lowest-order mode, $\nu = 0$. From formula (3.6-11) for coupling to the true radiation modes of the fiber, we obtain with the help of Eqs. (3.4-19) and (4.2-24) the following expression for the power loss coefficient:

$$2\alpha = \frac{b^2}{a^2}\frac{2[(n_1/n_2)-1]\gamma^2 J_0{}^2(\kappa a)}{\pi J_1{}^2(\kappa a)}\frac{|\beta|\,J_0{}^2(\sigma a)}{|\sigma J_1(\sigma a)\,H_0^{(1)}(\rho a)-\rho J_0(\sigma a)\,H_1^{(1)}(\rho a)|^2} \tag{4.4-6}$$

Using the coupling formula for the "free space" radiation modes, Eq. (3.6-12), we obtain similarly

$$2\alpha = b^2\frac{\pi nk[(n_1/n_2)-1]\gamma^2 J_0{}^2(\kappa a)}{J_1{}^2(\kappa a)}\frac{\beta^2 J_0{}^2(\sigma a)}{\beta^2+(nk)^2} \tag{4.4-7}$$

The parameter β is obtained from the propagation constant β_0 of the guided mode and the spatial frequency θ of the sinusoidal core boundary perturbation,

$$\beta = \beta_0 - \theta \tag{4.4-8}$$

Figure 4.4.2 shows $2\alpha a^3/b^2$ as a function of ρa,

$$\rho = (n_2{}^2k^2-\beta^2)^{1/2} \tag{4.4-9}$$

The variation of ρa may be considered as being caused by varying θ. The solid line represents Eq. (4.4-6), while the dashed line is a plot of Eq. (4.4-7). We have used $J_0(\sigma a)$ instead of $J_0(\rho a)$ in Eq. (4.4-7) [compare Eq. (3.6-12)]. The "free space" radiation modes apply to the case $n_1 \approx n_2$, so that the distinction between σ and ρ disappears. Using σ instead of ρ causes the maxima and zeros of the solid and dashed lines in Fig. 4.4.2 to coincide.

Figure 4.4.2 was plotted for $n_1 = 1.6$, $n_2 = 1.5$, $ka = 10$, so that $V = 5.57$. The solution of the eigenvalue Eq. (2.2-39) yields $\gamma a = 3.20$. The solid and dashed lines are remarkably close to each other. For small values of ρa, we expect the solid line to be more accurate, while the dashed line should be more reliable for large values of ρa. The oscillations of both curves are caused by interference between radiation originating at opposite sides of the sinusoidally

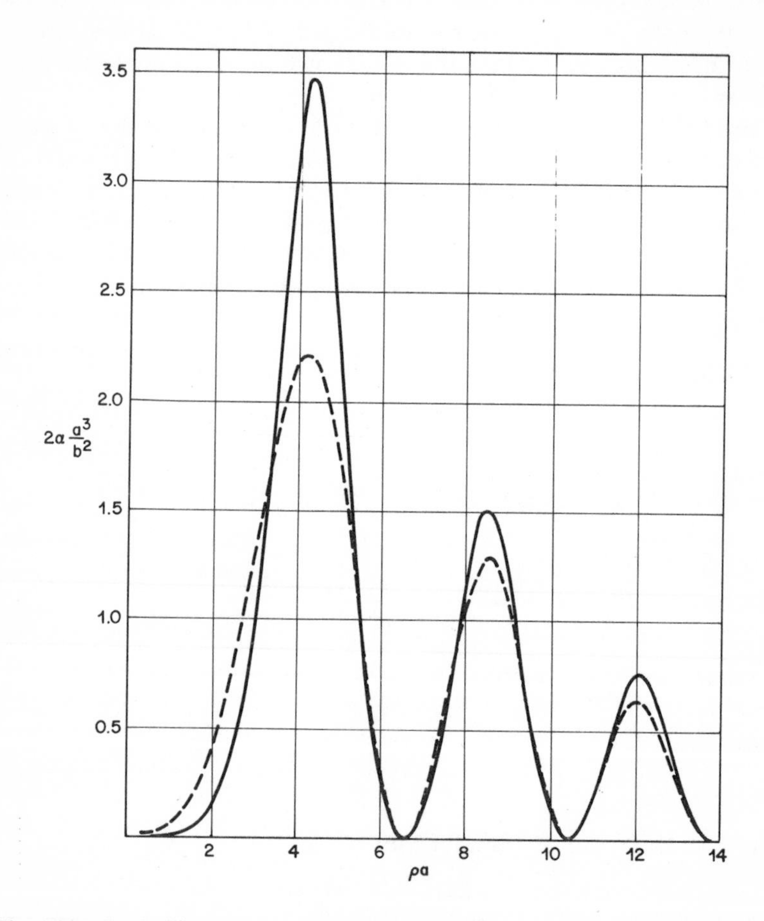

Fig. 4.4.2 *Power loss coefficient 2α as a function of the sinusoidal spatial frequency θ that enters via $\rho = [n_2{}^2k^2 - (\beta_\nu - \theta)^2]^{1/2}$. The solid line was drawn by using the radiation modes of the waveguide. The dotted line was drawn by using only the "free space" radiation modes.*

deformed core boundary. This mechanism works equally well for the true radiation modes of the fiber and for the "free space" radiation modes. The corresponding Fig. 4.2.2 for the slab waveguide also showed oscillations. However, they were caused by interference between a direct beam and the reflection from the other core boundary. We discussed in Sect. 4.2 that this phenomenon could also be interpreted as leaky wave "resonances." That leaky waves are not the cause of the oscillations of the curves in Fig. 4.4.2 is apparent from the fact that the dashed curve shows the same oscillations as the solid curve. The dashed curve is produced by coupling between the guided mode and the "free space" radiation modes. The latter do not contain any information about the core boundary and can thus not be related to leaky waves of the fiber.

The larger values for the peaks in Fig. 4.2.2 compared to the peak values of the curves in Fig. 4.4.2 are caused by the normalization. The slab width d, used to normalize the curves of Fig. 4.2.2, corresponds to $2a$ of the fiber. If we use $2a$ instead of a to normalize the curves of Fig. 4.4.2, we would raise the curves by a factor of 8. This would cause the highest peak in Fig. 4.4.2 to be roughly twice as large as the corresponding peak in Fig. 4.2.2. This larger value can be explained by the fact that the entire core boundary of the fiber is sinusoidally deformed, while only one side of the slab core carries the ripple.

The index ratio of $n_1/n_2 = 1.6/1.5 = 1.067$ is actually too large for the "weakly guiding" approximation to be accurate. We used this value to be able to compare the fiber theory with the slab waveguide theory, at least to order of magnitude.

Our evaluation of the radiation loss does not include the E_z component in the scalar product in Eq. (3.6-3). We obtain an estimate of the range of ρa values over which Fig. 4.4.2 is reasonably accurate by using Eq. (3.6-7). With our choice of parameters we have approximately $\gamma/\beta_g = 0.2$. We must keep the ratio $\rho/\beta_r < 0.5$ in order to ensure that the contribution of the E_z components remains less than 10%. This requirement limits ρa to the range $0 < \rho a < 7.7$. The first peak value in Fig. 4.4.2 is thus reasonably accurate, with decreasing accuracy for the remaining parts of the curve with $\rho a > 7.7$. The region of applicability of our approximate formula is larger for smaller values of $n_1/n_2 - 1$. Equation (4.4-2) and the solid line in Fig. 4.4.1 do not suffer from this problem, which is peculiar to our treatment of coupling to radiation modes.

4.5 Change of Polarization

The modes of the weakly guiding fiber that are listed in Sect. 2.2 are linearly polarized. In the perfect fiber the linear polarization would be maintained. However, it is observed that the polarization in an actual fiber changes rapidly, so that a field that has traveled some distance in the fiber emerges unpolarized. This change in the polarization state can be attributed to deformations of the core from its circular symmetry or to anisotropy of the fiber material. The present discussion is limited to polarization changes of the lowest-order guided mode, with $\nu = 0$, caused by elliptical core deformations.

The loss of linear polarization can be described as a coupling process between orthogonally polarized modes. We have seen in Sect. 2.2 that there are always two modes of the same kind but with transverse electric field components that are polarized perpendicular to each other. Core deformations or anisotropies of the waveguide material couple these orthogonally polarized modes to each other causing the polarization of the superposition field of all the modes to change. The form (3.3-20) of the coupling coefficient shows clearly that orthogonal electric fields can not couple to each other in an isotropic material. Core deformations in isotropic dielectric waveguides can cause coupling between perpendicularly polarized modes only *via* their longitudinal field components E_z. So far we have neglected the contribution of the E_z components to the coupling process, since it is much less than the contribution of the transverse field components, if it exists. In the absence of coupling of the transverse components, the contribution of the E_z components must be considered.

We return to the coupling coefficient (3.6-2) and (3.6-3) and replace the product of the two electric field components by the product of their E_z components. With $\nu = 0$, we obtain from Eqs. (2.2-8), (2.2-28) (using the upper set of trigonometric functions), (2.2-42), (3.6-2), and (3.6-3),

$$K_{00} = (\kappa^2 \gamma^2 f(z)/2i\pi ank\beta^2) \int_0^{2\pi} \sin\phi \cos\phi \cos(m\phi + \Psi)\, d\phi \tag{4.5-1}$$

This coupling coefficient applies to two modes whose transverse electric field components are polarized perpendicular to each other so that only their E_z components contribute to the coupling process. The refractive index n indicates an average value of n_1 and n_2.

For $\Psi = 0$, the integral in Eq. (4.5-1) vanishes for all values of m. We must thus set $\Psi = \pi/2$. A finite contribution is obtained only for $m = 2$. The core deformation (3.6-1) assumes an elliptical shape for $m = 2$, so that only elliptical core deformations couple the two perpendicular $\nu = 0$ modes to each other. We thus have

$$K_{00} = \kappa^2 \gamma^2 f(z)/4iank\beta^2 \tag{4.5-2}$$

An effective transfer of power can occur only for $f(z) = $ constant, since the propagation constants of the two modes are identical. In a section of elliptically deformed guide, the two modes can exchange their power completely according to the laws that are expressed in Eqs. (4.2-8)–(4.2-12). Complete transfer of power from one mode to its perpendicularly polarized counterpart happens in the distance

$$L = \pi/2|K_{00}| = 2\pi nak\beta^2/\kappa^2\gamma^2 f \tag{4.5-3}$$

Instead of the different signs appearing in Eqs (4.2-8) and (4.2-9) we now have the same sign but an imaginary value of $\hat{R}_{12}$ in this case. The factor p is unity, since we are dealing with modes traveling in the same direction. The coupling process (4.2-10) and (4.2-11) is unaffected by these changes. Instead of $\hat{R}_{12}$, it is now the absolute value of K_{00} that appears in the argument of the sine and cosine functions.

As an example we use the conditions that were applied to the study of the radiation loss caused by a sinusoidal core boundary ripple. We again use $V = 5.57$, with $\kappa a = 2.01$ for the lowest-order mode. With these values we compute $\beta a = 15.9$, $\gamma a = 5.2$, and $ka = 10$. We use $n = 1.55$ and assume that $f = a/100$. The waveguide core deformation is thus only 1% of the radius. With these values we obtain from Eq. (4.5-3) the relative length

$$L/a = 2.25 \times 10^4 \tag{4.5-4}$$

With a vacuum wavelength $\lambda = 1\ \mu\text{m}$, we have $a = 1.59 \times 10^{-4}$ cm. The absolute length for a change of polarization by 90° is thus

$$L = 3.6 \quad \text{cm} \tag{4.5-5}$$

Even for this slight ellipticity of the waveguide core and the fact that only the weak E_z components are responsible for the coupling process, we obtain complete power exchange and thus a change of polarization by 90° in less than 4 cm of length. This distance is even shorter for higher-order modes. In an actual waveguide we may expect to find random changes of the core diameter. The ellipticity is thus slowly changing in direction and orientation, which causes the polarization to change randomly. Furthermore, there is a very slight difference in the propagation constants of the two modes in an elliptically deformed dielectric waveguide. The two perpendicularly polarized modes will thus slowly get out of step with each other, so that we do not just have a change in the direction of linear polarization but obtain arbitrary elliptical polarization. In multimode waveguides the picture is further complicated by the fact that all the modes couple to their perpendicularly polarized counterparts, so that the polarization of the total wave field becomes random, resembling the polarization state of incoherent light. This tendency for the wave to depolarize is increased by slight anisotropies that may be caused by stresses in the waveguide core.

4.6 Fiber with More General Interface Deformations

The radiation losses of optical fibers with random core boundary variations can be calculated with the help of Eq. (3.4-19). However, the evaluation of this formula is possible only if the power spectrum of the core distortion function $f(z)$ [see Eq. (3.6-1)] is known. In most practical cases this function is unknown, and it is necessary to gain insight into the behavior of the radiation losses by considering statistical models.

It is also possible to turn the problem around and use Eq. (3.4-19) to determine the power spectrum $|F(\theta)|^2$. We have seen in Sect 4.2, Eq. (4.2-44), that the radiation caused by a sinusoidal core boundary ripple of spatial frequency θ escapes at an angle

$$\phi = \arccos[(\beta_i - \theta)/n_2 k] \tag{4.6-1}$$

Each Fourier component of a more general core distortion function is thus responsible for radiation escaping in a definite direction. Refraction and reflection at the outer cladding boundary can be avoided by using an index matching fluid with refractive index n_2. An experimental determination of the power spectrum of the distortion function requires the use of a single-mode fiber, since the same process that causes radiation losses would tend to couple power from the incident guided mode i to other guided modes. Each guided mode produces a radiation lobe whose direction—given by Eq. (4.6-1)—is different even for the same value of θ.

Assuming single-mode operation, we transform the integral in Eq. (3.4-19) to the integration variable ϕ using Eq. (4.2-43):

$$d\beta = -n_2 k \sin\phi \, d\phi = -\rho \, d\phi \tag{4.6-2}$$

We can then obtain the fractional amount of power radiated into the angular interval $\Delta\phi$ from Eqs. (3.4-19) and (4.6-2):

$$\Delta P = P_i L \left(\sum |\hat{K}_{\rho i}|^2\right) |\beta_s| \, |F(\theta)|^2 \, \Delta\phi \tag{4.6-3}$$

The power P_i of the incident guided mode is $|c_i|^2 P$, and β_s and θ are defined by Eqs. (4.2-43) and (4.6-1). The fractional power ΔP and the radiation angle can be measured. If some prior knowledge about the nature of the core boundary distortion exists, $|\hat{K}_{\rho i}|^2$ can be computed from Eqs. (3.6-11) and (3.6-12), so that the power spectrum $|F(\theta)|^2$ can be computed from Eq. (4.6-3). In most practical cases it is unlikely that we can calculate $|\hat{K}_{\rho i}|^2$, because we do not know to what types of radiation modes the incident mode is coupled. A rough idea about the azimuthal symmetry of the core boundary distortion may be available by observing the azimuthal dependence of the radiation. However, even if we do not know the nature of the core boundary distortion, we can still obtain an estimate of the shape of the power spectrum by assuming that $\sum |\hat{K}_{\rho i}|^2$ is approximately independent of the angle ϕ.

Rawson has made measurements of the angle dependence of the scattered radiation from a round optical fiber [Rn1]. He found that the radiation pattern is sharply peaked in the forward direction, indicating that the power spectrum of the distortion function favors low spatial frequencies. This result is understandable if we consider that the drawing process has the tendency of stretching any imperfection existing in the glass preform to long length. Imperfections of long stretched-out shapes correspond to low spatial frequencies and hence to small angles.

Interesting insight into the loss properties of imperfect optical fibers can be obtained by assuming statistical models of the core boundary deformation. If we take an ensemble average of Eq. (3.4-19), we obtain the following expression for the average radiation loss:

$$2\alpha = \sum \int_{-n_2 k}^{n_2 k} |\hat{K}_{\rho i}|^2 \langle |F(\beta_i - \beta)|^2 \rangle (|\beta|/\rho)\, d\beta \tag{4.6-4}$$

The symbol $\langle\ \rangle$ indicates an ensemble average. Instead of the power spectrum, we consider the autocorrelation function $R(u)$ of the function $f(z)$. For a stationary random process this function is defined as

$$R(u) = \langle f(z) f(z \pm u) \rangle \tag{4.6-5}$$

The assumption of a stationary process is expressed by considering R only as a function of the displacement u between the two points z and $z \pm u$ at which the functional values of $f(z)$ are sampled. The correlation function $R(u)$ has a maximum at $u = 0$, since at this point it consists of the average value of the square of the function. As the value of u is increased, $R(u)$ decreases. This decrease is caused by the fact that, for some members of the ensemble, $f(x)$ assumes different signs at $x = z$ and $x = z + u$. For large values of u, $R(u)$ approaches zero, since there is no longer any correlation between the function at the two sample points. We define two parameters that characterize the statistical process:

$$R(0) = \bar{\sigma}^2 \tag{4.6-6}$$

is called the variance of $f(z)$, since we always assume that

$$\langle f(z) \rangle = 0 \tag{4.6-7}$$

The distance $u = D$ at which the function $R(u)$ has decreased to a fraction of its maximum value—usually to $1/e$—is called the correlation length, since it is a measure for the length over which the displacement of the core boundary are correlated to each other. The autocorrelation function is useful since it has an intuitive meaning. The two parameters, correlation length and variance, can easily be visualized. The square root of the variance, $\bar{\sigma}$, is also called the rms deviation.

The connection of the autocorrelation function with our radiation loss theory is given by the well-known fact that the power spectrum is the Fourier transform of the correlation function [Me1] [the additional factor L occurring in Eq. (9.4-7) of [Me1] is caused by a difference in the definition of $F(\theta)$]:

$$\langle |F(\theta)|^2 \rangle = \int_{-\infty}^{\infty} R(u) e^{-i\theta u} du \tag{4.6-8}$$

It is not known what type of correlation function should be attributed to the core boundary deformations of optical fibers. The following autocorrelation function is sometimes used as a convenient model:

$$R(u) = \bar{\sigma}^2 \exp(-|u|/D) \tag{4.6-9}$$

From Eq. (4.6-8) we obtain, in this case,

$$\langle |F(\theta)|^2 \rangle = 2\bar{\sigma}^2/D[\theta^2 + (1/D^2)] \tag{4.6-10}$$

The Fourier transform of an exponential function is a Lorentzian function.

Figure 4.6.1 shows the normalized radiation loss $2\alpha a^3/\bar{\sigma}^2$ for several values of ka as a function of the normalized correlation length D/a. The curves apply to the radiation loss of the lowest-order mode with $v = 0$ that is caused by a pure diameter change, $m = 0$ in Eq. (3.6-1). The refractive index ratio is taken as $n_1/n_2 = 1.01$. The curves were obtained by numerical integration of Eq. (4.6-4) with Eq. (4.6-10). In the range $0.95 n_2 k \leqslant |\beta| \leqslant n_2 k$, the coupling coefficient for the true radiation modes, Eq. (3.6-11), was used, while we used

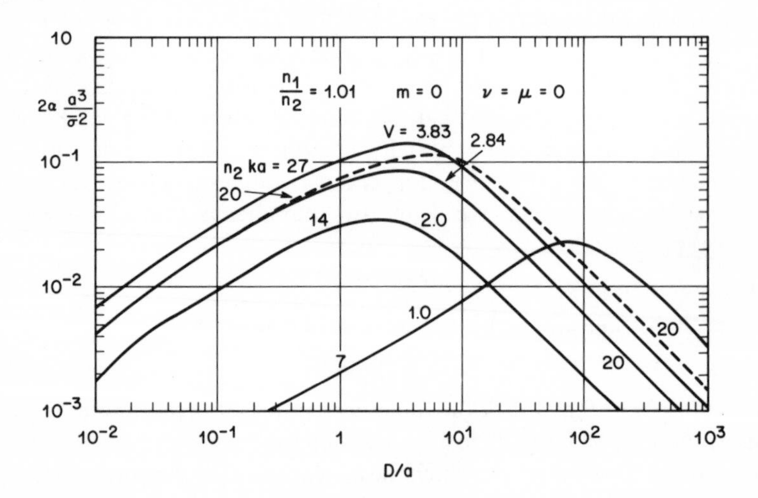

Fig. 4.6.1 *Power loss coefficient 2α as a function of the correlation length D of the core diameter distortion function. The dashed line was drawn by using only "free space" radiation modes.* (From *D. Marcuse, "Coupled mode theory of round optical fibers,"* Bell Syst. Tech. J. ***52**, 817–842 (1973). Copyright 1973, The American Telephone and Telegraph Co., reprinted by permission.)*

Eq. (3.6-12) outside this range. The dashed line in the figure was calculated by using the coupling coefficient Eq. (3.6-12) to the "free space" radiation modes only. It is apparent that this approximation yields good results only for small values of the correlation length. To understand this result we must remember that, according to Eq. (4.6-10), a short correlation length causes a wide Fourier spectrum. A wide power spectrum, in turn, contains high spatial frequencies, causing a large amount of radiation to escape at large angles. For large radiation angles the "free space" radiation modes are useful and provide a good approximation. For long correlation length the Fourier spectrum is narrow and consists mainly of low spatial frequencies. This causes predominantly forward directed radiation, which must be described by the true radiation modes of the fiber. It is noteworthy that the solid curves of Fig. 4.6.1 are in excellent agreement with corresponding curves of a far more complicated theory that is based on the exact guided and radiation modes of the fiber [Me2]. This agreement provides a good justification for the usefulness of the approximate modes derived in Chap. 2.

Figure 4.6.1 shows that the radiation loss peaks between $D/a = 1$ and $D/a = 10$ in most cases. Only for small values of V [see Eq. (4.4-5)] is the maximum shifted to large values of D/a. It has been found that the loss curve to the right of the peak value is strongly dependent on the statistical model. However, at the peak and for small values of D/a to the left of the maximum, we obtain the same loss curve almost independently of the assumed statistical model. For this reason we use only the maximum loss values to consider a few numerical examples. The loss values for large D/a are strongly dependent on the assumed correlation function and have no general validity.

We discuss the rms deviation $\bar{\sigma}$ that is required to produce 10 dB/km radiation loss provided that the statistical situation is such that the maximum value of each loss curve shown in Fig. 4.6.1 is reached. We assume that the fiber is operated at a vacuum wavelength of $\lambda = 1$ μm. Using the fact that 10 dB/km loss corresponds to $2\alpha = 2.3 \times 10^{-9}$ μm^{-1}, we obtain the values appearing in Table 4.6.1.

TABLE 4.6.1

Core radius a, normalized power loss coefficient 2α, and rms core deviation $\bar{\sigma}$ for four different values of ka.

ka	V	a (μm)	$2\alpha a^3/\bar{\sigma}^2$	$\bar{\sigma}$ (μm)
27	3.83	4.3	0.14	1.14×10^{-3}
20	2.84	3.18	0.086	9.28×10^{-4}
14	2.0	2.23	0.035	8.54×10^{-4}
7	1.0	1.11	0.023	2.71×10^{-4}

The table makes it clear that the tolerance requirements for core boundary perturbations are very stringent indeed. The small values of the rms deviation $\bar{\sigma}$ that cause 10 dB/km radiation loss become even more dramatic when we realize that 10^{-3} μm = 10 Å. However, it is necessary to keep in mind that the figures in Table 4.6.1 represent the worst possible case. Practical experience has shown that radiation losses of less than 10 dB/km are indeed achievable, so that we must conclude that the correlation length of practical fibers is probably very much longer than the worst possible value of 1–10 core radii. That the correlation length of practical fibers is very long is born out by the measurements of Rawson [Rn1], which indicate that the Fourier spectrum peaks at very low spatial frequencies corresponding to long correlation length.

Figure 4.6.2 shows curves that are very similar to those of Fig. 4.6.1, except that they represent the result of a random ellipticity of the core boundary corresponding to $m = 2$ in Eq. (3.6-1). The order of magnitude of the radiation losses caused by random elliptical core deformation is roughly the same as that caused by random diameter changes. The guided mode whose losses are show in the figure again has the ϕ symmetry $\nu = 0$, but core radius deviations with $m = 2$ cause coupling to radiation modes with the ϕ symmetry $\mu = 2$. It was again assumed that $n_1/n_2 = 1.01$. The radiation modes of the fiber as well as the "free space" radiation modes were again used to compute the curves in Fig. 4.6.2.

We close this section with a discussion of mode coupling caused by curvature

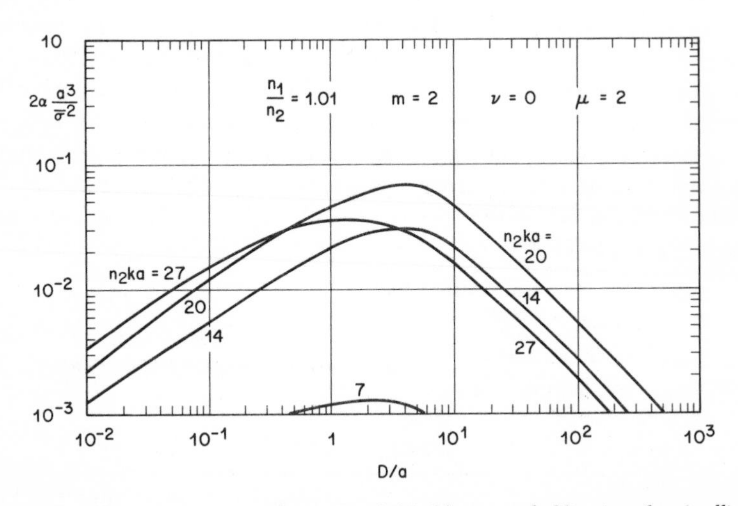

Fig. 4.6.2 *These curves are similar to Fig. 4.6.1. The core–cladding interface is elliptically distorted in this case.* (From *D. Marcuse, "Coupled mode theory of round optical fibers,"* Bell Syst. Tech. J. **52**, *817–842 (1973). Copyright 1973, The American Telephone and Telegraph Co., reprinted by permission.)*

of the waveguide axis. We have seen in Chap. 3 that the coupling coefficients contain either the distortion function $f(z)$ or its derivative. The form that the coupling coefficients assume depends on the modes that are used for the field expansion (3.2-18) and (3.2-19). The coupling coefficients for modes of the ideal waveguide are proportional to $f(z)$. This causes coupling among these modes even if $f(z)$ is constant. The ideal modes are coupled even if the waveguide is perfectly uniform but has a constant core radius that is slightly different from that of the ideal waveguide whose modes are used for the field expansion. The coupled mode theory can be used in this case to calculate the normal modes of the uniform waveguide by requiring that the z dependence of the expansion coefficients is of the form

$$a_\nu(z) = A_\nu e^{-i\beta z} \tag{4.6-11}$$

The propagation constant β does not depend on the mode index ν. Substitution of Eq. (4.6-11) into Eqs. (3.2-42) and (3.2-43) results in an algebraic eigenvalue problem. The solutions of this equation system yield the modes and eigenvalues (propagation constant β) of the modes of the waveguide. This method can, in principle, be used to determine the modes of an arbitrary uniform waveguide by means of a field expansion in terms of modes of a simpler waveguide.

However, for purposes of determining mode coupling to nonuniform waveguides, the ideal mode expansion may be inconvenient if the inhomogeneous waveguide contains long sections of uniform cross section. In this case a field expansion in terms of local normal modes is preferable. Its coupling coefficients depend on df/dz and vanish on uniform portions of the waveguide.

However, we are now going one step further and consider a dielectric waveguide whose direction changes in space. If we consider two straight, uniform waveguide sections that are connected by a bend, we have $f(z) \neq 0$ and $df(z)/dz \neq 0$ everywhere, perhaps with the exception of one of the two straight waveguide sections. Not only the ideal modes but also the local normal modes are coupled on at least one of the straight waveguide sections connected to the bend. We realize that this awkward behavior is caused by an inappropriate choice of modes. The ideal modes and the local normal modes both belong to waveguides whose directions are fixed in space. We have good reason to expect that we could choose local normal modes of a different type that belong to a waveguide whose direction is locally tangential to the direction of the waveguide under consideration. The coupling coefficients of these new local normal modes must be proportional to d^2f/dz^2 so that no coupling results if the waveguide does not change its direction. We have seen in Sect. 3.4, and expressed in Eq. (3.5-22), that $f(z)$ is equivalent to

$$f(z) \to -[i/(\beta_\nu - \beta_\mu)]\,(df/dz) \tag{4.6-12}$$

The transition (4.6-12) corresponds to changing from modes of the ideal waveguide to local normal modes of straight waveguides that adjust locally to the cross section of the actual waveguide. The remaining portion of the coupling coefficients remains unchanged, provided that the radius deformations of the nonuniform waveguide are very slight. We are now prepared to use the equivalence

$$f(z) \to -[1/(\beta_\nu - \beta_\mu)^2](d^2 f/dz^2) \tag{4.6-13}$$

We expect that substitution of Eq. (4.6-13) into Eq. (3.6-2) and use of Eqs. (3.6-10)–(3.6-12) yields coupling coefficients that are suitable to describe mode coupling caused by waveguide curvature. The same procedure should, of course, also work for slab waveguides and could be used in Eqs. (3.5-27)–(3.5-37).

There is an additional effect that needs to be considered when we work with curved dielectric waveguides. In addition to mode coupling caused by changes in curvature, curved waveguides exhibit radiation losses that are intrinsic to uniform bends. Curvature losses have been discussed in [Me1] and [Mi2]; they are not included in perturbation solutions of the coupled mode theory.

For use in connection with statistical treatments of random mode coupling in multimode dielectric waveguides, to be discussed in Chap. 5, we briefly consider here coupling between guided modes caused by waveguide curvature. Restricting ourselves to the far from cutoff approximation, we obtain from Eqs. (3.6-2), (3.6-16), and (4.6-13) the coupling coefficient

$$K_{\mu\nu} = -[1/(\beta_\mu - \beta_\nu)^2][e_{\mu\nu 1}/(e_\nu e_\mu)^{1/2}](\kappa_\mu \kappa_\nu / 2inka)(d^2 f/dz^2) \tag{4.6-14}$$

We have used $m = 1$, since according to Eq. (3.6-1) this corresponds to a change in the direction of the waveguide axis. Indeed, Eq. (3.6-1) describes a short, circular bend if we use $m = 1$ and $f(z) = cz^2$. In the spirit of our extension of the coupling coefficients of ideal waveguide modes to more general local normal modes, we must use Eq. (4.6-13) in conjunction with $m = 1$. Using

$$f(z) = b \cos \theta z \tag{4.6-15}$$

we define

$$K_{\mu\nu} = -2i\hat{R}_{\mu\nu} \cos \theta z \tag{4.6-16}$$

and obtain from Eq. (4.6-14)

$$\hat{R}_{\mu\nu} = [e_{\mu\nu 1}/(e_\nu e_\mu)^{1/2}] \kappa_\mu \kappa_\nu b / 4nka \tag{4.6-17}$$

Equations (4.6-16) and (4.6-17) correspond to Eqs. (4.2-2) and (4.2-3). If we assume that one of the modes has $\nu = 0$, we find from Table 3.6.1 that this mode can be coupled only to a mode with $\mu = 1$, so that we have

$$e_{101}/(e_0 e_1)^{1/2} = \sqrt{2} \tag{4.6-18}$$

Thus Eq. (4.6-17) becomes

$$\hat{R}_{\mu\nu} = \kappa_\mu \kappa_\nu b/2\sqrt{2}nka \tag{4.6-19}$$

This formula holds for modes that are far from cutoff. A comparison with the slab waveguide coupling formula (4.2-3) is possible if we assume $\beta_\mu \approx \beta_\nu \approx nk$, $\gamma_\mu d \gg 1$, and $\delta_\mu d \gg 1$. Associating the full width of the slab with the diameter $2a$ of the fiber, we see that the slab formula differs from the fiber coupling coefficient (4.6-19) only by a factor of $1/\sqrt{2}$. The slab formula (4.2-3) does not actually apply to a waveguide with sinusoidal deformation of its axis, since only one of the two core boundaries is deformed. However, deformation of the other interface would only contribute a factor of 2 to Eq. (4.2-3), so that the corresponding formula for a slab with sinusoidally deformed axis (more accurately we would say axial plane) would be very similar to Eq. (4.2-3) and hence to Eq. (4.6-19). The order of magnitude of the conversion effects to be expected from the fiber with sinusoidally deformed axis are thus the same as those of the slab discussed in Sect. 4.2.

4.7 Rayleigh Scattering

The theory of coupled modes enables us to calculate the power loss to a plane wave caused by Rayleigh scattering. Rayleigh scattering is of fundamental importance since it sets the lower limit to scattering losses in dielectric materials. In the absence of all other loss mechanisms, waves always lose power by scattering from inhomogeneities in the dielectric materials. Maxwell's theory includes the interaction between electromagnetic fields and materials by means of the refractive index. The index of refraction and its square, the dielectric constant, are usually regarded as constant with respect to the spatial coordinates or as slowly varying functions. This continuum approach ignores the fact that all matter is composed of atoms or molecules. On the molecular level, electromagnetic radiation is scattered from the individual molecules of the material. This scattering manifests itself as a phase retardation slowing the progress of the wave. The light that is scattered out of the original direction of wave propagation is randomly phased, and the superposition of all the randomly scattered wavelets cancels almost completely. The forward scattered light superimposes itself coherently. The net effect of this coherent superposition is described very accurately by the phenomenological material constants, the refractive index and the dielectric constant. However, the incoherent light scattering caused by the individual molecules does not cancel completely. A small amount of radiation is actually scattered out of the light beam and causes Rayleigh scattering losses. These losses are the fundamental loss limit of dielectric optical waveguides. Absorption losses and losses caused by the random core–cladding interface deviations can, in principle, be reduced to arbitrarily small values by purification of the dielectric materials and by careful manufacture of the dielectric waveguide. The scattering caused by the random position of the molecules in amorphous dielectric materials cannot be influenced by the experimenter except by choosing materials whose Rayleigh scattering is naturally smaller than that of other dielectrics.

The theory of Rayleigh scattering has been described in [Me10] on the basis of a molecular scattering theory. In this book we derive the Rayleigh scattering formula from the coupled mode theory, assuming that the refractive index is subject to random fluctuations.

We are considering Rayleigh scattering losses of a plane wave in an infinitely extended medium. For purposes of calculation, we assume that the fluctuations are present only inside a region with the very large volume V, while the refractive index outside this region is perfectly homogeneous. The region of volume V is a cube whose sides have the length L, so that $V = L^3$.

The modes of the structure are the plane waves of the infinitely extended homogeneous medium with refractive index n_0:

$$\mathscr{E}_\nu = (1/2\pi)(2\omega\mu_0 P/|\beta_\nu|)^{1/2}\mathbf{e}_\nu \exp[-i(\kappa x + \sigma y)] \tag{4.7-1}$$

$$\mathscr{H}_\nu = (1/2\pi)(2\omega\mu_0 P/|\beta_\nu|)^{1/2}(\mathbf{k}_\nu \times \mathbf{e}_\nu/\omega\mu_0)\exp[-i(\kappa x + \sigma y)] \tag{4.7-2}$$

$\mathbf{e}_\nu$ is a unit vector that determines the polarization of $\mathscr{E}_\nu$ at right angles to the propagation vector $\mathbf{k}_\nu$:

$$\mathbf{e}_\nu \cdot \mathbf{k}_\nu = 0 \tag{4.7-3}$$

The propagation vector is defined as ($\mathbf{e}_x$, $\mathbf{e}_y$, and $\mathbf{e}_z$ are unit vectors in x, y, and z direction)

$$\mathbf{k}_\nu = \kappa\mathbf{e}_x + \sigma\mathbf{e}_y + \beta_\nu\mathbf{e}_z = n_0 k\mathbf{k}_\nu/|\mathbf{k}_\nu| \tag{4.7-4}$$

The parameters κ and σ are the components of the plane wave propagation vector $\mathbf{k}_\nu$ in x and y direction. The label ν is an abbreviation for the number pair κ, σ. The factor $\exp(-i\beta_\nu z)$ is not included in the expression for the mode fields since they are solutions of Eqs. (3.3-15) and (3.3-16). The coupling coefficient (3.2-44) is used in the approximate form

$$K_{\mu\nu} = (\omega\varepsilon_0/4iP)\iint_{-\infty}^{\infty}(n^2 - n_0^2)\mathscr{E}_\mu^* \cdot \mathscr{E}_\nu\, dx\, dy \tag{4.7-5}$$

This approximation is valid if $n(x, y, z)$ is only very slightly different from n_0. Substitution of Eq. (4.7-1) leads to

$$K_{\mu\nu} = [k^2/8\pi^2(|\beta_\nu \beta_\mu|)^{1/2}](\mathbf{e}_\mu \cdot \mathbf{e}_\nu) \int\int_{-\infty}^{\infty} (n^2 - n_0^2) \cdot \exp\{i[(\kappa_\mu - \kappa_\nu)x + (\sigma_\mu - \sigma_\nu)y]\}\, dx\, dy \tag{4.7-6}$$

We write the incident plane wave in the form

$$\mathscr{E}_0 = L^{-1}(2\omega\mu_0 P/n_0 k)^{1/2}\mathbf{e}_0 \tag{4.7-7}$$

This wave propagates in z direction with the propagation vector

$$\mathbf{k}_0 = n_0 k \mathbf{e}_z \tag{4.7-8}$$

The unit vector $\mathbf{e}_z$ points in z direction and, as usual, we use

$$k = \omega(\varepsilon_0 \mu_0)^{1/2} = 2\pi/\lambda \tag{4.7-9}$$

The wave (4.7-7) carries the finite amount of power P through the (very large) cross section with area L^2. Comparison with the mode field (4.7-1) shows that the amplitude of the incident wave is

$$c_0 = 2\pi/L \tag{4.7-10}$$

The amplitudes of the scattered waves are obtained by first-order perturbation theory in close analogy to the procedure that leads to Eq. (3.4-8) ($L \to \infty$ is implied):

$$\begin{aligned} c_\nu &= c_0 \int_{-L/2}^{L/2} K_{\nu 0} \exp[i(\beta_\nu - n_0 k)z]\, dz \\ &= [k^2(\mathbf{e}_\nu \cdot \mathbf{e}_0)/4\pi L(n_0 k|\beta_\nu|)^{1/2}] \int\int\int_{-\infty}^{\infty} (n^2 - n_0^2) \exp[i(\mathbf{k}_\nu - \mathbf{k}_0) \cdot \mathbf{r}]\, dx\, dy\, dz \end{aligned} \tag{4.7-11}$$

The vector $\mathbf{r}$ is defined as

$$\mathbf{r} = x\mathbf{e}_x + y\mathbf{e}_y + z\mathbf{e}_z \tag{4.7-12}$$

The total amount of scattered power is obtained from Eq. (3.4-16):

$$P_s = P \int\int_{-\infty}^{\infty} |c_\nu|^2\, d\kappa\, d\sigma \tag{4.7-13}$$

Introducing a spherical coordinate system we can express the components of the propagation vector (4.7-4) in the form

$$\kappa = n_0 k \sin\theta \cos\phi, \qquad \sigma = n_0 k \sin\theta \sin\phi, \qquad \beta_\nu = n_0 k \cos\theta \tag{4.7-14}$$

The integral can thus be transformed to the coordinates θ and ϕ with the help of

$$d\kappa\, d\sigma = \begin{vmatrix} \partial\kappa/\partial\theta & \partial\kappa/\partial\phi \\ \partial\sigma/\partial\theta & \partial\sigma/\partial\phi \end{vmatrix} d\theta\, d\phi = (n_0 k)^2 \cos\theta \sin\theta\, d\theta\, d\phi$$
$$= n_0 k \beta_\nu\, d\Omega \tag{4.7-15}$$

We have used Eq. (4.7-14) and define the element of solid angle as

$$d\Omega = \sin\theta\, d\theta\, d\phi \tag{4.7-16}$$

The total scattered power can now be expressed as

$$P_s = P \int W\, d\Omega \tag{4.7-17}$$

The relative amount of power scattered into the unit element of solid angle is defined by

$$W = n_0 k \beta_\nu |c_\nu|^2$$
$$= (k^4/16\pi^2 L^2)(\mathbf{e}_\nu \cdot \mathbf{e}_0)^2 \left| \int\!\!\int\!\!\int_{-\infty}^{\infty} (n^2 - n_0{}^2) \exp[i(\mathbf{k}_\nu - \mathbf{k}_0)\cdot \mathbf{r}]\, dx\, dy\, dz \right|^2 \tag{4.7-18}$$

We take an ensemble average of Eq. (4.7-17) [and consequently of Eq. (4.7-18)]. In analogy with Eq. (4.6-8), we now have

$$\left\langle \left| \int\!\!\int\!\!\int_{-\infty}^{\infty} (n^2 - n_0{}^2) \exp(i\mathbf{K}\cdot\mathbf{r})\, dx\, dy\, dz \right|^2 \right\rangle$$
$$= L^3 \int\!\!\int\!\!\int_{-\infty}^{\infty} R(u, v, w) \exp(i\mathbf{K}\cdot\mathbf{u})\, du\, dv\, dw \tag{4.7-19}$$

The vector **u** has the components u, v, and w. The autocorrelation function is defined as

$$R(u, v, w) = \langle [n^2(x, y, z) - n_0{}^2][n^2(x+u, y+v, z+w) - n_0{}^2] \rangle \tag{4.7-20}$$

The refractive index fluctuations that cause Rayleigh scattering are very rapid compared to the wavelength of light. The exact shape of the autocorrelation function is thus not important. We use as a statistical model the function

$$R(u, v, w) = \begin{cases} \langle (n^2 - n_0{}^2)^2 \rangle, & \text{inside } S \\ 0, & \text{outside } S \end{cases} \tag{4.7-21}$$

The region S is a cube of volume D^3. The linear dimension D is the correlation length. With our assumption that D is much shorter than the wavelength of the scattered radiation, we obtain, from Eqs. (4.7-18)–(4.7-21):

$$L^{-1}W = (k^4/16\pi^2)\,D^3\langle(n^2-n_0{}^2)^2\rangle[(\mathbf{e}_{v1}\cdot\mathbf{e}_0)^2+(\mathbf{e}_{v2}\cdot\mathbf{e}_0)^2] \tag{4.7-22}$$

Equation (4.7-22) is the important Rayleigh scattering formula. It represents the relative amount of power that is scattered out of the plane wave per unit length into the unit element of solid angle. We have incorporated the contributions of the two possible orthogonal polarizations of the scattered waves indicated by $\mathbf{e}_{v1}$ and $\mathbf{e}_{v2}$. The quantity $\langle(n^2-n_0{}^2)^2\rangle$ is the variance of the square of the refractive index in the volume D^3.

The power loss factor 2α caused by Rayleigh scattering can be defined as

$$2\alpha = L^{-1}\int W\,d\Omega \tag{4.7-23}$$

The integral over the scalar products of the polarization vectors was evaluated on pp. 369–370 of [Me10] with the result

$$\int[(\mathbf{e}_{v1}\cdot\mathbf{e}_0)^2+(\mathbf{e}_{v2}\cdot\mathbf{e}_0)^2]\,d\Omega = 8\pi/3 \tag{4.7-24}$$

The power loss coefficient is thus

$$2\alpha = (k^4/6\pi)\,D^3\langle(n^2-n_0{}^2)^2\rangle \tag{4.7-25}$$

Any further evaluation of the Rayleigh loss formula (4.7-25) requires knowledge about the statistics of the dielectric material. The result is different for different types of materials. For ideal gases the fluctuations of the gas density and hence of the refractive index are known. We use the formula

$$n = 1+(n_0-1)(\rho/\rho_0) \tag{4.7-26}$$

The density ρ is slightly different from the average density ρ_0. Since $n-n_0$ is always much less than unity, we can use the approximation

$$n^2-n_0{}^2 \approx 2n_0(n-n_0) = 2n_0(n_0-1)[(\rho/\rho_0)-1] \tag{4.7-27}$$

The ensemble average of the square of this quantity is

$$\langle(n^2-n_0{}^2)^2\rangle = 4n_0{}^2(n_0-1)^2\langle[(\rho/\rho_0)-1]^2\rangle \tag{4.7-28}$$

From the theory of statistical mechanics [Tn1] we obtain the result

$$\langle[(\rho/\rho_0)-1]^2\rangle = 1/\overline{N} \tag{4.7-29}$$

where $\overline{N}$ indicates the number of molecules contained inside the volume whose average density fluctuations are being considered. In our case this volume is

D^3. The number of molecules per unit volume is therefore

$$N = \overline{N}/D^3 \tag{4.7-30}$$

Collecting these results yields the following formula for the Rayleigh scattering loss in gases:

$$2\alpha = (2n_0{}^2k^4/3\pi N)(n_0 - 1)^2 \tag{4.7-31}$$

We have thus rederived Eq. (8.3-61) of [Me10, p. 370]. (Note that the power loss coefficient was defined by the symbol α in [Me10]). The calculations based on scattering by individual molecules thus lead to the same result as the theory that is based on density fluctuations of the gas. For $\lambda = 0.6\ \mu$m and air at atmospheric pressure, the Rayleigh scattering loss is approximately 0.03 dB/km. The Rayleigh scattering loss increases as the fourth power of the light frequency or decreases as the inverse fourth power of its wavelength.

Equation (4.7-25) is identical with Eq. (9.1.4) of [Kr1, p. 489] where further evaluation of this equation with the help of thermodynamic methods is also shown.

5

Coupled Power Theory

5.1 Introduction

This chapter is devoted to multimode waveguides. In Chap. 3 we presented an exact coupled mode theory that is capable, at least in principle, of handling any kind of mode conversion and radiation effect that may occur in a multimode or single-mode optical dielectric waveguide. However, the infinite set of coupled integrodifferential Eq. (3.2-42) is very hard to solve. We have been able to utilize the coupled wave equations for the derivation of radiation loss coefficients and for the description of two modes coupled by sinusoidal waveguide imperfections. The problem of a randomly deformed multimode waveguide is too complex to be treated directly with the coupled wave theory. The complexity of the coupled wave equations is caused by the fact that they contain too much information. The system of coupled Eqs. (3.2-42) contains a detailed description of the phase and amplitude of all the modes at any point z along the waveguide. However, we are never interested in the phases of the individual modes and only rarely in the exact amplitudes of each mode. For most practical purposes, it would suffice to know the average amount of power carried by each mode or groups of modes. We are interested to know how the total power carried by the waveguide is distributed among the modes and how this distribution is changing as a result of coupling and loss processes. One should expect that this more modest information should be obtainable from simpler equations. This expectation is indeed justified.

The title of the present chapter shows that we are now dealing with coupled equations for the power carried by the modes. Actually we should call it "coupled equations for the average mode power." For the sake of brevity we omit the explicit reference to the statistical treatment by dropping the word average and speaking loosely of coupled power equations, leaving it understood that it is always the average power that is meant.

The exact coupled mode equations of Chap. 3 can be used to derive coupled equations for the average power carried by each mode. The coupled power equations are not exact. In this book we derive coupled power equations from the exact coupled wave equations by a method that is based on perturbation theory. The resulting coupled power equations hold only for relatively weak coupling. However, this weak coupling case is most often encountered in practical applications. Weak coupling means that changes in the power distribution take place over distances that are very long compared to the wavelength of light. Coupling that requires a distance of 1000 wavelengths for a complete exchange of power between two modes is considered to be weak in this context. However, compared to the dimensions of our macroscopic world a distance of 1000 optical wavelengths is approximately only 1 mm long. The approximate theory is thus applicable to situations where considerable amounts of power are transferred among the modes over distances of millimeters or centimeters. Undesired, random coupling caused by waveguide imperfections is always weak and is thus accessible to the approximate coupled power theory.

The coupled power equations are a set of a finite number of first-order differential equations with constant and symmetric coefficients. These equations can always be solved at least with the aid of a computer. In certain special cases it is even possible to transform the system of coupled first-order differential equations into one partial differential equation. This has the advantage of offering the possibility of obtaining solutions in closed form for waveguides with such a large set of guided modes that it can be regarded as a continuum.

Why are we interested in multimode waveguides? If a source of coherent, single-mode light is available (that is, if a single-mode laser can be used), it is possible to excite a single-mode fiber efficiently. However, at the present state of laser development it is not clear if cheap lasers with long lifetimes may become available. However, cheap, incoherent light sources are already available. Luminescent diodes, or LEDs for short, deliver several milliwatts of power, have long lifetimes, and can be operated with low direct current, low voltage power sources. Since multimode incoherent light cannot efficiently be launched into a single-mode waveguide, multimode fibers are needed to transmit the light of luminescent diodes. It appears that communications systems based on the use of luminescent diodes and multimode glass fibers may be very economical, offering the possibility of transmitting signals of 10–100 MHz bandwidth or data rates of 10^7 to 10^8 bits. Such a multimode transmission medium requires the analytical methods of handling problems of wave propagation in multimode waveguides that are the subject of this chapter.

5.2 Derivation of Coupled Power Equations

In order to derive coupled differential equations for the average power carried by each mode of a dielectric optical waveguide, we introduce the quantity

$$P_\mu = \langle |a_\mu|^2 \rangle = \langle |c_\mu|^2 \rangle \tag{5.2-1}$$

The rapidly varying mode amplitude a_μ is related to the slowly varying amplitude c_μ by Eq. (3.2-34). P_μ is the average power of mode μ. According to Eq. (3.4-15) we obtain the actual power carried by each mode by multiplying the absolute square of the expansion coefficient a_μ or c_μ with the normalization coefficient P. So far, the mode amplitudes have always been considered dimensionless quantities. However, the coupled wave Eqs. (3.2-49) are homogeneous in the expansion coefficients c_μ. We can multiply both sides of Eqs. (3.2-49) or (3.3-8) with $\sqrt{P}$. The product of the mode amplitudes c_μ with $\sqrt{P}$ is again called c_μ. It satisfies the same coupled wave Eq. (3.2-49) as the original mode amplitudes but has the advantage of allowing us to consider Eq. (5.2-1) as the actual power carried by the mode.

The ensemble average is taken over statistically similar waveguides. We no longer consider only one individual dielectric waveguide but assume that we have a very large number of waveguides that are all built according to the same method but are, nevertheless, not all identical. All these waveguides have imperfections coupling their modes among each other. These waveguide imperfections are all similar in nature. For example, they are all pure diameter changes, elliptical core deformations, or random bends of the waveguide axis. The z dependence of the imperfections is, however, not the same. In particular, we assume that the phases of periodic wiggles are randomly distributed throughout the members of the statistical ensemble.

We take the z derivative of the average power (5.2-1):

$$\frac{dP_\mu}{dz} = \left\langle \frac{dc_\mu}{dz} c_\mu^* \right\rangle + \left\langle c_\mu \frac{dc_\mu^*}{dz} \right\rangle = \left\langle \frac{dc_\mu}{dz} c_\mu^* \right\rangle + \text{cc} \tag{5.2-2}$$

The notation +cc indicates that the complex conjugate of the term already appearing in the equation is to be added.

We limit our discussion to forward traveling waves. This allows us to drop the plus or minus superscripts in Eq. (3.2-49) and dispense with Eq. (3.2-50) altogether. Coupling to forward traveling guided modes occurs for distortion functions $f(z)$ whose Fourier spectra are limited to spatial frequencies in the range

$$0 < \theta < 2n_2 k \tag{5.2-3}$$

Rawson's measurements [Rn1] and other corroborating evidence have shown that most optical fibers have spatial Fourier spectra that satisfy restriction (5.2-3). There is, however, always a background of a nearly flat Fourier spectrum that causes Rayleigh scattering from a given guided mode to all radiation modes and guided modes in forward as well as backward direction. It is easy to include the radiation losses of this Rayleigh background in our coupled power theory. The small amount of coupling to backward traveling guided modes is now neglected. We shall consider coupling among two modes traveling in opposite directions in Sect. 5.7. The coupled wave Eqs. (3.2-49) are now written with the help of Eq. (3.4-7):

$$dc_\mu/dz = \sum_{\nu=1}^{N} \hat{K}_{\mu\nu} f(z) c_\nu(z) \exp[i(\beta_\mu - \beta_\nu)z] \tag{5.2-4}$$

The summation is restricted to the N guided modes of the waveguide. We substitute Eq. (5.2-4) into Eq. (5.2-2) and obtain

$$dP_\mu/dz = \sum_{\nu=1}^{N} \hat{K}_{\mu\nu} \langle c_\nu c_\mu^* f(z) \rangle \exp[i(\beta_\mu - \beta_\nu)z] + \text{cc} \tag{5.2-5}$$

The triple product of the two mode amplitudes and the distortion function $f(z)$ requires special attention. Since the mode amplitudes c_ν and c_μ are both taken at the same point z that also enters the argument of $f(z)$, it is clear that these three quantities are not statistically independent. However, we have discussed in Sect. 4.6 that the statistics of the random function $f(z)$ can be expressed in terms of an autocorrelation function that has appreciable values only over a limited range of its argument. We thus assume that $f(z)$ is a stationary random process that can be described by an autocorrelation function with a finite correlation length D. Since the waves approach the point z from the left (which means from smaller values of z), we can make use of the fact that, owing to the finite correlation length of the function $f(z)$, the field amplitude $c_\nu(z')$ and $f(z)$ are uncorrelated if

$$z - z' \gg D \tag{5.2-6}$$

Because of the assumed lack of correlation we can write

$$\langle c_\nu(z')\, c_\mu^*(z')\, f(z)\rangle = \langle c_\nu(z')\, c_\mu^*(z')\rangle \langle f(z)\rangle \tag{5.2-7}$$

This discussion leads to the logical conclusion that we must express the field amplitudes at point z in terms of the field amplitudes at point z'. This is achieved with the help of the approximate solution of the coupled wave Eq. (5.2-4). If we can assume that the field amplitudes change only slightly over distances that are large compared to the correlation length D, we can consider the field amplitudes c_n as nearly constant and obtain the following perturbation solution of Eq. (5.2-4):

$$c_m(z) = c_m(z') + \sum_{n=1}^{N} \hat{K}_{mn}\, c_n(z') \int_{z'}^{z} f(x) \exp[i(\beta_m - \beta_n)x]\, dx \tag{5.2-8}$$

It is at this point that the weak coupling assumption and ideas of perturbation theory enter the analysis. Substitution of Eq. (5.2-8) into Eq. (5.2-5) leads to

$$\begin{aligned} dP_\mu/dz = \sum_{\nu,n} \Big\{ &\hat{K}_{\mu\nu} \hat{K}_{\nu n} \langle c_n(z')\, c_\mu^*(z')\rangle \exp[i(\beta_\mu - \beta_\nu)z] \\ &\cdot \int_{z'}^{z} \langle f(z) f(x)\rangle \exp[i(\beta_\nu - \beta_n)x]\, dx \\ &+ \hat{K}_{\mu\nu} \hat{K}_{\mu n}^* \langle c_\nu(z')\, c_n^*(z')\rangle \exp[i(\beta_\mu - \beta_\nu)z] \\ &\cdot \int_{z'}^{z} \langle f(z) f(x)\rangle \exp[-i(\beta_\mu - \beta_n)x]\, dx + \text{cc} \Big\} \end{aligned} \tag{5.2-9}$$

A number of additional assumptions are incorporated in Eq. (5.2-9). We have used Eq. (5.2-7), as discussed earlier, but we have also assumed that the average of $f(z)$ vanishes:

$$\langle f(z)\rangle = 0 \tag{5.2-10}$$

Contributions of products of the first term on the right-hand side of Eq. (5.2-8) thus do not appear in Eq. (5.2-9). Products of the second term are proportional to $\hat{K}_{\nu\mu} \hat{K}_{mn}$. In the spirit of perturbation theory and the weak coupling assumption, such second-order terms have been neglected. Only cross products of the first and the second terms in Eq. (5.2-8) are included in Eq. (5.2-9). The principle that is expressed in Eq. (5.2-7) was also applied to quadruple products of two-mode amplitudes and the product of $f(z) f(x)$. The average of this quadruple product can be expressed as the product of two averages by the same argument that was advanced above. Next we introduce the autocorrelation function

$$R(u) = R(-u) = \langle f(z) f(z-u)\rangle \tag{5.2-11}$$

Since we have already used the fact that $R(u)$ contributes only over a range on the order of the correlation length, we can change the lower integration limit from z' to $-\infty$ and obtain

$$\exp[i(\beta_\mu-\beta_\nu)z]\int_{z'}^{z}\langle f(z)f(x)\rangle \exp[i(\beta_\nu-\beta_n)x]\,dx$$

$$= \exp[i(\beta_\mu-\beta_n)z]\int_{0}^{\infty} R(u)\exp[-i(\beta_\nu-\beta_n)u]\,du \qquad (5.2\text{-}12)$$

and

$$\exp[i(\beta_\mu-\beta_\nu)z]\int_{z'}^{z}\langle f(z)f(x)\rangle \exp[-i(\beta_\mu-\beta_n)x]\,dx$$

$$= \exp[i(\beta_n-\beta_\nu)z]\int_{0}^{\infty} R(u)\exp[i(\beta_\mu-\beta_n)u]\,du \qquad (5.2\text{-}13)$$

Equation (5.2-9) contains a double summation. However, we argue now that only diagonal elements contribute to the summations. This argument is based on two assumptions that reinforce each other. One the one hand, we expect that the phases of the complex field amplitudes are sufficiently random, so that we have

$$\langle c_n c_\mu{}^*\rangle = \langle |c_\mu|^2\rangle \delta_{n\mu} \qquad (5.2\text{-}14)$$

This equation removes one of the summations from Eq. (5.2-9). However, even if Eq. (5.2-14) were not valid, we could argue that only nonoscillatory terms contribute appreciably to the equation system, since oscillatory terms have the tendency to cancel the contributions from any integration required to solve the differential Eq. (5.2-9). The oscillatory terms stem from Eqs. (5.2-12) and (5.2-13). Only if $\beta_\mu = \beta_n$ does the oscillatory term in Eq. (5.2-12) cease to oscillate. We thus obtain the same result, already contained in Eq. (5.2-14), that only the terms with $n = \mu$ contribute to the first summation term on the right-hand side of Eq. (5.2-9). A similar argument applies to the second term, so that we obtain from Eqs. (5.2-1), (5.2-9), and (5.2-14):

$$dP_\mu(z)/dz = \sum_{\nu=1}^{N}\Big\{[\hat{K}_{\mu\nu}\hat{K}_{\nu\mu}P_\mu(z') + |\hat{K}_{\mu\nu}|^2 P_\nu(z')]$$

$$\cdot\int_{0}^{\infty} R(u)\exp[i(\beta_\mu-\beta_\nu)u]\,du + \text{cc}\Big\} \qquad (5.2\text{-}15)$$

Equation (5.2-15) is almost in the desired form. It is a system of coupled first-order differential equations containing only the average power of the modes but not their amplitudes. The equation system has the flaw of containing P_μ on the left-hand side with the argument z, while the average power on the right-hand side of the equation is taken at $z = z'$. However, when we derived the equation system we had to assume that the slowly varying field amplitudes, and hence the power, is nearly constant over the range from z' to z. We are thus justified in replacing the argument z' on the right-hand side with z.

There is an important property of the coupling coefficients $\hat{K}_{\mu\nu}$ that we can use to simplify Eq. (5.2-15). This property is a consequence of conservation of power and can be derived from the coupled wave Eq. (5.2-4). The requirement that the total power of all the modes must be independent of z in a lossless system can be expressed as follows:

$$\begin{aligned}\frac{d}{dz}\sum_{\mu}|c_\mu|^2 &= \sum_{\mu}\left(\frac{dc_\mu}{dz}c_\mu{}^* + c_\mu\frac{dc_\mu{}^*}{dz}\right)\\ &= \sum_{\mu\nu}(\hat{K}_{\mu\nu}+\hat{K}^*_{\nu\mu})f(z)\,c_\nu c_\mu{}^*\exp[i(\beta_\mu-\beta_\nu)z]\\ &= 0\end{aligned} \tag{5.2-16}$$

Equation (5.2-16) must hold for any combination of mode amplitudes. The amplitudes are quite arbitrary, since we have considerable freedom in choosing initial conditions. Equation (5.2-16) can be satisfied for all conceivable initial conditions only if

$$\hat{K}_{\mu\nu} = -\hat{K}^*_{\nu\mu} \tag{5.2-17}$$

With the use of Eq. (5.2-17) we see that the expression inside the square bracket, on the right-hand side of Eq. (5.2-15), is real. The complex conjugate terms thus contribute the complex conjugate of the integral already explicitly stated in Eq. (5.2-15). The sum of the two integrals is

$$\begin{aligned}&\int_0^\infty R(u)\exp[-i(\beta_\mu-\beta_\nu)u]\,du + \int_0^\infty R(u)\exp[i(\beta_\mu-\beta_\nu)u]\,du\\ &\qquad = \int_{-\infty}^{\infty} R(u)\exp[-i(\beta_\mu-\beta_\nu)u]\,du\\ &\qquad = \langle|F(\beta_\mu-\beta_\nu)|^2\rangle\end{aligned} \tag{5.2-18}$$

In the second integral on the left-hand side, we changed u to $-u$ and used $R(u) = R(-u)$. The last expression on the right-hand side was inserted using Eq. (4.6-8).

We can now write the coupled power equations in their final form

$$dP_\mu/dz = -2\alpha_\mu P_\mu + \sum_{\nu=1}^{N} h_{\mu\nu}(P_\nu - P_\mu) \tag{5.2-19}$$

with the following expression for the power coupling coefficients:

$$h_{\mu\nu} = |\hat{K}_{\mu\nu}|^2\langle|F(\beta_\mu-\beta_\nu)|^2\rangle \tag{5.2-20}$$

There is an additional term on the right-hand side of Eq. (5.2-19). We introduced the power loss coefficient $2\alpha_\mu$ in order to extend the validity of the coupled power equations to the lossy case. The loss is a combination of radiation and heat losses. It is possible to introduce the loss by allowing the propagation constants β_μ to have a slight imaginary component. However, this additional feature would complicate the derivation of the coupled power equations unnecessarily. The additional loss term is so logical that it does not require any explanation.

The coupled power equations have a simple intuitive meaning. If we assume that at some point z only mode μ carries power, we would have

$$dP_\mu/dz = -\left(2\alpha_\mu + \sum_{\nu=1}^{N} h_{\mu\nu}\right)P_\mu \tag{5.2-21}$$

The loss term is now augmented by the sum over all the coupling coefficients indicating that the guided mode μ loses power to all the other guided modes. If, on the other hand, mode μ should carry no power at a particular point z, we have

$$dP_\mu/dz = \sum_{\nu=1}^{N} h_{\mu\nu} P_\nu \tag{5.2-22}$$

The z derivative of P_μ is now positive, indicating that this mode gains power by being coupled to all the other guided modes.

The coupled power equations are so simple and have such clear intuitive meaning that we could have written them down without any derivation. However, our derivation has served two purposes. It has shown that the intuitive form (5.2-19) is indeed valid, at least to the approximation implied by our derivation. In addition we have obtained the explicit expression (5.2-20) for the power coupling coefficients. The correct form of the power coupling coefficients would have been harder to guess.

The coupled power equations are a simple system of first-order differential equations with constant coefficients. In addition to being constant, we see from Eq. (5.2-17) that $h_{\mu\nu}$ is also symmetric:

$$h_{\mu\nu} = h_{\nu\mu} \tag{5.2-23}$$

The symmetry of the power spectrum, also contained in Eq. (5.2-20), follows from the fact that $R(u)$ is an even function of u.

To appreciate the simplicity of the coupled power equations, we take one more look at the coupled wave Eq. (5.2-4) and recall that $f(z)$ is a random function of z. There is practically no hope to solve Eq. (5.2-4) for a general case. Such solutions are hard to obtain even on a computer. The problem of solving the coupled power Eq. (5.2-19) can always be reduced to an algebraic eigenvalue problem that can be solved on a computer. Methods for solving the coupled power equations are discussed in the next section.

5.3 cw Operation of Multimode Waveguides

The coupled power Eqs. (5.2-19) do not contain the time coordinate t. They describe continuous wave (cw) operation of multimode dielectric waveguides. We solve this system of coupled first-order differential equations by substitution of

$$P_\nu(z) = A_\nu e^{-\sigma z} \tag{5.3-1}$$

into Eq. (5.2-19). The eigenvalue σ is independent of the mode label. The system of differential equations is now converted into a homogeneous system of N equations with N unknowns A_ν:

$$\sum_{\nu=1}^{N} [h_{\mu\nu} + (\sigma - 2\alpha_\nu - b_\nu)\,\delta_{\mu\nu}]\,A_\nu = 0 \tag{5.3-2}$$

with the abbreviation

$$b_\mu = \sum_{\nu=1}^{N} h_{\mu\nu} \tag{5.3-3}$$

Equation (5.3-2) is an algebraic eigenvalue problem. Its solutions are the N eigenvectors $\mathbf{A}^{(n)}$ and the eigenvalues $\sigma^{(n)}$. The symbol $\mathbf{A}^{(n)}$ indicates a column vector with N elements A_ν. Since a system of N homogeneous equations has N eigenvectors and N eigenvalues, we need a system of labels to indicate the νth element of the nth eigenvector. We thus write this element as $A_\nu^{(n)}$. The homogeneous equation system (5.3-2) has solutions only if its determinant vanishes. We thus obtain the eigenvalues $\sigma^{(n)}$ as solutions of the eigenvalue equation

$$\det[h_{\mu\nu} + (\sigma - 2\alpha_\nu - b_\nu)\,\delta_{\mu\nu}] = 0 \tag{5.3-4}$$

The notation $\det[g_{\mu\nu}]$ indicates the determinant of the matrix with elements $g_{\mu\nu}$. Equation (5.3-4) is a polynomial of order N with N solutions $\sigma^{(n)}$. Not all the $\sigma^{(n)}$ need to be different from each other. If two of them coincide we speak of degenerate eigenvalues or of a case of degeneracy.

The N eigenvectors $\mathbf{A}^{(n)}$ are mutually orthogonal. To prove this assertion we write the eigenvalue problem Eq. (5.3-2) in the form

$$\sum_{\nu=1}^{N} H_{\mu\nu} A_\nu^{(n)} = \sigma^{(n)} A_\mu^{(n)} \tag{5.3-5}$$

with the matrix elements

$$H_{\mu\nu} = h_{\mu\nu} - (2\alpha_\nu + b_\nu)\,\delta_{\mu\nu} \tag{5.3-6}$$

The matrix elements are symmetric:

$$H_{\mu\nu} = H_{\nu\mu} \tag{5.3-7}$$

We multiply Eq. (5.3-5) with $A_\mu^{(m)}$ and sum over μ, obtaining

$$\sum_{\nu,\mu} H_{\mu\nu} A_\mu^{(m)} A_\nu^{(n)} = \sigma^{(n)} \sum_\mu A_\mu^{(m)} A_\mu^{(n)} \tag{5.3-8}$$

We also use Eq. (5.3-5) with m instead of n, multiply with $A_\mu^{(n)}$, sum, and have

$$\sum_{\nu,\mu} H_{\mu\nu} A_\mu^{(n)} A_\nu^{(m)} = \sigma^{(m)} \sum_\mu A_\mu^{(n)} A_\mu^{(m)} \tag{5.3-9}$$

By interchanging the summation indices ν and μ in Eq. (5.3-9), we see with the help of Eq. (5.3-7) that the left-hand sides of Eqs. (5.3-8) and (5.3-9) are identical. For $\sigma^{(n)} \neq \sigma^{(m)}$, we obtain from the difference of Eqs. (5.3-8) and (5.3-9),

$$\sum_{\mu=1}^{N} A_\mu^{(m)} A_\mu^{(n)} = \delta_{mn} \tag{5.3-10}$$

The normalization implied by Eq. (5.3-10) does not follow from this derivation. However, since the solutions of the homogeneous equation system (5.3-5) are determined only up to an arbitrary multiplicative constant, it is convenient to choose the eigenvectors so that they are normalized as indicated by Eq. (5.3-10). Eigenvectors belonging to a degenerate eigenvalue are not automatically orthogonal to each other, but it is always possible to use linear combinations of such eigenvectors, which are still solutions Eq. (5.3-5) and are also mutually orthogonal. In this way it is possible to obtain a complete orthogonal system of N eigenvectors even if some of the eigenvalues are degenerate.

The orthogonality relation (5.3-10) involves summation over the index μ labeling the vector components. There is also another orthogonality relation involving summation over the label m that specifies the eigenvector. This second orthogonality relation follows from the fact that any arbitrary vector **B** with N components B_μ can be expressed as a superposition of eigenvectors:

$$B_\mu = \sum_{n=1}^{N} c_n A_\mu^{(n)} \tag{5.3-11}$$

The expansion coefficients can be expressed in terms of the eigenvectors and the vector **B** by multiplying Eq. (5.3-11) with $A_\mu^{(m)}$, summing over μ, and using the orthogonality relation Eq. (5.3-10):

$$c_m = \sum_{\mu=1}^{N} A_\mu^{(m)} B_\mu \tag{5.3-12}$$

We now choose a special vector **B** that has the components $\delta_{\mu\nu}$. The expansion coefficients for this special vector are

$$c_m = A_\nu^{(m)} \tag{5.3-13}$$

Substitution of Eq. (5.3-13) into Eq. (5.3-11) leads to the expansion for the special vector $B_\mu = \delta_{\mu\nu}$:

$$\sum_{n=1}^{N} A_\mu^{(n)} A_\nu^{(n)} = \delta_{\mu\nu} \tag{5.3-14}$$

This is the orthogonality relation of the second kind.

We have found a solution of the coupled power equations. The nth solution of the eigenvalue problem allows us to write Eq. (5.3-1) in the form

$$P_\nu^{(n)}(z) = A_\nu^{(n)} \exp(-\sigma^{(n)} z) \tag{5.3-15}$$

Just as any arbitrary N component vector could be expressed as an expansion of eigenvectors in form (5.3-11), we can now express the most general solution of the coupled power equations as

$$P_\nu(z) = \sum_{n=1}^{N} c_n A_\nu^{(n)} \exp(-\sigma^{(n)} z) \tag{5.3-16}$$

The distribution of power versus mode number ν depends on the initial power distribution at $z = 0$ which, in principle, can be determined by the experimenter. Setting $z = 0$ in Eq. (5.3-16), we obtain the expansion coefficients c_n, with the help of Eq. (5.3-12),

$$c_n = \sum_{\nu=1}^{N} A_\nu^{(n)} P_\nu(0) \tag{5.3-17}$$

Equations (5.3-16) and (5.3-17) represent the complete (formal) solution of the time independent coupled power Eq. (5.2-19). The problem of determining the distribution of power versus mode number and its change along the waveguide is thus reduced to the solution of the algebraic eigenvalue problem, Eqs. (5.3-2) or (5.3-5).

It might appear as though not much has been gained by the formal solution (5.3-16) and (5.3-17). However, this general solution allows us to reach an important conclusion. It is apparent from Eq. (5.3-17) that the distribution of power versus mode number ν depends on the initial power distribution $P_\nu(0)$. The eigenvalues $\sigma^{(n)}$ are all different from each other except, perhaps, for an occasional degeneracy. Let us assume that at least the first eigenvalue $\sigma^{(1)}$ is nondegenerate and that we label the eigenvalues such that

$$\sigma^{(1)} < \sigma^{(2)} \leqslant \sigma^{(3)} \leqslant \cdots \sigma^{(N)} \tag{5.3-18}$$

holds. It follows from physical arguments that the eigenvalues can not be negative. If any of the eigenvalues were negative, $P_\nu(z)$ would become infinitely large at $z \to \infty$.

Since all eigenvalues must be positive, it is clear that the terms under the summation sign in Eq. (5.3-16) decrease at a different rate as z increases. The ratio of the first and second term is

$$(c_1 A_v^{(1)}/c_2 A_v^{(2)}) \exp(\sigma^{(2)} - \sigma^{(1)}) z \qquad (5.3\text{-}19)$$

This ratio increases indefinitely as z increases. The first term of the series is thus dominant over all other terms for sufficiently large values of z. This is a significant observation. It means that the distribution of power versus mode number becomes independent of the initial power distribution and depends only on the shape of the first eigenvector $A_v^{(1)}$. We say that the power distribution reaches a steady state after a distance $z = L_s$ given by

$$L_s = (\ln K)/(\sigma^{(2)} - \sigma^{(1)}) \qquad (5.3\text{-}20)$$

The value of K depends on the criterion that we want to apply to the steady state. If we are satisfied if the first term in the series expansion (5.3-16) is 100 times larger than the second term, we set $K = 100$ or $\ln K = 4.61$. The steady state power distribution is given by

$$P_{vs}(z) = \left[\sum_{v=1}^{N} A_v^{(1)} P_v(0)\right] A_v^{(1)} \exp(-\sigma^{(1)} z) \qquad (5.3\text{-}21)$$

The steady state distribution applies, of course, only to the average power of the modes. Fluctuations about this average value do occur. However, the steady state power distribution allows us to define a definite loss value for a multimode waveguide. As long as the steady state is not yet reached, the decrease of the power in the multimode fiber depends on the power distribution versus mode number. A unique loss can not be assigned to this transient condition of the waveguide. However, once the steady state is reached, the average power carried by each mode decreases at the same rate, so that the total power distribution decreases without change of its shape according to the loss coefficient $\sigma^{(1)}$. The first eigenvalue and eigenvector thus have important physical meanings. The first eigenvalue represents the steady state power loss coefficient, while the first eigenvector determines the shape of the power versus mode number distribution once the steady state is reached. The other eigenvalues and eigenvectors do not have a physical significance. They are needed to adjust the actual power distribution at $z = 0$, and in the transition region $0 \leqslant z < L_s$, to the initial power distribution that is determined by the experimental conditions. The difference between $\sigma^{(2)}$ and $\sigma^{(1)}$ determines the length of the transition region, but it is important to keep in mind that the eigenvectors $A_v^{(n)}$, with $n > 1$, do not represent physically realizable power distributions. Because of its physical meaning as the steady state power distribution, all elements of the first eigenvector $A_v^{(1)}$ must be positive:

$$A_v^{(1)} \geqslant 0, \qquad \text{for all } v \qquad (5.3\text{-}22)$$

However, since the sum of the products of all other eigenvectors with $A_\nu^{(1)}$ must vanish according to Eq. (5.3-10), at least some of their elements must be negative. Since negative power has no physical meaning, we see that none of the eigenvectors, except the first one, have an independent physical reality.

Another interesting fact can be deduced from Eq. (5.3-18). Because of its meaning as the steady state power loss coefficient, we must have $\sigma^{(1)} = 0$ in the lossless case. Inequality (5.3-18) implies that

$$\sigma^{(n)} > 0, \qquad \text{for} \quad n \neq 1 \tag{5.3-23}$$

even if there are no losses at all. The strength of the eigenvalues, with the exception of the first one, determines only the speed with which the power redistributes itself in the transition region $0 \leqslant z < L_s$.

In the lossless case we have

$$A_\nu^{(1)} = \sqrt{N}^{-1}, \qquad \sigma^{(1)} = 0 \tag{5.3-24}$$

To prove this assertion we only need to show that the eigenvalue Eq. (5.3-2), with $\alpha_\nu = 0$, is satisfied by the solution (5.3-24). (Actually it is sufficient if $\alpha_\nu = \alpha =$ const. and $\sigma^{(1)} = 2\alpha$.) We obtain from Eqs. (5.3-2) and (5.3-24):

$$\left[\left(\sum_{\nu=1}^{N} h_{\mu\nu}\right) - b_\mu\right]\sqrt{N}^{-1} = 0$$

That this equation holds follows from Eq. (5.3-3). Solutions of homogeneous algebraic equation systems are unique except for a constant multiplier. Equation (5.3-24) is thus the correct solution for the first eigenvector of the equation system (5.3-2) in the absence of losses. The form of the solution (5.3-24) satisfies the normalization condition (5.3-10). Equations (5.3-21) and (5.3-24) indicate that any kind of mode coupling tends to distribute the power evenly over all the modes provided that it does not introduce radiation losses. Equations (5.3-10) and (5.3-24) lead to

$$\sum_{\mu=1}^{N} A_\mu^{(n)} = 0, \qquad \text{for} \quad n \neq 1 \tag{5.3-25}$$

in the lossless case. This equation shows that the unphysical power distributions represented by the higher-order eigenvectors do not carry power in the lossless case. Because of Eqs. (5.3-24) and (5.3-25), we obtain from Eqs. (5.3-16) and (5.3-17) for the total power carried by all the modes in the lossless case

$$P(z) = \sum_{\nu=1}^{N} P_\nu(z) = \sqrt{N}\, c_1 = \sum_{\nu=1}^{N} P_\nu(0) \tag{5.3-26}$$

This equation shows that the formalism works, for we must certainly expect that the total power is equal to the power injected into all the modes at $z = 0$ if there is no loss.

Two Coupled Modes

The general solution of the steady state power distribution, Eq. (5.3-21), can not be evaluated further unless the specific form of the coupling coefficients is given. Only for the case of vanishing losses (or equal losses for all the modes) did we find the simple solution, Eq. (5.3-24), that indicates that the power is evenly distributed over all the modes once the steady state is reached. In order to obtain insight into the behavior of the solutions of the coupled power equations, we limit ourselves to the case of only two guided modes. The algebraic eigenvalue problem (5.3-2) assumes the simple form

$$(\sigma - 2\alpha_1 - h) A_1 + hA_2 = 0 \tag{5.3-27}$$

$$hA_1 + (\sigma - 2\alpha_2 - h) A_2 = 0 \tag{5.3-28}$$

with

$$h_{12} = h_{21} = h \tag{5.3-29}$$

The requirement of a vanishing system determinant leads to the eigenvalue equation

$$\sigma^2 - 2(\alpha_1 + \alpha_2 + h)\sigma + 4\alpha_1\alpha_2 + 2(\alpha_1 + \alpha_2)h = 0 \tag{5.3-30}$$

The two eigenvalues are the solutions of this second-order equation:

$$\sigma^{\binom{1}{2}} = (\alpha_1 + \alpha_2 + h) \mp [(\alpha_2 - \alpha_1)^2 + h^2]^{1/2} \tag{5.3-31}$$

The upper and lower superscripts 1 and 2 are associated with the upper and lower sign of the square root.

In the lossless case we have $\alpha_1 = \alpha_2 = 0$, so that Eq. (5.3-31) specializes to

$$\sigma^{(1)} = 0 \tag{5.3-32}$$

and

$$\sigma^{(2)} = 2h \tag{5.3-33}$$

The first eigenvalue is thus zero in accordance with the general result, and the second eigenvalue does not vanish even though there are no losses.

If the coupling vanishes, $h = 0$, we find from Eq. (5.3-31)

$$\sigma^{(1)} = 2\alpha_1 \tag{5.3-34}$$

and

$$\sigma^{(2)} = 2\alpha_2 \tag{5.3-35}$$

The identification of $2\alpha_1$ with the first eigenvalue is possible if we label the modes so that we have $\alpha_1 < \alpha_2$. In the absence of coupling the eigenvalues become identical with the power loss coefficients of the individual modes.

The distance (5.3-20) that is required for the transient to die out is now

$$L_s = (\ln K)/2[(\alpha_2-\alpha_1)^2+h^2]^{1/2} \tag{5.3-36}$$

The steady state is reached more quickly if the coupling is strong or if the differential loss is high.

The two eigenvectors are obtained from either one of Eqs. (5.3-27) or (5.3-28) and (5.3-31). The first eigenvector, in its properly normalized form, has the components

$$A_1^{(1)} = h/\sqrt{2}\{[(\alpha_2-\alpha_1)^2+h^2]^{1/2}-(\alpha_2-\alpha_1)\}^{1/2}[(\alpha_2-\alpha_1)^2+h^2]^{1/4} \tag{5.3-37}$$

and

$$A_2^{(1)} = \{[(\alpha_2-\alpha_1)^2+h^2]^{1/2}-(\alpha_2-\alpha_1)\}^{1/2}/\sqrt{2}[(\alpha_2-\alpha_1)^2+h^2]^{1/4} \tag{5.3-38}$$

The components of the second eigenvector are

$$A_1^{(2)} = A_2^{(1)} \tag{5.3-39}$$

and

$$A_2^{(2)} = -A_1^{(1)} \tag{5.3-40}$$

These two eigenvectors satisfy both orthogonality conditions (5.3-10) and (5.3-14). We also see that both components of the first eigenvector are positive, while the components of the second eigenvector have opposite signs.

In the lossless case, $\alpha_1 = \alpha_2 = 0$, we have for the first eigenvector

$$A_1^{(1)} = A_2^{(1)} = 1/\sqrt{2} \tag{5.3-41}$$

in agreement with Eq. (5.3-24), and for the second eigenvector

$$A_1^{(2)} = -A_2^{(2)} = 1/\sqrt{2} \tag{5.3-42}$$

Equation (5.3-41) confirms that both modes carry equal average power in the lossless case.

In the absence of coupling, $h = 0$, the first eigenvector has the components

$$A_1^{(1)} = 1 \tag{5.3-43}$$

and

$$A_2^{(1)} = 0 \tag{5.3-44}$$

This is an interesting result. It shows that in the steady state all the power is carried by the mode with the lower loss if coupling between the modes does not exist. The physical explanation for this result is simple. Since there is no coupling the lossier mode dies away more rapidly, so that for large values of z only the lower loss mode, mode 1, carries power.

The second eigenvector has the following form in the absence of coupling, $h = 0$:

$$A_1^{(2)} = 0 \tag{5.3-45}$$

and

$$A_2^{(2)} = -1 \tag{5.3-46}$$

There is no physical reality associated with this second eigenvector.

Finally we consider the strong coupling case $h \gg |\alpha_2 - \alpha_1|$ and obtain, by expanding the square root of Eq. (5.3-31),

$$\sigma^{(1)} = (2\alpha_1 + 2\alpha_2)/2 - (\alpha_2 - \alpha_1)^2/2h \tag{5.3-47}$$

This expression shows that the steady-state loss approaches the average value of the power loss coefficients of the two modes in case of strong coupling.

The two eigenvalues $\sigma^{(1)}$ and $\sigma^{(2)}$ are degenerate only in the trivial case of no coupling, $h = 0$, and equal mode losses, $\alpha_1 = \alpha_2$. In general, $\sigma^{(1)}$ is not identical with $\sigma^{(2)}$.

The Multimode Case

If there are only two guided modes, a complete solution of the coupled power equations is possible for any arbitrary coupling process. In the multimode case, no closed form solution is available. In the special case that only neighboring modes are coupled to each other, it is possible to convert the system of coupled differential equations to a partial differential equation, provided that the number of guided modes is very large. This conversion of the coupled equation system (5.2-19) to a partial differential equation is discussed in Sect. 5.6.

We provide an insight into the multimode behavior of dielectric waveguides by discussing computer solutions of the eigenvalue problem (5.3-2). In [Me5] it was assumed that the autocorrelation function of the random coupling process has Gaussian shape:

$$R(u) = \bar{\sigma}^2 \exp[-(u/D)^2] \tag{5.3-48}$$

As a consequence, the power spectrum is also Gaussian. From Eqs. (4.6-8) and (5.3-48) we obtain

$$\langle |F(\theta)|^2 \rangle = \sqrt{\pi}\, \bar{\sigma}^2 D \exp[-(\theta D/2)^2] \tag{5.3-49}$$

In [Me5] the theory was applied to the TE modes of a symmetric slab waveguide. It was assumed that both core interfaces are distorted with the same power spectrum but with no correlation between them. This causes the contribution from the two interfaces to add in their contribution to the power coupling coefficient (5.2-20). We thus have from Eqs. (3.5-22) and (3.5-27), with $\hat{K}_{\mu\nu} = \hat{R}_{\mu\nu}$,

$$|\hat{K}_{\mu\nu}|^2 = 2\kappa_\mu{}^2\kappa_\nu{}^2/\beta_\mu\beta_\nu[d+(2/\gamma_\mu)][d+(2/\gamma_\nu)] \qquad (5.3\text{-}50)$$

The factor 2 in the numerator accounts for the fact that both interfaces of the slab contribute in an uncorrelated fashion. Since we are considering a symmetric slab, we have used $\gamma_\nu = \delta_\nu$. We compare Eq. (5.3-50) with the corresponding expression for the round fiber whose far from cutoff approximation follows from Eq. (3.6-16):

$$|\hat{K}_{\mu\nu}|^2 = (e_{\mu\nu m}^2/e_\nu e_\mu)\,\kappa_\mu{}^2\kappa_\nu{}^2/4n^2k^2a^2 \qquad (5.3\text{-}51)$$

The two expressions are strikingly similar. Far from cutoff we have $\gamma_\nu d \gg 1$. In the weak guidance approximation the relation $\beta_\mu\beta_\nu \approx n^2k^2$ can be substituted in Eq. (5.3-50). If we had $e_{\mu\nu m}^2/e_\nu e_\mu = 2$, the agreement between Eqs. (5.3-50) and (5.3-51) would be complete, since d corresponds to $2a$. It is certainly possible to have $e_{\mu\nu m} = 2$, $e_\nu = 2$, and $e_\mu = 1$, so that the two coupling coefficients are indeed identical. However, even if we consider guided modes and coupling processes such that the numerical factors of Eqs. (5.3-50) and (5.3-51) do not coincide completely, it is clear that the results of the slab waveguide theory are in very close agreement with the results that apply to the round optical fiber. The results differ at most by a constant factor that is easy to include in any actual numerical comparison.

The process that causes coupling among the guided modes also couples the guided modes to the radiation modes so that radiation losses result. The radiation losses can be included in the theory through the loss coefficients α_μ appearing in Eq. (5.2-19). An exact calculation of the radiation losses from Eq. (3.4-19) is not possible for the power spectrum (5.3-49). An estimate of the radiation losses was attempted in [Me5].

Figure 5.3.1 shows the first two normalized eigenvalues $\sigma^{(i)}a/\bar{\sigma}^2k^2$, for $i = 1$ and $i = 2$, as a function of the normalized correlation length D/a, for $n_1 = 1.5$, $n_1/n_2 = 1.01$. Since our result was calculated for the symmetric slab waveguide, a must be associated with $d/2$.

The first eigenvalue represents the steady state losses of the 10 (solid lines)- and 21 (dashed lines)-mode slab waveguide. The second eigenvalue is plotted in Fig. 5.3.1 in order to provide information about the distance L_s, defined by Eq. (5.3-20), that is required until the steady state is reached. The steady state loss curves of Fig. 5.3.1 resemble the single-mode radiation loss curves, Fig. 4.6.1. However, the autocorrelation function was chosen differently in both cases.

In order to obtain a feeling for the rms deviation required to produce 10 dB/km steady state loss, we again use the peak values of the curves in Fig. 5.3.1 and obtain, for a free space wavelength of $\lambda = 1$ μm for $ka = 82$, the value

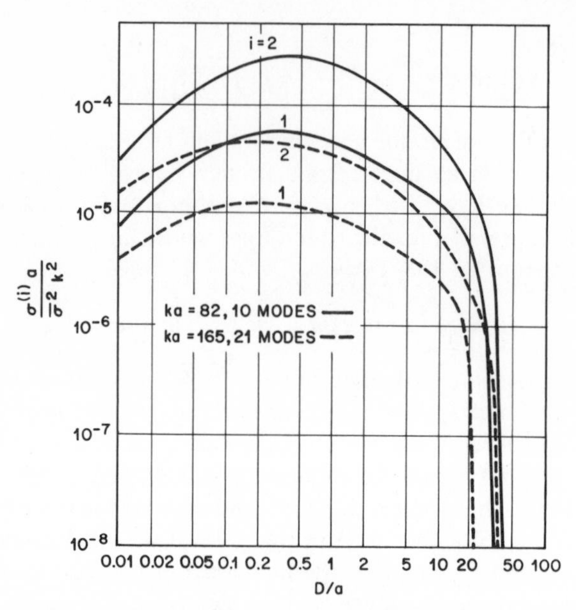

Fig. 5.3.1 *The first two normalized eigenvalues $\sigma^{(1)}$ and $\sigma^{(2)}$ are shown as functions of the correlation length of a Gaussian autocorrelation function. The first eigenvalue $\sigma^{(1)}$ represents the power loss coefficient.* (From *D. Marcuse, "Power distribution and radiation loss in multimode dielectric slab waveguides,"* Bell Syst. Tech. J. ***51***, *429–454 (1972). Copyright 1972, The American Telephone and Telegraph Co., reprinted by permission.*)

$a = 13\ \mu\text{m}$. We find in Fig. 5.3.1 the value $\sigma^{(1)}a/\bar{\sigma}^2k^2 = 5 \times 10^{-5}$ for the 10-mode case, for $D/a = 0.4$. A loss of 10 dB/km corresponds to $\sigma^{(1)} = 2.3 \times 10^{-9}$ μm^{-1}. This loss is caused by an rms deviation of $\bar{\sigma} = 4 \times 10^{-3}\ \mu\text{m}$. For the 21-mode case with $ka = 165$, we have $a = 26.3\ \mu\text{m}$, $\sigma^{(1)}a/\bar{\sigma}^2k^2 = 1.3 \times 10^{-5}$, so that we find for 10 dB/km loss, $\bar{\sigma} = 0.01\ \mu\text{m}$. The larger waveguide is more tolerant of random core boundary imperfections. However, the relative tolerance requirement is almost the same in both cases. For the 10-mode guide we have $(\bar{\sigma}/a)100 = 0.03\%$. The corresponding number for the 21-mode case is $(\bar{\sigma}/a)100 = 0.04\%$. The relative tolerance requirement for low (scattering)-loss operation is very stringent indeed. However, it must be pointed out once more that our tolerance estimates are based on the worst possible case. There is evidence that the correlation length in actual low-loss fibers is very much longer than the value we assumed. Because of the very rapid decrease of the curves of Fig. 5.3.1 with increasing value of D/a, the tolerance requirement becomes far less stringent for larger values of D/a. Equation (5.3-49) shows that $2/D$ corresponds to the spatial bandwidth of the Fourier spectrum of the distortion function $f(z)$. A narrow Fourier spectrum corresponds to long correlation length. The scattering experiment made by Rawson

[Rn1] indicates that the Fourier spectrum is very narrow, so that the actual correlation length of optical fibers is very long.

Table 5.3.1 contains a listing of the eigenvalues and the components of the first eigenvector for three different values of the ratio D/a for the 10-mode case. The labels i and ν, labeling the eigenvalues and the components of the first eigenvector, are different. The label i is used to indicate the 10 different

TABLE 5.3.1

Eigenvalues and the First Eigenvector[a]

D/a	i or ν	$(a/\bar{\sigma}^2k^2)\sigma^{(i)}$	$A_\nu^{(1)}$
	1	7.624×10^{-6}	9.999×10^{-1}
	2	3.044×10^{-5}	8.486×10^{-5}
	3	6.845×10^{-5}	7.160×10^{-5}
	4	1.214×10^{-4}	6.789×10^{-5}
0.01	5	1.891×10^{-4}	6.630×10^{-5}
	6	2.711×10^{-4}	6.546×10^{-5}
	7	3.666×10^{-4}	6.497×10^{-5}
	8	4.743×10^{-4}	6.466×10^{-5}
	9	5.909×10^{-4}	6.444×10^{-5}
	10	7.049×10^{-4}	6.424×10^{-5}
	1	5.175×10^{-6}	7.284×10^{-1}
	2	1.402×10^{-5}	3.799×10^{-1}
	3	4.576×10^{-5}	3.212×10^{-1}
	4	1.008×10^{-4}	2.770×10^{-1}
20	5	1.951×10^{-4}	2.376×10^{-1}
	6	3.329×10^{-4}	2.011×10^{-1}
	7	5.217×10^{-4}	1.648×10^{-1}
	8	7.596×10^{-4}	1.248×10^{-1}
	9	1.046×10^{-3}	7.475×10^{-2}
	10	4.345×10^{-3}	3.129×10^{-3}
	1	6.415×10^{-9}	3.296×10^{-1}
	2	2.552×10^{-7}	3.294×10^{-1}
	3	1.112×10^{-6}	3.293×10^{-1}
	4	4.251×10^{-6}	3.291×10^{-1}
35	5	7.789×10^{-6}	3.290×10^{-1}
	6	1.608×10^{-5}	3.287×10^{-1}
	7	3.149×10^{-5}	3.279×10^{-1}
	8	5.837×10^{-5}	3.248×10^{-1}
	9	1.036×10^{-4}	3.094×10^{-1}
	10	1.765×10^{-4}	2.067×10^{-1}

[a] *From* D. Marcuse, "Power distribution and radiation losses in multimode dielectric slab waveguides," *Bell. Syst. Tech. J.* **51**, 429–454 (1972). Copyright 1972, The American Telephone and Telegraph Co., reprinted by permission.

solutions of the eigenvalue problem, while ν labels the 10 components of the eigenvector $\mathbf{A}^{(1)}$.

The first eigenvector determines the shape of the power *versus* mode number distribution of the steady state. We see from the last column of Table 5.3.1 that the steady state distribution favors the lowest-order mode for small values of D/a; practically all the power is contained in mode 1. For $D/a = 20$, the power decreases gradually with increasing mode number. Finally, for $D/a = 35$, all the modes, with the exception of the last, carry nearly the same amount of power.

The shape of the steady state power distribution is determined by the interplay of radiation losses and intermode coupling. For small values of D/a, all modes are coupled directly to the radiation mode continuum, since the wide Fourier spectrum contains a sufficient number of spatial frequencies so that Eq. (4.2-20) is satisfied for every guided mode. However, the radiation loss formula, Eq. (4.2-25), for a sinusoidal imperfection shows that higher-order modes lose more power because the radiation loss coefficient is proportional to κ_ν^2, and κ_ν increases linearly with the mode number ν, as Eq. (1.3-89) indicates. The loss process causes more power loss than the intermode coupling process can counteract, so that only the lowest loss mode, $\nu = 1$, carries an appreciable amount of power. This situation corresponds to the weak coupling case (5.3-43) and (5.3-44).

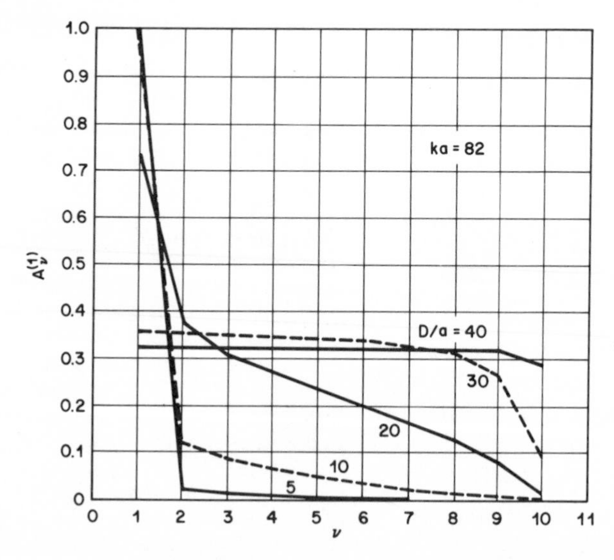

Fig. 5.3.2 *Steady state power distributions for different values of the correlation length D. The waveguide carries 10 modes.* (From *D. Marcuse, "Power distribution and radiation losses in multimode dielectric slab waveguides,"* Bell Syst. Tech. J. *51, 429–454 (1972). Copyright 1972, The American Telephone and Telegraph Co., reprinted by permission.*)

For very long correlation length, the Fourier spectrum is narrow, so that only the highest-order mode is coupled to the radiation field. The intermode coupling is thus able to distribute power evenly over all the modes with the exception of the last one, $\nu = 10$, which loses power at a somewhat higher rate than can be supplied by its neighbors. For very narrow Fourier spectra, only nearest neighbors couple to each other. In fact, it is possible, at least in principle, to shape the Fourier spectrum so that all guided modes are coupled to each other *via* the nearest neighbor coupling mechanism, while no radiation losses occur. This case will be discussed in more detail in Sect. 5.5.

Figure 5.3.2 is a graphical representation of the components of the first eigenvector corresponding to the shape of the steady state power distribution *versus* mode number. The trend from a distribution favoring only the lowest-order mode to one that puts equal power in all the modes is clearly discernible. The plot should actually be a bar graph since only integer values of ν have physical meaning. In order to be able to show several graphs in one figure, we connected the isolated points of the figure by straight lines. Figure 5.3.2 again describes the 10-mode case. The same trend is, however, true for any mode number. The different average levels of the curves in Fig. 5.3.2 are caused by the normalization. The square of the power contained in all ten modes adds up to unity. The refractive indices are again chosen as $n_1 = 1.5$ and $n_1/n_2 = 1.01$.

5.4 Power Fluctuations

The coupled power theory presented in the two preceding sections deals with the average power carried by the guided modes. This theory is simple in comparison to the coupled wave theory of Chap. 3 and gives interesting and valuable insight into the multimode behavior of dielectric optical waveguides. However, it is important to remember that it is only the average and not the actual power that is described by this theory. When the theory predicts equal power in all the guided modes, we must realize that the actual power distribution that is observed at the output of one individual waveguide may deviate considerably from the predicted ensemble average. In order to gain insight into the expected departure from the average value, we study power fluctuation in this section.

Our approach to the power fluctuation problem is very similar to the derivation of the coupled power equations in Sect. 5.2. We derive a system of coupled differential equations for the variance of the power by means of perturbation theory. The resulting equation system can be solved for a special case that is of practical value.

The variance of the power of mode μ is defined as

$$(\Delta P_\mu)^2 = \langle (Q_\mu - P_\mu)^2 \rangle = \langle Q_\mu{}^2 \rangle - P_\mu{}^2 \tag{5.4-1}$$

Since the symbol P_μ is already defined as the average power, we use the symbol Q_μ for the actual power of the mode, and $\langle\ \rangle$ indicates the ensemble average. With Eq. (5.2-1) and

$$Q_\mu = c_\mu c_\mu^* \tag{5.4-2}$$

we obtain from Eq. (5.4-1)

$$\tfrac{1}{2}\, d(\Delta P_\mu)^2/dz = [\langle c_\mu c_\mu^* (dc_\mu/dz) c_\mu^* \rangle + \mathrm{cc}] - P_\mu (dP_\mu/dz) \tag{5.4-3}$$

The symbol cc indicate again that the complex conjugate expression must be added. Substitution of Eq. (5.2-4) into Eq. (5.4-3) results in

$$\tfrac{1}{2}\, d(\Delta P_\mu)^2/dz = \left[\sum_\nu \hat{K}_{\mu\nu} \langle c_\mu c_\mu^* c_\nu c_\mu^* f(z)\rangle \exp[i(\beta_\mu - \beta_\nu) z] + \mathrm{cc}\right] - P_\mu (dP_\mu/dz) \tag{5.4-4}$$

Proceeding in close analogy to the derivation of the coupled power equations in Sect. 5.2, we replace all mode amplitudes with the perturbation solution (5.2-8). The mode amplitudes are now evaluated at the point z' that is several correlation lengths removed from the point z at which the derivative of the variance is taken. Because $f(z)$ is uncorrelated with the field amplitudes at $z' < z$, the ensemble average of the product of mode amplitudes and products of $f(z)$ with itself can be expressed as the product of the ensemble averages. It may seem disturbing that this process includes $f(x)$, with $z' < x < z$. However, $f(x)$ appears in the combination $f(x) f(z)$. When x approaches z', $f(x)$ is certainly correlated with the field amplitudes, but $f(z)$ is uncorrelated to all the other terms under the ensemble average sign, so that the result vanishes. No error results from combining $f(x)$ with $f(z)$ under one average sign. In addition to Eq. (5.2-8) we use Eq. (5.2-10) and obtain from Eq. (5.4-4)

$$\begin{aligned} \tfrac{1}{2}\, d(\Delta P_\mu)^2/dz = \Bigg\{ \sum_{n,\nu} \Bigg[& \hat{K}_{\mu\nu} \hat{K}_{\mu n} \langle c_n c_\mu^* c_\nu c_\mu^* \rangle \exp[i(2\beta_\mu - \beta_\nu - \beta_n) z] \\ & \cdot \int_{z'}^{z} \langle f(z) f(x) \rangle \exp[i(\beta_\mu - \beta_n)(x - z)]\, dx \\ & + 2\hat{K}_{\mu\nu} \hat{K}_{\mu n}^* \langle c_\mu c_n^* c_\nu c_\mu^* \rangle \exp[i(\beta_n - \beta_\nu) z] \\ & \cdot \int_{z'}^{z} \langle f(z) f(x) \rangle \exp[-i(\beta_\mu - \beta_n)(x - z)]\, dx \\ & + \hat{K}_{\mu\nu} \hat{K}_{\nu n} \langle c_\mu c_\mu^* c_n c_\mu^* \rangle \exp[i(\beta_\mu - \beta_n) z] \\ & \cdot \int_{z'}^{z} \langle f(z) f(x) \rangle \exp[i(\beta_\nu - \beta_n)(x - z)]\, dx \Bigg] + \mathrm{cc} \Bigg\} \\ & - P_\mu \sum_\nu h_{\mu\nu} (P_\nu - P_\mu) \end{aligned} \tag{5.4-5}$$

Terms of higher than second order have been neglected. The derivative of the average power was replaced with the help of Eq. (5.2-19). However, the loss term has been omitted since we are concentrating on the lossless case for the moment.

We are now invoking the random phase argument that was also used in the derivation of the coupled power equations. In the spirit of the random phase approximation, we use

$$\langle c_n c_\mu^* c_\nu c_\mu^* \rangle = \langle (c_\mu c_\mu^*)^2 \rangle \, \delta_{n\mu} \, \delta_{\nu\mu} \tag{5.4-6}$$

$$\langle c_\mu c_n^* c_\nu c_\mu^* \rangle = \langle c_\mu c_\mu^* c_\nu c_\nu^* \rangle \, \delta_{n\nu} \tag{5.4-7}$$

and

$$\langle c_\mu c_\mu^* c_n c_\mu^* \rangle = \langle (c_\mu c_\mu^*)^2 \rangle \, \delta_{n\mu} \tag{5.4-8}$$

Using Eqs. (5.2-17), (5.2-18), (5.2-20), (5.4-2), and (5.4-6)–(5.4-8), we obtain

$$d(\Delta P_\mu)^2/dz = 2 \sum_{\nu=1}^{N} h_{\mu\nu}(2\langle Q_\mu Q_\nu \rangle - \langle Q_\mu^2 \rangle + P_\mu^2 - P_\mu P_\nu) \tag{5.4-9}$$

The first term in Eq. (5.4-5) does not appear in Eq. (5.4-9). This term would contain $h_{\mu\mu}$. We omitted it under the assumption that the self-coupling term in Eq. (5.2-4) vanishes, so that we use $K_{\mu\mu}=0$. Self-coupling can only modify the propagation constant of the wave. For this reason the self-coupling term $K_{\mu\mu}$ can be absorbed in β_μ. The value of $h_{\mu\mu}$ does not affect the coupled power Eq. (5.2-19) since $P_\nu - P_\mu = 0$, for $\nu = \mu$. In Eq. (5.4-9) Q_ν is actually taken at $z = z'$. However, using the argument that the perturbation theory is valid only if $Q_\nu(z)$ is very nearly constant over the region between z' and z, we replace z' with z in Eq. (5.4-9).

With the help of Eqs. (5.3-3) and (5.4-1), we can rewrite Eq. (5.4-9) in the following way:

$$d(\Delta P_\mu)^2/dz = -2b_\mu(\Delta P_\mu)^2 + 2 \sum_{\nu=1}^{N} h_{\mu\nu}(2\langle Q_\mu Q_\nu \rangle - P_\mu P_\nu) \tag{5.4-10}$$

The term containing $b_\mu = \sum_\nu h_{\mu\nu}$ in Eq. (5.2-21) could be interpreted as loss, since it caused a decrease in the average power level in case only one mode was excited. The same interpretation can also be given to the b_μ term in Eq. (5.4-10). If only mode μ carries power the sum vanishes (since we assumed $h_{\mu\mu} = 0$), so that the variance decreases at a rate proportional to its value.

The derivation of Eq. (5.4-10) was based on the assumption that the modes do not lose power to radiation or by a heat dissipation mechanism. The coupled power Eq. (5.2-19) was also derived for a lossless waveguide and the loss term with $2\alpha_\mu$ was inserted phenomenologically. A similar procedure can also be applied to Eq. (5.4-10). In the absence of mode coupling, the right-hand side of Eq. (5.4-10) vanishes. However, we now assume that each mode suffers a loss that is determined by the power loss coefficient $2\alpha_\mu$, so that we have

$$Q_\mu(z) = Q_\mu(z') \exp[-2\alpha_\mu(z-z')] \tag{5.4-11}$$

We assume that a given mode suffers the same loss in every waveguide of the ensemble. This assumption has the consequence that the average value decreases at the same rate,

$$P_\mu = P_\mu(z') \exp[-2\alpha_\mu(z-z')] \tag{5.4-12}$$

From Eq. (5.4-1) we now derive

$$d(\Delta P_\mu)^2/dz = d\{[\Delta P_\mu(z')]^2 \exp[-4\alpha_\mu(z-z')]\}/dz = -4\alpha_\mu(\Delta P_\mu)^2 \tag{5.4-13}$$

In the absence of coupling, Eq. (5.4-10) must assume the form (5.4-13). For weak coupling and low losses we expect that these two effects are additive. Using the cross correlation

$$\langle (Q_\mu - P_\mu)(Q_\nu - P_\nu)\rangle = \langle Q_\mu Q_\nu \rangle - P_\mu P_\nu \tag{5.4-14}$$

and introducing the loss term (5.4-13) yields the following set of differential equations for the variance of the mode power in lossy waveguides:

$$d(\Delta P_\mu)^2/dz = -a_\mu(\Delta P_\mu)^2 + 2\sum_{\nu=1}^{N} h_{\mu\nu}[2\langle (Q_\mu - P_\mu)(Q_\nu - P_\nu)\rangle + P_\mu P_\nu] \tag{5.4-15}$$

with

$$a_\mu = 4\alpha_\mu + 2\sum_{\nu=1}^{N} h_{\mu\nu} \tag{5.4-16}$$

The derivative of the variance can be positive or negative. The variance can increase or decrease depending on the conditions of the power fluctuations at the point z and on the interplay between coupling and loss.

Differential Eq. (5.4-15) can easily be integrated, with the result

$$[\Delta P_\mu(z)]^2 = [\Delta P_\mu(0)]^2 \exp(-a_\mu z) + 2\exp(-a_\mu z) \cdot \int_0^z \exp(a_\mu x) \sum_{\nu=1}^{N} h_{\mu\nu}[2\langle (Q_\mu - P_\mu)(Q_\nu - P_\nu)\rangle + P_\mu P_\nu]\, dx \tag{5.4-17}$$

If we knew the average power and the cross correlation for all the modes, we could calculate the variance from Eq. (5.4-17). The average power can be obtained from solutions of the coupled power equations, but the cross correlation is not known. In spite of this difficulty we can extract useful information from Eq. (5.4-17). The cross correlation has the property that it vanishes if the power carried by the modes is completely random, which means uncorrelated. Because of the existing correlation in the coupling function we can not expect that the cross correlation vanishes exactly. However, it seems reasonable to assume that the cross correlation may be small. Since we have no other information, we neglect the cross correlation term in Eq. (5.4-17) in the hope that the power carried in each mode is only slightly correlated with the power carried in other modes. For a guide supporting only two guided modes, this assumption is clearly wrong. As one mode gains power the other mode must lose an equal amount of power, so that the fluctuations of the two modes are perfectly correlated. However, one may expect that the modes of a multimode waveguide with a large number of guided modes are only slightly correlated. In general, we expect that the cross-correlation term is negative. If neighboring modes are coupled preferentially, $h_{\mu\nu}$ is large only for ν values close to μ. As one mode gains power its neighbors are likely to have decreased in power so that, as one mode increases above average, the neighboring modes may be expected to decrease below their average values. As a result we see that the two terms in the cross-correlation product are more likely to be of opposite signs, so that the cross correlation is negative. If this is the case, the cross correlation would tend to counteract the influence of the positive term $P_\mu P_\nu$ in Eq. (5.4-17), so that the variance remains smaller than in the absence of cross correlation. Neglecting the cross-correlation term thus leads to a pessimistic result exaggerating the variance.

We have seen in Sect. 5.3 that the average power settles down to a steady state distribution after an initial transient has died out. The steady state distribution (5.3-21) is constant except for a common loss term. If we assume that the waveguides of the ensemble are excited so that the power averages correspond to the steady state distribution from the beginning, we can replace the average power terms in Eq. (5.4-17) with

$$P_\mu = P_{\mu 0} \exp(-\sigma^{(1)} z) \tag{5.4-18}$$

With Eq. (5.4-18) and without the cross correlation (note that $h_{\mu\mu} = 0$ also removes the term with $\nu = \mu$) we obtain from Eq. (5.4-17), with $\Delta P_\mu(0) = 0$,

$$[\Delta P_\mu(z)]^2 = 2\{[\exp(-2\sigma^{(1)}z) - \exp(-a_\mu z)]/(a_\mu - 2\sigma^{(1)})\} P_{\mu 0} \sum_{\nu=1}^{N} h_{\mu\nu} P_{\nu 0} \tag{5.4-19}$$

We derive an expression for the relative power fluctuations by dividing Eq. (5.4-19) by the square of Eq. (5.4-18) and taking the square root:

$$\frac{\Delta P_\mu(z)}{P_\mu(z)} = \left[2 \frac{1 - \exp[-(a_\mu - 2\sigma^{(1)})z]}{a_\mu - 2\sigma^{(1)}} \sum_{\nu=1}^{N} h_{\mu\nu} \frac{P_{\nu 0}}{P_{\mu 0}}\right]^{1/2} \tag{5.4-20}$$

The difference $a_\mu - 2\sigma^{(1)}$ is always positive. For the lossless case, we have $\sigma^{(1)} = 0$, so that $a_\mu - 2\sigma^{(1)} > 0$ follows immediately. For high losses $\sigma^{(1)}$ approaches $2\alpha_1$. Since $\alpha_1 < \alpha_\nu$, with $\nu \neq 1$, we see from Eq. (5.4-16) that $a_\mu - 2\sigma^{(1)} > 0$ also in this case. We obtained expression (5.4-20) for the relative power fluctuations by assuming that the variance vanishes at $z = 0$. This means that the steady state distribution is actually launched into every waveguide of the ensemble. Equation (5.4-20) tells us that the relative power fluctuation does not remain zero even though it is zero initially. It grows monotonically until it reaches the asymptotic value

$$\frac{\Delta P_\mu}{P_\mu} = \left(\frac{2}{a_\mu - 2\sigma^{(1)}} \sum_{\nu=1}^{N} h_{\mu\nu} \frac{P_{\nu 0}}{P_{\mu 0}}\right)^{1/2} \tag{5.4-21}$$

In some cases of multimode operation it is desirable to have strong coupling among the guided modes. Strong coupling in this context means that the coupling term is considerably larger than the loss term in Eq. (5.4-16). In case of strong coupling, Eq. (5.4-21) becomes

$$\Delta P_\mu / P_\mu = \left\{\sum_{\nu=1}^{N} h_{\mu\nu}(P_{\nu 0}/P_{\mu 0}) / \sum_{\nu=1}^{N} h_{\mu\nu}\right\}^{1/2} \tag{5.4-22}$$

In this case Eq. (5.3-24) applies, indicating that all modes carry equal amounts of power. We thus obtain finally

$$\Delta P_\mu / P_\mu = 1 \tag{5.4-23}$$

Our derivation makes it clear that Eq. (5.4-23) represents the largest value that the relative fluctuations can assume. Any existing cross correlation may be expected to reduce this value. It is also less if the losses play an important role in determining the steady state distribution.

Our theory has provided us with important information about the power fluctuations. We have learned that the fluctuations build up steadily until they reach an asymptotic value that remains constant. The distance required until this asymptotic value is essentially reached follows from Eq. (5.4-20). If we use the criterion that the exponential function in Eq. (5.4-20) decreases to 10^{-2}, we obtain for the distance that is needed for the asymptotic fluctuation value to be reached

$$L_\mu = 2.3/(2\alpha_\mu - \sigma^{(1)} + \sum_{\nu=1}^{N} h_{\mu\nu}) \tag{5.4-24}$$

In general, the asymptotic distance is different for every mode. Comparison with Eq. (5.3-20) indicates that the asymptotic distance, which is characteristic for power fluctuations, is not identical with the distance that is required until the steady state power distribution is reached.

The relative power fluctuations can become quite large. According to Eq. (5.4-23), the rms value of the power fluctuation can become equal to the average value of the power. The fluctuations may thus become as large as 100%. Such large fluctuations are not unusual for physical processes. The short-term time-averaged power of electrical thermal noise with Gaussian amplitude distribution is known to fluctuate this way [Bn1]. The probability distribution for the power fluctuations in multimode waveguides may thus be expected to follow the same law as thermal noise power fluctuations in electrical circuits. We thus assume that the probability distribution for Q_μ is given by

$$W(Q_\mu) = P_\mu^{-1} \exp(-Q_\mu/P_\mu) \tag{5.4-25}$$

With this probability distribution, we find for the variance

$$(\Delta P_\mu)^2 = \left[\int_0^\infty W(Q_\mu)\, Q_\mu{}^2\, dQ_\mu\right] - P_\mu{}^2 = P_\mu{}^2 \tag{5.4-26}$$

in agreement with Eq. (5.4-23). The fact that the probability distribution (5.4-25) leads to the correct value for the variance does not prove its validity. However, it does appear plausible that the mode amplitudes may be Gaussian random variables, at least to some approximation, so that the probability distribution (5.4-25) would be obtained for the average power of the modes.

Our discussion of power fluctuations in multimode waveguides may have created the impression that the coupled power theory is of little value, since the actual power values fluctuate so much around the average value. The situation is really not as bad as it may appear at first glance. When the coupled power theory predicts equal power in all the modes for the strong coupling case, we know that we cannot expect to find all the power preferentially in the lowest-order mode. On the other hand, if we obtain a steady state power distribution, predicting that almost all the power resides in the lowest-order mode, we do not expect much power in any of the other modes even if the fluctuations around the predicted average value are as large as 100%. In this case, the losses are instrumental in determining the shape of the steady state distribution, so that the power fluctuations may be expected to be considerably less than 100%. However, even in the worst possible case of 100% fluctuations, we still gain valuable insight into the power distribution to be expected from an actual experiment, knowing the distribution of power versus mode number for the average power.

The situation improves even further if we consider not a single mode but a large group of modes. In optical multimode waveguides the number of guided modes can be very large. It is thus impossible to single out any particular mode for detection. Most experiments that are designed to select some of the guided modes instead of the total output power arriving in all the modes can be expected to detect an average value over a fairly large number of modes. It is thus necessary to investigate the fluctuations of groups of modes. Defining the power carried by a group of M modes as

$$Q_M = \sum_{\nu=n}^{n+M-1} Q_\nu \tag{5.4-27}$$

we obtain the average value

$$P_M = \sum_{\nu=n}^{n+M-1} P_\nu \tag{5.4-28}$$

and the variance

$$(\Delta P_M)^2 = \langle Q_M{}^2 \rangle - P_M{}^2 = \sum_{\nu=n}^{n+M-1} \sum_{\mu=n}^{n+M-1} (\langle Q_\nu Q_\mu \rangle - P_\nu P_\mu) \tag{5.4-29}$$

Introducing the cross correlation, Eq. (5.4-14), we have

$$(\Delta P_M)^2 = \sum_\nu \sum_\mu \langle (Q_\mu - P_\mu)(Q_\nu - P_\nu) \rangle \tag{5.4-30}$$

With the assumption that there is no correlation between the modes, only the diagonal elements of Eq. (5.4-30) remain:

$$(\Delta P_M)^2 = \sum_{\nu=n}^{n+M-1} \langle (Q_\nu - P_\nu)^2 \rangle = \sum_{\nu=n}^{n+M-1} (\Delta P_\nu)^2 \tag{5.4-31}$$

Finally we assume that all modes of the group have the same average power and obey Eq. (5.4-23). We thus have

$$\Delta P_M / P_M = \left(P_\nu{}^2 \sum_{\nu=n}^{n+M-1} 1 \middle/ P_\nu{}^2 \sum_{\nu=n}^{n+M-1} 1 \sum_{\mu=n}^{n+M-1} 1 \right)^{1/2} = \sqrt{M}^{-1} \tag{5.4-32}$$

The relative fluctuation of the power contained in M equally excited modes is thus reduced by $M^{-1/2}$ compared to the relative fluctuation of the power of one mode. The power that is carried by 100 modes thus fluctuates only by 10%. This discussion shows that the experimentally observed power fluctuations may be quite modest if we detect not one signal mode but groups of modes.

The power of all the modes may still fluctuate slightly from one waveguide to the next since the loss mechanism may be slightly different for each waveguide. However, for statistically similar guides the fluctuation of the total power is expected to be very small.

5.5 Pulse Propagation in Multimode Waveguides

The coupled mode theory presented in the preceding four sections did not contain the time variable, so that it applied only to the cw case. For the purposes of optical communications it is important to know the properties of pulse propagation in multimode dielectric optical waveguides. This theory will again be based on a description of the average power carried by each mode. A time-dependent theory of coupled average power is obtained from the coupled power equations of Sect. 5.2 by a straightforward generalization.

The propagation of a pulse of arbitrary shape is described by a function of the following kind:

$$P_\mu(z,t) = G_\mu(z-v_\mu t) \qquad (5.5\text{-}1)$$

where P_μ is the average power of mode μ, and v_μ is its group velocity. Single-mode dispersion of each mode is neglected in our present treatment. The power pulse (5.5-1) is the solution of the differential equation

$$(\partial P_\mu/\partial z) + (1/v_\mu)(\partial P_\mu/\partial t) = 0 \qquad (5.5\text{-}2)$$

We can easily incorporate waveguide losses by using the partial differential equation

$$(\partial P_\mu/\partial z) + (1/v_\mu)(\partial P_\mu/\partial t) = -2\alpha_\mu P_\mu \qquad (5.5\text{-}3)$$

The solution of this equation is

$$P_\mu = \exp(-2\alpha_\mu z)\, G_\mu(z-v_\mu t) \qquad (5.5\text{-}4)$$

Differential Eqs. (5.5-2) and (5.5-3) apply to a stationary observer who sees the pulse passing by. An observer moving with the pulse would have the impression of a continuous process. To him the pulse is not a passing phenomenon but appears like a cw signal viewed by a stationary observer. A moving observer would describe the pulse by dropping the partial time derivative term if he uses the z coordinate as his only variable. For the moving observer, Eq. (5.5-3) has the same form as Eq. (5.2-19) in the absence of coupling, $h_{\mu\nu} = 0$. If the modes are coupled it is natural to expect that the moving observer would describe the pulse by Eq. (5.2-19), since to him it appears very much like a cw process. The stationary observer would write the same differential equation in the form

$$(\partial P_\mu/\partial z) + (1/v_\mu)(\partial P_\mu/\partial t) = -2\alpha_\mu P_\mu + \sum_{\nu=1}^{N} h_{\mu\nu}(P_\nu - P_\mu) \qquad (5.5\text{-}5)$$

We use the partial differential Eq. (5.5-5) as the basis of our description of pulse propagation in multimode optical waveguides. We do not claim that Eq. (5.5-5) is an exact description of a pulse. The coupled power Eqs. (5.2-19) are not exact but hold only for reasonably weak coupling. If the group velocities are considerably different for different modes, the pulses spread apart and even fail to overlap. The pulse does not appear as a cw process even to the moving observer. The applicability of the time-dependent coupled power equations is thus clearly limited to long pulses and nearly identical group velocities. However, Eq. (5.5-5) describes pulse propagation in the multimode waveguide correctly in the absence of coupling, $h_{\mu\nu} = 0$, for arbitrary group velocities. We show in this section that coupling has the tendency to keep the pulse from spreading even if the group velocities of the modes are different. It is thus possible that the time dependent coupled power Eqs. (5.5-5) provide a good description of pulse propagation even for large differences in the group velocities of the individual modes in case of reasonably strong coupling. This brief discussion shows that the range of applicability and the accuracy of the time-dependent coupled power equations is not known. The results that are obtained from the time-dependent coupled power theory are plausible and inspire confidence that the theory is indeed a reasonably accurate approximation to the—as yet unknown—exact theory.

We find a formal solution of the coupled power Eqs. (5.5-5) by using the trial solution

$$P_\mu = B_\mu e^{-\rho z} e^{i\omega t} \tag{5.5-6}$$

The time dependence of Eq. (5.5-6) does not mean that the average power of the mode oscillates with the angular frequency ω. In fact, Eq. (5.5-6) does not have a physical meaning all by itself. Only the superposition of infinitely many such solutions in the form of a Fourier integral expansion and a superposition of eigensolutions of an algebraic eigenvalue problem assumes a physical reality. Substitution of Eq. (5.5-6) into Eq. (5.5-5) results in the following algebraic eigenvalue problem:

$$\sum_{\nu=1}^{N} \{h_{\mu\nu} + [\rho - (i\omega/v_a) - i\omega V_\nu - 2\alpha_\nu - b_\nu]\, \delta_{\mu\nu}\}\, B_\nu = 0 \tag{5.5-7}$$

with the abbreviations

$$V_\mu = (1/v_\mu) - (1/v_a) \tag{5.5-8}$$

and

$$b_\mu = \sum_{\nu=1}^{N} h_{\mu\nu} \tag{5.5-9}$$

The term V_μ was introduced as a small perturbation to emphasize that the group velocities v_μ of the different modes are nearly equal to their average velocity v_a. The assumption $V_\mu \ll 1/v_a$ allows us to solve the equation system (5.5-7) by perturbation theory.

For $V_\mu = 0$, Eq. (5.5-7) is identical with Eq. (5.3-2). To zero order of approximation we find the same eigenvectors as in Sect. 5.3. The zero-order eigenvalue, $\rho_0^{(i)}$, can be expressed in terms of the eigenvalue $\sigma^{(i)}$ of the time-independent problem,

$$\rho_0^{(j)} = \sigma^{(j)} + (i\omega/v_a) \tag{5.5-10}$$

and the zero-order eigenvectors are identical to the eigenvectors of the time-independent problem,

$$\mathbf{B}_0^{(j)} = \mathbf{A}^{(j)} \tag{5.5-11}$$

In the general case with $V_\mu \neq 0$, the eigenvalue problem (5.5-7) is different from Eq. (5.3-2). However, it is possible to solve the eigenvalue problem (5.5-7), since its matrix is also symmetrical. We again find N eigenvalues $\rho^{(j)}(\omega)$ and the corresponding N eigenvectors $\mathbf{B}^{(j)}(\omega)$. The notation indicates that the eigenvalues and eigenvectors are functions of ω. The general solution of Eq. (5.5-5) is a superposition of all the eigensolutions of the algebraic eigenvalue problem for all possible values of the parameter ω:

$$P_\mu(z,t) = \sum_{j=1}^{N} \int_{-\infty}^{\infty} c_j(\omega)\, B_\mu^{(j)}(\omega) \exp[-\rho^{(j)}(\omega) z]\, e^{i\omega t}\, d\omega \tag{5.5-12}$$

The eigenvectors satisfy the orthogonality relations (5.3-10):

$$\sum_{\mu=1}^{N} B_\mu^{(i)} B_\mu^{(j)} = \delta_{ij} \tag{5.5-13}$$

We determine the expansion coefficient $c_j(\omega)$ in two steps. First we invert the Fourier transformation and obtain from Eq. (5.5-12), with $z = 0$,

$$\sum_{j=1}^{N} c_j(\omega)\, B_\mu^{(j)}(\omega) = (1/2\pi) \int_{-\infty}^{\infty} P_\mu(0,t)\, e^{-i\omega t}\, dt \tag{5.5-14}$$

Next we multiply Eq. (5.5-14) with $B_\mu^{(k)}$ and sum over μ. Using Eq. (5.5-13) and replacing k again with j, we have

$$c_j(\omega) = (1/2\pi) \sum_{\mu=1}^{N} B_\mu^{(j)}(\omega) \int_{-\infty}^{\infty} P_\mu(0,t)\, e^{-i\omega t}\, dt \tag{5.5-15}$$

Equations (5.5-12) and (5.5-15) represent the complete solution of the time-dependent coupled power Eq. (5.5-5). The expansion coefficient $c_j(\omega)$ appearing in Eq. (5.5-12) is determined by Eq. (5.5-15) in terms of the power distribution of all the modes as a function of time at $z = 0$. However, this formal solution is expressed in terms of the solutions of the algebraic eigenvalue problem (5.5-7). Solutions of this problem are hard to obtain analytically if the number of modes is large. Computer solutions could be obtained, but since the eigenvectors and eigenvalues are complex functions of the parameter ω, numerical methods would be very time consuming and costly. For this reason we proceed to describe an approximate evaluation of Eq. (5.5-5) by means of perturbation theory.

Perturbation Solution

The eigenvalues and eigenvectors of the algebraic eigenvalue problem (5.5-7) depend on V_μ. For $V_\mu = 0$, we have $\rho^{(j)} = \rho_0^{(j)}$ and $\mathbf{B}^{(j)} = \mathbf{B}_0^{(j)}$ with $\rho_0^{(j)}$ and $\mathbf{B}_0^{(j)}$ of Eqs. (5.5-10) and (5.5-11). For small values of V_μ, the solutions are only slightly different from their value for $V_\mu = 0$. For this reason we use the perturbation expansions

$$\rho^{(j)} = \rho_0^{(j)} + i\omega\rho_1^{(j)} + \omega^2\rho_2^{(j)} \tag{5.5-16}$$

and

$$\mathbf{B}^{(j)} = \mathbf{B}_0^{(j)} + \mathbf{B}_1^{(j)} + \mathbf{B}_2^{(j)} \tag{5.5-17}$$

It is immediately apparent that substitution of Eq. (5.5-17) into Eq. (5.5-12) results in three terms of decreasing magnitude. The first term with $\mathbf{B}^{(j)} = \mathbf{B}_0^{(j)}$ is by far the largest and most important. We make only a slight error when we limit expansion (5.5-17) to its first term. However, the eigenvalue enters the exponent of the z-dependent exponential function in Eq. (5.5-12). Small changes of the eigenvalue result in very significant changes of the pulse shape since we must allow z to become very large. The expansion of the eigenvalue is thus of considerable importance. Before we discuss the explicit evaluation of the first- and second-order perturbations of the eigenvalue, we study the effect of the perturbation expansion (5.5-16) on the behavior of light pulses.

Our problem becomes relatively simple if we use a Gaussian shape for the input pulse at $z = 0$:

$$P_\mu(0, t) = G_\mu \exp[-(2t/\tau)^2] \tag{5.5-18}$$

where τ is the full width at the $1/e$ points of the Gaussian distribution. Substitution of Eq. (5.5-18) into Eq. (5.5-15) results in the following expression for the expansion coefficient:

$$c_j(\omega) = (k_j\tau/4\sqrt{\pi})\exp[-(\omega\tau/4)^2] \tag{5.5-19}$$

with

$$k_j = \sum_{\mu=1}^{N} A_\mu^{(j)} G_\mu \tag{5.5-20}$$

We have used only the zero-order term of the perturbation expansion of the eigenvector Eq. (5.5-17) and replaced $\mathbf{B}_0^{(j)}$ with $\mathbf{A}^{(j)}$ according to Eq. (5.5-11).

Next we substitute Eqs. (5.5-16) and (5.5-19) into the general solution (5.5-12) of the pulse propagation problem:

$$P_\mu(z,t) = (\tau/4\sqrt{\pi}) \sum_{j=1}^{N} \left(k_j A_\mu^{(j)} \exp(-\sigma^{(j)} z) \cdot \int_{-\infty}^{\infty} \exp\{i\omega[t-(z/v_a)-\rho_1^{(j)}z]\} \exp\{-\omega^2[(\tau^2/16)+\rho_2^{(j)}z]\}\, d\omega \right) \tag{5.5-21}$$

The terms $\sigma^{(j)}$ and v_a enter Eq. (5.5-21) through relation (5.5-10).

The integral in Eq. (5.5-21) extends over ω from $-\infty$ to ∞. These infinite integration limits seem to contradict the requirement that $\omega\rho_1^{(j)}$ and $\omega^2\rho_2^{(j)}$ in Eq. (5.5-16) ought to be small quantities. However, the integrand of the integral in Eq. (5.5-21) drops off rapidly for large values of z. Regardless of the actual size of $\rho_2^{(j)}$, the effective integration range becomes arbitrarily narrow as z grows sufficiently large. The perturbation theory is thus always applicable for large values of z. Close to the input end of the waveguide, for small values of z, we have to require that the width τ of the input pulse must be sufficiently wide if the perturbation theory is to be applicable. However, even if the input pulse is too narrow for the perturbation theory to hold for small values of z, its applicability is assured if we only make z large enough, since the function $\exp(-\omega^2\rho_2^{(j)}z)$ becomes arbitrarily narrow for large values of z. This feature of the perturbation theory is quite advantageous, since we are most interested in the pulse shape at the end of long optical waveguides.

After the Fourier integral of the Gaussian function is evaluated with the help of integration tables, we obtain from Eq. (5.5-21) the solution of our problem:

$$P_\mu(z,t) = \sum_{j=1}^{N} \left[\frac{\tau}{T_j} k_j A_\mu^{(j)} \exp(-\sigma^{(j)} z) \exp\left(-\frac{\{t-[(1/v_a)+\rho_1^{(j)}]z\}^2}{(T_j/2)^2} \right) \right] \tag{5.5-22}$$

with

$$T_j = (\tau^2 + 16\rho_2^{(j)} z)^{1/2} \tag{5.5-23}$$

We have seen in our discussion of the steady state power distribution in Sect. 5.3 that only the first term of series (5.3-16) need to be considered for large values of z. Since our perturbation solution is more accurate for large z values, it seems advisable to study the results of the perturbation theory only for large values of z where the steady state power distribution has established itself.

The steady state (or equilibrium) solution of our pulse propagation problem is thus

$$P_\mu(z,t) = \frac{\tau}{T_1} k_1 A_\mu^{(1)} \exp(-\sigma^{(1)}z) \exp\left(-\frac{\{t-[(1/v_a)+\rho_1^{(1)}]z\}^2}{(T_1/2)^2}\right) \tag{5.5-24}$$

with

$$T_1 = 4(\rho_2^{(1)}z)^{1/2} \tag{5.5-25}$$

Since it was assumed that z is very large, we have neglected τ in Eq. (5.5-23).

Before we discuss the many interesting consequences of our result we return to the problem of determining the first- and second-order perturbations of the eigenvalues.

Determination of the Perturbed Eigenvalues

We consider the eigenvalue problem

$$\sum_{\nu=1}^{N} (H_{\mu\nu} + i\omega V_{\mu\nu}) B_\nu^{(k)} = \rho^{(k)} B_\mu^{(k)} \tag{5.5-26}$$

The matrices H and V are assumed to be real and symmetric. The matrix elements $V_{\mu\nu}$ are supposed to be small quantities of first order of magnitude. The matrix H determines the zero-order eigenvalue problem whose solutions are assumed to be known:

$$\sum_{\nu=1}^{N} H_{\mu\nu} B_{\nu 0}^{(k)} = \rho_0^{(k)} B_{\mu 0}^{(k)} \tag{5.5-27}$$

To keep the discussion general, we are considering the possibility that at least some of the eigenvalues are degenerate, in which case there are several eigenvectors that belong to the same eigenvalue. In fact, any combination of these eigenvectors is also an eigenvector of Eq. (5.5-27). We express these linear combinations of eigenvectors, belonging to the same degenerate eigenvalue, as follows:

$$\hat{B}_{\mu 0}^{(k\gamma)} = \sum_\sigma a_\sigma^{(k\gamma)} B_{\mu 0}^{(k\sigma)} \tag{5.5-28}$$

The label σ is used to distinguish the different eigenvectors that belong to the same eigenvalue $\rho_0^{(k)}$. The summation over σ extends over the subspace of degenerate eigenvectors. There are, of course, as many linearly independent combinations as there are degenerate eigenvectors. The label γ is used to distinguish between them. The vector space $\hat{B}_0^{(k\gamma)}$ is of the same dimension as the vector space that is spanned by $B_0^{(k\sigma)}$. The expansion coefficients $a_\sigma^{(k\gamma)}$ will be determined by the first-order perturbation theory.

We use the perturbation expansion

$$\rho^{(k)} = \rho_0^{(k)} + i\omega\rho_1^{(k\gamma)} + \omega^2\rho_2^{(k\gamma)} \tag{5.5-29}$$

for the eigenvalues. The γ label that is attached to the first- and second-order perturbations of the eigenvalue expresses our expectation that the degeneracy will be broken by the perturbation, so that we have N values of $\rho_1^{(k\gamma)}$ and N values of $\rho_2^{(k\gamma)}$. In case of degeneracy, the number of different values of $\rho_0^{(k)}$ is less than N. A similar perturbation expansion is used for the eigenvectors

$$B_\mu^{(k)} = \hat{B}_{\mu 0}^{(k\gamma)} + B_{\mu 1}^{(k\gamma)} + B_{\mu 2}^{(k\gamma)} = B_\mu^{(k\gamma)} \tag{5.5-30}$$

The first terms in Eqs. (5.5-29) and (5.5-30) are of zero order, the second terms of first order, and the third terms of second order of magnitude. We substitute Eqs. (5.5-29) and (5.5-30) into Eq. (5.5-26) and equate equal orders of magnitude separately. This procedure results in the following equations:

$$\sum_{\nu=1}^{N} H_{\mu\nu}\hat{B}_{\nu 0}^{(k\gamma)} = \rho_0^{(k)}\hat{B}_{\mu 0}^{(k\gamma)} \tag{5.5-31}$$

$$\sum_{\nu=1}^{N} (H_{\mu\nu}B_{\nu 1}^{(k\gamma)} + i\omega V_{\mu\nu}\hat{B}_{\nu 0}^{(k\gamma)}) = \rho_0^{(k)}B_{\mu 1}^{(k\gamma)} + i\omega\rho_1^{(k\gamma)}\hat{B}_{\mu 0}^{(k\gamma)} \tag{5.5-32}$$

$$\sum_{\nu=1}^{N} (H_{\mu\nu}B_{\nu 2}^{(k\gamma)} + i\omega V_{\mu\nu}B_{\nu 1}^{(k\gamma)}) = \rho_0^{(k)}B_{\mu 2}^{(k\gamma)} + i\omega\rho_1^{(k\gamma)}B_{\mu 1}^{(k\gamma)} + \omega^2\rho_2^{(k\gamma)}\hat{B}_{\mu 0}^{(k\gamma)} \tag{5.5-33}$$

The zero-order Eq. (5.5-31) is satisfied by our choice of the zero-order solution. In order to solve the second- and third-order equations, we expand the first-order perturbation of the eigenvector in terms of the zero-order eigenvectors $\hat{B}_0^{(k\gamma)}$:

$$B_{\mu 1}^{(k\gamma)} = \sum_{j\sigma} b_{j\sigma}^{(k\gamma)}\hat{B}_{\mu 0}^{(j\sigma)} \tag{5.5-34}$$

Substitution of Eq. (5.5-34) into Eq. (5.5-32) yields, with the help of Eq. (5.5-31),

$$\sum_{j\sigma} \rho_0^{(j)}b_{j\sigma}^{(k\gamma)}\hat{B}_{\mu 0}^{(j\sigma)} + i\omega\sum_{\nu=1}^{N} V_{\mu\nu}\hat{B}_{\nu 0}^{(k\gamma)} = \rho_0^{(k)}\sum_{j\sigma} b_{j\sigma}^{(k\gamma)}\hat{B}_{\mu 0}^{(j\sigma)} + i\omega\rho_1^{(k\gamma)}\hat{B}_{\mu 0}^{(k\gamma)} \tag{5.5-35}$$

In order to determine the first-order perturbation of the eigenvalue, we multiply Eq. (5.5-35) with $B_{\mu 0}^{(k\sigma')}$, sum over μ, and use the orthogonality relations

$$\sum_{\mu=1}^{N} B_{\mu 0}^{(k\sigma')}\hat{B}_{\mu 0}^{(j\gamma)} = 0, \qquad \text{if} \quad k \neq j \tag{5.5-36}$$

and

$$\sum_{\mu=1}^{N} B_{\mu 0}^{(k\sigma')}B_{\mu 0}^{(j\sigma)} = \delta_{kj}\delta_{\sigma\sigma'} \tag{5.5-37}$$

With the help of Eq. (5.5-28), we obtain

$$\sum_{\mu=1}^{N}\sum_{\nu=1}^{N}\sum_{\sigma} B_{\mu 0}^{(k\sigma')} V_{\mu\nu} B_{\nu 0}^{(k\sigma)} a_{\sigma}^{(k\gamma)} = \rho_{1}^{(k\gamma)} a_{\sigma'}^{(k\gamma)} \tag{5.5-38}$$

The zero-order eigenvalues have cancelled. It is convenient to introduce the abbreviated notation

$$V_{\sigma'\sigma}^{(k)} = \sum_{\mu=1}^{N}\sum_{\nu=1}^{N} B_{\mu 0}^{(k\sigma')} V_{\mu\nu} B_{\nu 0}^{(k\sigma)} \tag{5.5-39}$$

Equation (5.5-38) defines another algebraic eigenvalue problem

$$\sum_{\sigma} (V_{\sigma'\sigma}^{(k)} - \rho_{1}^{(k\gamma)} \delta_{\sigma'\sigma}) a_{\sigma}^{(k\gamma)} = 0 \tag{5.5-40}$$

This equation system determines the expansion coefficients in Eq. (5.5-28). The first-order perturbations $\rho_{1}^{(k\gamma)}$ of the eigenvalues are obtained from the condition that the system determinant of Eq. (5.5-40) must vanish:

$$\det(V_{\sigma'\sigma}^{(k)} - \rho_{1}^{(k\gamma)} \delta_{\sigma'\sigma}) = 0 \tag{5.5-41}$$

The number of solutions of this equation is equal to the degree of degeneracy. If all solutions are different from each other the degeneracy, existing in the zero-order eigenvalue problem, is broken. In case that no degeneracy exists to zero order, there is only one term in the sum in Eq. (5.5-40), and condition (5.5-41) simplifies to

$$\rho_{1}^{(k)} = V^{(k)} \tag{5.5-42}$$

The labels γ and σ are not necessary in this case.

Because $V_{\sigma'\sigma}^{(k)}$ defines a symmetric matrix, the vectors with components $a_{\sigma}^{(k\gamma)}$ are mutually orthogonal. As a consequence of this fact and of Eq. (5.5-37) we have the orthogonality relations

$$\sum_{\mu=1}^{N} \hat{B}_{\mu 0}^{(k\gamma')} \hat{B}_{\mu 0}^{(j\gamma)} = \delta_{kj}\, \delta_{\gamma'\gamma} \tag{5.5-43}$$

We have now determined the first-order perturbations of the eigenvalues of the eigenvalue problem Eq. (5.5-26) and have also found the expansion coefficients for the proper combinations of the zero-order eigenvectors Eq. (5.5-28). With the help of these results we may now proceed to determine the expansion coefficients $b_{j\sigma}^{(k\gamma)}$ of the first-order perturbations of the eigenvectors, Eq. (5.5-34). We multiply Eq. (5.5-35) with $\hat{B}_{\mu 0}^{(i\gamma')}$, sum over μ, and use the orthogonality relation (5.5-43):

$$b_{j\gamma'}^{(k\gamma)} = -i\omega \sum_{\mu\nu} \hat{B}_{\mu 0}^{(j\gamma')} V_{\mu\nu} \hat{B}_{\nu 0}^{(k\gamma)} / (\rho_{0}^{(j)} - \rho_{0}^{(k)}), \qquad \text{for} \quad j \neq k \tag{5.5-44}$$

The index i has again been replaced with j. We simplify the notation by introducing

$$V_{j\gamma',k\gamma} = \sum_{\mu=1}^{N} \sum_{\nu=1}^{N} \hat{B}_{\mu 0}^{(j\gamma')} V_{\mu\nu} \hat{B}_{\nu 0}^{(k\gamma)} \tag{5.5-45}$$

The vector $\hat{B}_0^{(k\gamma)}$ is obtained from Eq. (5.5-28) in case of degeneracy, but it is simply $\hat{B}_0^{(k\gamma)} = B_0^{(k)}$ if no degeneracy exists. The expansion coefficient (5.5-44) assumes, with the help of Eq. (5.5-45), the simpler form

$$b_{j\gamma'}^{(k\gamma)} = -i\omega V_{j\gamma',k\gamma}/(\rho_0^{(j)} - \rho_0^{(k)}), \qquad \text{for} \quad j \neq k \tag{5.5-46}$$

The expansion coefficients (5.5-46) are defined only for $j \neq k$. The coefficients $b_{k\gamma}^{(k\gamma)}$ are obtained from the normalization condition. The eigenvector (5.5-30) must be normalized:

$$\sum_{\mu=1}^{N} (B_{\mu}^{(k\gamma)})^2 = \sum_{\mu=1}^{N} [(\hat{B}_{\mu 0}^{(k\gamma)})^2 + 2\hat{B}_{\mu 0}^{(k\gamma)} B_{\mu 1}^{(k\gamma)}] = 1 \tag{5.5-47}$$

Second-order terms are neglected in Eq. (5.5-47) since we are considering only first-order perturbation theory for the moment. The zero-order eigenvectors are also normalized, so that we obtain, from Eqs. (5.5-47) and (5.5-34),

$$\sum_{\mu=1}^{N} \hat{B}_{\mu 0}^{(k\gamma)} B_{\mu 1}^{(k\gamma)} = b_{k\gamma}^{k\gamma} = 0 \tag{5.5-48}$$

Eigenvectors belonging to the same value of k but having different γ labels must be orthogonal to each other. This condition results in

$$\sum_{\mu=1}^{N} B_{\mu}^{(k\gamma)} B_{\mu}^{(k\gamma')} = \sum_{\mu=1}^{N} (\hat{B}_{\mu 0}^{(k\gamma)} \hat{B}_{\mu 0}^{(k\gamma')} + \hat{B}_{\mu 0}^{(k\gamma)} B_{\mu 1}^{(k\gamma')} + \hat{B}_{\mu 0}^{(k\gamma')} B_{\mu 1}^{(k\gamma)}) = 0 \tag{5.5-49}$$

Since the zero-order eigenvectors are mutually orthogonal, we obtain, from Eqs. (5.5-49) and (5.5-34),

$$b_{k\gamma}^{(k\gamma')} + b_{k\gamma'}^{(k\gamma)} = 0 \tag{5.5-50}$$

Except for condition (5.5-50) the expansion coefficients $b_{k\gamma}^{(k\gamma')}$, with $\gamma' \neq \gamma$, remain arbitrary to first order of perturbation theory. These coefficients are determined from the second-order eigenvalue Eq. (5.5-33).

For our purposes the second-order perturbation of the eigenvalue is of most interest. It is obtained by multiplication of Eq. (5.5-33) with $\hat{B}_{\mu 0}^{(k\gamma)}$, summation over μ, and application of the orthogonality relation (5.5-43). We obtain, with the help of Eq. (5.5-48) and with

$$\sum_{\mu=1}^{N} \sum_{\nu=1}^{N} \hat{B}_{\mu 0}^{(k\gamma)} H_{\mu\nu} B_{\nu 2}^{(k\gamma)} = \sum_{\mu=1}^{N} \sum_{\nu=1}^{N} B_{\mu 2}^{(k\gamma)} H_{\mu\nu} \hat{B}_{\nu 0}^{(k\gamma)}$$
$$= \rho_0^{(k)} \sum_{\mu=1}^{N} \hat{B}_{\mu 0}^{(k\gamma)} B_{\mu 2}^{(k\gamma)} \tag{5.5-51}$$

the following expression for the second-order perturbation of the eigenvalue:

$$\omega^2 \rho_2^{(k\gamma)} = i\omega \sum_{\mu=1}^{N} \sum_{\nu=1}^{N} \hat{B}_{\mu 0}^{(k\gamma)} V_{\mu\nu} B_{\nu 1}^{(k\gamma)}$$

$$= i\omega \sum_{\mu\nu} \sum_{j\gamma'} b_{j\gamma'}^{(k\gamma)} \hat{B}_{\mu 0}^{(k\gamma)} V_{\mu\nu} \hat{B}_{\nu 0}^{(j\gamma')} \qquad (5.5\text{-}52)$$

The last step was accomplished with the help of Eq. (5.5-34). It is easy to prove that

$$\sum_{\mu\nu} \hat{B}_{\mu 0}^{(k\gamma)} V_{\mu\nu} \hat{B}_{\nu 0}^{(k\gamma')} = \rho_1^{(k\gamma)} \delta_{\gamma\gamma'} \qquad (5.5\text{-}53)$$

We use expansion (5.5-28) and obtain, with the help of Eqs. (5.5-38)–(5.5-40):

$$\sum_{\sigma'\sigma} a_{\sigma'}^{(k\gamma)} V_{\sigma'\sigma}^{(k)} a_{\sigma}^{(k\gamma')} = \rho_1^{(k\gamma')} \sum_{\sigma'} a_{\sigma'}^{(k\gamma)} a_{\sigma'}^{(k\gamma')} = \rho_1^{(k\gamma')} \delta_{\gamma\gamma'} \qquad (5.5\text{-}54)$$

The left-hand side of this equation is identical with the left-hand side of Eq. (5.5-53). Its right-hand side follows from the orthogonality of the eigenvectors of the eigenvalue problem (5.5-40). This orthogonality relation follows directly from the symmetry of the matrix elements $V_{\sigma'\sigma}^{(k)}$ by the same method that was used to prove Eq. (5.3-10). The symmetry of $V_{\sigma'\sigma}^{(k)}$ is a consequence of the symmetry of $V_{\mu\nu}$, according to its definition (5.5-39).

Because of Eqs. (5.5-48) and (5.5-53), all terms in the sum over γ' with $j = k$ vanish in Eq. (5.5-52). The case $j = k$ can thus be excluded from the summation. The undetermined coefficients $b_{k\gamma'}^{(k\gamma)}$ are not needed for the evaluation of $\rho_2^{(k\gamma)}$. Substitution of Eqs. (5.5-45) and (5.5-46) into Eq. (5.5-52) yields the following expression for the second-order perturbation of the eigenvalue:

$$\rho_2^{(k\gamma)} = \sum_{\substack{j\gamma' \\ j \neq k}} V_{k\gamma, j\gamma'}^2 / (\rho_0^{(j)} - \rho_0^{(k)}) \qquad (5.5\text{-}55)$$

The symmetry of the matrix element $V_{j\gamma', k\gamma}$—defined by Eq. (5.5-45)—was used in Eq. (5.5-55).

For the sake of completeness, we derive the—as yet undetermined—expansion coefficient $b_{k\gamma'}^{(k\gamma)}$. Even though this coefficient belongs to the expansion of the first-order perturbation of the eigenvector (5.5-34), it is obtained from the second-order terms, Eq. (5.5-33), of the eigenvalue equation. We multiply Eq. (5.5-33) with $\hat{B}_{\mu 0}^{(k\gamma')}$, sum over μ, and use Eqs. (5.5-31), (5.5-34), and (5.5-43) to obtain

$$\sum_{\mu\nu} \sum_{\substack{j\sigma \\ j \neq k}} b_{j\sigma}^{(k\gamma)} \hat{B}_{\mu 0}^{(k\gamma')} V_{\mu\nu} \hat{B}_{\nu 0}^{(j\sigma)} = \rho_1^{(k\gamma)} b_{k\gamma'}^{(k\gamma)} \qquad (5.5\text{-}56)$$

The condition $j \neq k$ follows from Eqs. (5.5-48) and (5.5-53). With the help of

Eqs. (5.5-45) and (5.5-46), we can express Eq. (5.5-56) as follows:

$$b_{k\gamma'}^{(k\gamma)} = (1/\rho_1^{(k\gamma)}) \sum_{\substack{j\sigma \\ j \neq k}} V_{j\sigma, k\gamma} V_{k\gamma', j\sigma} / (\rho_0^{(j)} - \rho_0^{(k)}), \qquad \text{for} \quad \gamma \neq \gamma' \tag{5.5-57}$$

The first-order perturbation of the eigenvector is now completely determined by means of Eqs. (5.5-34), (5.5-46), (5.5-48), (5.5-50), and (5.5-57).

We have thus obtained the eigenvalues of equation system (5.5-26) up to the second order of perturbation theory. The zero-order eigenvalue must be computed by solving equation system (5.5-27). The first-order perturbation of the eigenvalue can then be computed directly from Eqs. (5.5-42) and (5.5-39), in case of nondegenerate eigenvalues, or as the solution of another eigenvalue problem from Eq. (5.5-41). The second-order perturbation of the eigenvalues follows by direct calculation from Eq. (5.5-55). The zero-order eigenvector results as the solution of the eigenvalue problem (5.5-27), and its first-order perturbation follows from Eq. (5.5-34), with the coefficients given by Eqs. (5.5-46), (5.5-48), (5.5-50), and (5.5-57). The second-order correction of the eigenvector has not been calculated, since it is of no interest to our present discussion.

General Discussion of the Pulse Propagation Problem

In the absence of coupling, the eigenvalue problem, Eq. (5.5-7), has a simple solution. With $h_{\mu\nu} = 0$, we obtain for the eigenvalues

$$\rho^{(j)} = 2\alpha_j + (i\omega/v_j) \tag{5.5-58}$$

and for the components of the corresponding eigenvectors

$$B_\mu^{(j)} = A_\mu^{(j)} = \delta_{\mu j} \tag{5.5-59}$$

These solutions are exact. We see by comparison with Eqs. (5.5-10) and (5.5-16) that we now have

$$\sigma^{(j)} = 2\alpha_j \tag{5.5-60}$$

$$\rho_1^{(j)} = (1/v_j) - (1/v_a) \tag{5.5-61}$$

and

$$\rho_2^{(j)} = 0 \tag{5.5-62}$$

The general solution of the pulse propagation problem without mode coupling follows from Eqs. (5.5-22) and (5.5-23):

$$P_\mu(z, t) = k_\mu \exp(-2\alpha_\mu z) \exp\left\{-\frac{[t-(z/v_\mu)]^2}{(\tau/2)^2}\right\} \tag{5.5-63}$$

This solution is also exact. It states that in the absence of coupling, each mode carries the Gaussian pulse, which was launched at $z = 0$, independently of all the other modes. Each mode suffers its characteristic attenuation $2\alpha_\mu$ and travels at a different group velocity. If the pulses overlap, we receive a long pulse instead of a sequence of individual short pulses. The signal is distorted by the group delay differences of the modes. For weakly guiding optical fibers, we calculated the delay difference in Sect. 2.3. From Eqs. (2.3-23) and (2.3-24) we obtain, for large V,

$$\Delta\tau_{uc} = (L/c)(n_1 - n_2) \tag{5.5-64}$$

where $\Delta\tau_{uc}$ is the width (in time) of the distorted pulse arriving at the end of the fiber carried by all the uncoupled modes. The pulse width is seen to be proportional to the length L of the fiber. The difference between the refractive index of the core, n_1, and the cladding index n_2 is assumed to be small. The constant c is the velocity of light in vacuum.

If the modes are coupled among each other, the pulse performance of the multimode waveguide is changed. There is a transient state near the input end of the guide where the pulse shape is described as a superposition of Gaussian functions, Eq. (5.5-22). After the transient has died down, only the first term in the sum (5.5-22) contributes, so that an equilibrium pulse is reached. Equation (5.5-24) shows that the shape of this equilibrium pulse is a Gaussian function of time. It might appear that this Gaussian pulse shape is related to the fact that we started with a Gaussian pulse at $z = 0$. However, if we let the width of the input pulse shrink to zero, $\tau = 0$, Eq. (5.5-25) is no longer only an approximation of Eq. (5.5-23), and Eq. (5.5-24) describes the impulse response of the multimode waveguide. Since the impulse response is a Gaussian function, it is clear that we always obtain a Gaussian pulse, provided the equilibrium pulse has spread so much that its width is much larger than the width of the input pulse of arbitrary shape. The Gaussian pulse shape is caused by the ω^2 term in Eq. (5.5-16). The impulse response is Gaussian if higher-order terms can be neglected in the perturbation expansion (5.5-16).

The most important feature of the equilibrium pulse is the dependence of its width on the square root of the waveguide length z. Equation (5.5-64) shows that the width of a signal carried by uncoupled modes is proportional to the length of the waveguide. Mode coupling has the beneficial effect of reducing the spreading of the pulses. The pulses still spread, even if they are carried by coupled modes, but their width is proportional only to the square root of the waveguide length instead of being proportional to the length itself. The actual width of the equilibrium pulse depends on the coupling strength in a complicated way through Eqs. (5.5-25) and (5.5-55). In order to gain insight into the dependence of the pulse width on the coupling strength, we study the two-mode case.

The zero-order eigenvalue problem for the two-mode case has already been solved. We need only evaluate the first- and second-order perturbation of the eigenvalues. In the absence of degeneracy we obtain the first-order perturbation of the eigenvalue $\rho^{(1)}$ from Eqs. (5.5-39) (by dropping the subscripts and superscripts σ), (5.5-42), and (5.3-10):

$$\rho_1^{(1)} = \left(\frac{1}{v_1} - \frac{1}{v_a}\right) A_1^{(1)^2} + \left(\frac{1}{v_2} - \frac{1}{v_a}\right) A_2^{(1)^2} = \frac{A_1^{(1)^2}}{v_1} + \frac{A_2^{(1)^2}}{v_2} - \frac{1}{v_a} \tag{5.5-65}$$

We used Eq. (5.5-11) and

$$V_{\mu\nu} = [(1/v_\mu) - (1/v_a)]\, \delta_{\mu\nu} \tag{5.5-66}$$

The average group velocity is arbitrary. We can always choose it so that $\rho_1^{(1)} = 0$. We thus use Eq. (5.5-65) to define v_a. We are free to choose $\rho_1^{(1)}$ arbitrarily, since it is only the sum $1/v_a + 1/\rho_1^{(1)}$ that enters Eq. (5.5-24).

The second-order perturbation of the first eigenvalue follows from Eqs. (5.5-55) and (5.5-45). Using Eqs. (5.5-10) and (5.5-66), we obtain

$$\rho_2^{(1)} = \frac{1}{\sigma^{(2)} - \sigma^{(1)}} \left\{ \frac{1}{v_1} A_1^{(1)} A_1^{(2)} + \frac{1}{v_2} A_2^{(1)} A_2^{(2)} \right\}^2 = \frac{[(1/v_2) - (1/v_1)]^2}{\sigma^{(2)} - \sigma^{(1)}} A_1^{(1)^2} A_1^{(2)^2} \tag{5.5-67}$$

The average velocity v_a vanished from this expression because of the orthogonality of the eigenvectors. The orthogonality relation (5.3-10) was also used to convert the expression to the form that is shown on the extreme right-hand side of Eq. (5.5-67). Use of Eqs. (5.3-31), (5.3-37), and (5.3-38) allows us to obtain from (5.5-67) and (5.5-25) the following expression for the width of the equilibrium pulse in the two-mode case:

$$T_1 = \left(\frac{1}{v_2} - \frac{1}{v_1}\right) \frac{h(2L)^{1/2}}{[(\alpha_2 - \alpha_1)^2 + h^2]^{3/4}} \tag{5.5-68}$$

We used $z = L$ to indicate the length of the waveguide. Since

$$\Delta\tau_{uc} = [(1/v_2) - (1/v_1)]\, L \tag{5.5-69}$$

is the separation of the two uncoupled pulses, we can define the ratio of the coupled to uncoupled pulse width

$$R = T_1/\Delta\tau_{uc} \tag{5.5-70}$$

and find from Eqs. (5.5-68) and (5.5-69) the expression

$$R = \sqrt{2}\, h/[(\alpha_2 - \alpha_1)^2 + h^2]^{3/4} \sqrt{L} \tag{5.5-71}$$

for the relative improvement of the pulse distortion caused by mode coupling. Equation (5.5-71) shows that the ratio of coupled to uncoupled pulse width decreases with increasing waveguide length.

The width of the pulse has an interesting dependence on the coupling strength h. If the coupling is weak compared to the loss difference, $h \ll |\alpha_2 - \alpha_1|$, we obtain from Eq. (5.5-68)

$$T_1 = \sqrt{2}\, h \Delta\tau_{uc} / (\alpha_2 - \alpha_1)^{3/2} \sqrt{L} \tag{5.5-72}$$

In the loss-dominated case the pulse width increases with increasing coupling strength. This surprising result is explained by a look at Eqs. (5.3-43) and (5.3-44) for the eigenvector that represent the steady state power distribution *versus* mode number in case of high differential loss. We see that all the power resides in the first mode. Coupling of the two modes tends to put more power into the second mode. It is thus clear that the pulse width would increase in this situation. In the absence of coupling the equilibrium pulse width is zero (for an impulse), since all the power travels in the first mode. Coupling between the modes forces power into the slower mode, increasing the pulse width.

For strong coupling, $h \gg |\alpha_2 - \alpha_1|$, we have

$$T_1 = (2/hL)^{1/2} \Delta\tau_{uc} \tag{5.5-73}$$

In case of strong coupling the width of the equilibrium pulse decreases with increasing coupling strength. The steady state power distribution, Eqs. (5.3-41) and (5.3-42), shows equal power in both modes in this case. Increased coupling strength facilitates the transfer of power from the slow mode to the fast mode and *vice versa*, so that averaging occurs and break up of the power into two pulses is avoided. At the same time we get a narrower pulse with increasing coupling strength.

In many multimode applications the mode power losses $2\alpha_\mu$ increase with mode number. We shall see shortly that sometimes only the highest-order mode suffers high loss, while the remaining modes have relatively low losses. Mode coupling transfers power from the low-loss modes to the high-loss mode, so that the steady state loss increases with increasing coupling strength. The desirable pulse-width reduction caused by coupling is—at least partially—offset by increasing losses. We say that we pay a loss penalty for the pulse distortion reduction. This phenomenon can be studied even in the two-mode case.

We assume that the first mode has negligible loss, $\alpha_1 = 0$, and distinguish between strong and weak coupling depending on whether we have

$$h \gg \alpha_2 \tag{5.5-74}$$

or

$$h \ll \alpha_2 \tag{5.5-75}$$

In case of strong coupling, Eq. (5.5-74), we obtain from Eq. (5.3-31)

$$\sigma^{(1)}L = \alpha_2[1-(\alpha_2/2h)]L \qquad (5.5\text{-}76)$$

To the same approximation ($\alpha_2{}^2/h^2$ is neglected), we have from Eq. (5.5-71)

$$R^2 = 2/hL \qquad (5.5\text{-}77)$$

Equation (5.5-76) is the power loss coefficient for a waveguide of length L, and Eq. (5.5-77) is the relative pulse-width improvement for a guide of the same length. We can eliminate L from the two equations and express the pulse-width improvement R in terms of the loss that occurs due to mode coupling:

$$R^2(\sigma^{(1)}L) = (2\alpha_2/h)[1-(\alpha_2/2h)] \qquad (5.5\text{-}78)$$

Formula (5.5-78) is misleading in this form. It gives the impression that the loss penalty can be decreased for a given value of the improvement factor R by increasing the coupling. However, the power loss $2\alpha_2$ is itself a result of coupling between the guided mode 2 and the continuum of radiation modes. If the losses had a different origin, both guided modes would have very nearly the same loss, contrary to our assumption. We must thus assume that

$$\alpha_2 = a_2 h \qquad (5.5\text{-}79)$$

Equation (5.5-78) now assumes the form

$$R^2(\sigma^{(1)}L) = 2a_2[1-(a_2/2)] \qquad (5.5\text{-}80)$$

showing that the loss penalty is independent of coupling strength. The loss is determined only by the improvement R that we want to achieve and is independent of the amount of coupling and the length of the waveguide. The term "loss penalty" is very appropriate for this situation, since the amount of relative improvement in the pulse width that can be achieved depends only on the amount of loss that we are willing to tolerate.

For weak coupling, Eq. (5.5-75), we have from Eq. (5.3-31)

$$\sigma^{(1)}L = h[1-(h/2\alpha_2)]L \qquad (5.5\text{-}81)$$

and from Eq. (5.5-71)

$$R^2 = 2h^2/\alpha_2{}^3 L \qquad (5.5\text{-}82)$$

Multiplication of Eqs. (5.5-81) and (5.5-82) results in

$$R^2\sigma^{(1)}L = (2h^3/\alpha_2{}^3)[1-(h/2\alpha_2)] = (2/a_2{}^3)[1-(1/2a_2)] \qquad (5.5\text{-}83)$$

Equation (5.5-79) was used on the right-hand side of this expression. We see again that the loss penalty is independent of the coupling strength. However, our definition of the improvement factor R is not very useful in this case. We see from Eq. (5.5-72) that the pulse width of the impulse response goes to zero in the absence of coupling, whereas the constant width, Eq. (5.5-69), was used for the definition of R. The improvement factor compares the width of the equilbrium pulse to the maximum width of the uncoupled signal, assuming that all modes are actually carrying power. Instead of an improvement a broadening of the pulse results if the coupling strength is increased, provided that relation (5.5-75) applies.

Reduced Delay Distortion by Intentional Mode Coupling

Multimode pulse dispersion in optical waveguides can be improved, to some extent, by mode coupling. As we discussed in Sect. 5.1, multimode operation is desirable if only an incoherent light source is available. We do not wish to allow only one mode to survive in the steady state, since most of the power initially launched into other modes would then be lost in the transient. Once we have decided on multimode operation, with most or all of the modes carrying significant amounts of power, we are interested in counteracting multimode pulse dispersion. In spite of the fact that some amount of mode coupling may exist unintentionally, we proceed to discuss methods of improving multimode pulse dispersion by intentional mode coupling.

The power coupling coefficient (5.2-20) consists of a constant term $|K_{\mu\nu}|^2$ that depends on the geometrical nature of the core–cladding interface deformation. In addition, $h_{\mu\nu}$ contains the power spectrum of the distortion function $f(z)$. If we plan to design intentional mode coupling mechanisms, we can utilize both factors of Eq. (5.5-20). We have seen in Sect. 3.6 that core–cladding interface deformations of the type (3.6-1) couple modes of symmetry ν only to modes with $\mu = |\nu \pm m|$. Using the far from cutoff approximation (2.2-71),

$$\kappa a = (\nu + 2N)\pi/2 \tag{5.5-84}$$

[we used N to avoid confusion with m of Eq. (3.6-1), $\frac{1}{2}$ was neglected] for the parameter (2.2-10), we can approximate the propagation constant of the fiber in the following way:

$$\beta = (n_1{}^2k^2 - \kappa^2)^{1/2} = n_1 k - (\kappa^2/2n_1 k) = n_1 k - [(\nu+2N)^2\pi^2/8n_1 ka^2] \tag{5.5-85}$$

For each mode with a given set of values ν and N, there is another mode with a set of values $\mu = \nu + m$ and $N' = N - m/2$ (N' is understood to be the nearest integer) with nearly the same propagation constant. For a given type of core interface imperfection, which means for a given value of m, we obtain a set of coupled modes whose separation in β space is given by

$$\Delta\beta = (\pi^2/8n_1 ka^2)[(\nu+2N+1)^2 - (\nu+2N)^2] \approx (\pi^2/4n_1 ka^2)(\nu+2N) \tag{5.5-86}$$

The spacing in β space between guided modes is important since—as we have seen in Sects. 3.4 and 4.2, Eq. (4.2-4)—it is the Fourier component at the spatial frequency

$$\theta = \Delta\beta \tag{5.5-87}$$

of the distortion function $f(z)$ that is responsible for coupling between two guided modes. Equation (5.5-86) shows that the mode spacing increases with increasing mode number. If we design the core boundary distortion intentionally with a distortion function $f(z)$ whose power spectrum—as shown in Fig. 5.5-1—has a cutoff frequency θ_c, only modes with

$$\nu + 2N \leqslant (4n_1 ka^2/\pi^2)\theta_c \tag{5.5-88}$$

are coupled to each other. All modes satisfying Eq. (5.5-88) find themselves coupled to another guided mode by a suitable Fourier component of the

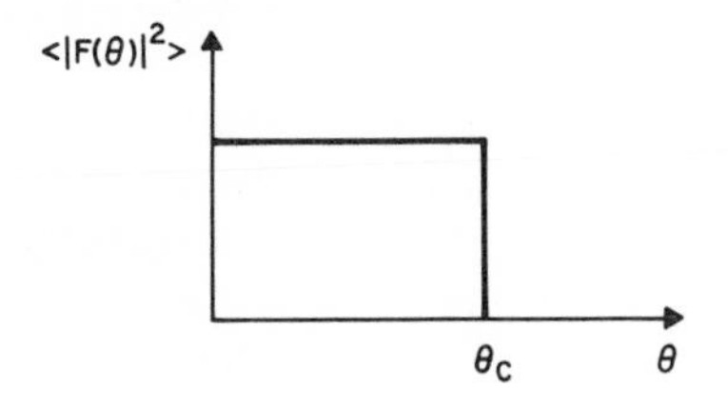

Fig. 5.5.1 *Ideal shape of the "power spectrum" of the core–cladding interface irregularities.* (From D. *Marcuse, "Pulse propagation in multimode dielectric waveguides,"* Bell Syst. Tech. J. ***51***, *1199–1232 (1972). Copyright 1972, The American Telephone and Telegraph Co., reprinted by permission.*)

power spectrum. Modes with larger mode numbers remain uncoupled, however, since there is no Fourier component available for mode coupling. The possibility of coupling only lower-order modes among each other, but leaving the high-order modes uncoupled, is provided by the fact that the mode spacing increases linearly with increasing mode number according to Eq. (5.5-86). If it were possible to design a coupling function $f(z)$ with a predetermined Fourier spectrum, we could achieve coupling among most of the guided modes without causing coupling of these modes to the continuous spectrum of radiation modes. Coupling among guided modes without a radiation loss penalty thus appears feasible.

It may seem ill advised to leave a few high-order modes uncoupled. If these modes are allowed to travel independently, their lower group velocity would cause the type of multimode pulse dispersion that we are trying to counteract. However, some amount of mode coupling is unavoidable because of random core–cladding interface roughness and random index inhomogeneities. The radiation loss formula (4.2-25) shows that the radiation losses caused by a single Fourier component of $f(z)$ increase as κ_ν^2, that is, proportional to the square of the mode number. The same law can also be deduced from the corresponding formula for the optical fiber, Eqs. (4.4-6) and (4.4-7), with the help of approximation (3.6-15). It is thus apparent that higher-order modes suffer more radiation losses than lower-order modes. Random core boundary perturbations cause more radiation losses to high-order modes than to low-order modes. Leaving high-order modes uncoupled (by intentional means) from the remaining guided modes is thus not necessarily detrimental, since these modes disappear on account of their higher radiation losses.

However, there is another consideration that must be taken into account. We have based our discussion of mode coupling and radiation losses entirely on the results of first-order perturbation theory. It can be shown [Me6, Me7] that there are higher-order effects that couple guided modes according to the law

$$|\Delta\beta| = M\theta \tag{5.5-89}$$

with integer values of M.

Mode coupling can be explained in terms of diffraction gratings. A sinusoidal core boundary ripple acts on the plane wave components of a guided wave in a slab waveguide like a phase grating. It is well known that a phase grating causes diffraction by producing radiation lobes of different order [Me1]. The first-order perturbation theory used in this book corresponds to considering only the grating lobes of first order. Coupling between two guided modes, for example, occurs if the first-order grating lobe produced by the light field of one mode scatters into the direction that is typical for the plane wave components of another guided mode. Higher-order grating lobes serve the same purpose. The integer M appearing in Eq. (5.5-89) indicates the grating order responsible for mode coupling. M is also the order of perturbation theory that is required for the calculation of the coupling coefficient in question. In Eq. (5.5-87), we have $M = 1$ corresponding to the first-order grating lobe or to first-order perturbation theory. The coupling strength depends on the grating order or the order of perturbation theory required to calculate the coupling coefficient. For core boundary ripples with small amplitudes, the first-order grating lobes are by far the most important. Higher-order effects can safely be neglected for the slight core boundary imperfections occurring unintentionally. Intentional core boundary ripples with larger amplitudes may require the consideration of higher-order scattering effects. It was shown in [Me7] that second-order grating lobes can cause significant radiation losses. However, even if the core boundary ripples are strong enough to cause significant higher-order effects, we can still use specially designed power spectra of the type shown in Fig. 5.5.1 to achieve guided mode coupling

TABLE 5.5.1

Parameters for the TE Modes of the Slab Waveguide[a,b]

$kd/2$	ν	$\kappa_\nu d/2$	$\beta_\nu d/2$	$(\beta_\nu - \beta_{\nu+1})d/2$
35	1	1.3821	52.48180	0.05429
(5-mode case)	2	2.7580	52.42751	0.08936
	3	4.1193	52.33815	0.12187
	4	5.4507	52.21628	0.14679
	5	6.7096	52.06949	—
	1	1.4708	104.98970	0.03089
	2	2.9047	104.95881	0.05140
	3	4.4086	104.90741	0.07281
	4	5.8733	104.83560	0.09199
70	5	7.3332	104.74361	0.11186
(10-mode case)	6	8.7861	104.63175	0.13115
	7	10.2286	104.50060	0.14939
	8	11.6544	104.35121	0.16531
	9	13.0498	104.18590	0.17295
	10	14.3634	104.01295	—
	1	1.5210	217.49468	0.01595
	2	3.0418	217.47873	0.02658
	3	4.5624	217.45214	0.03721
	4	6.0826	217.41493	0.04783
	5	7.6023	217.36710	0.05844
	6	9.1214	217.30865	0.06904
	7	10.6396	271.23961	0.07962
	8	12.1568	217.15999	0.09018
145	9	13.6728	217.06981	0.10070
(20-mode case)	10	15.1873	216.96911	0.11119
	11	16.7000	216.85793	0.12161
	12	18.2105	216.73631	0.13197
	13	19.7182	216.60434	0.14222
	14	21.2226	216.46213	0.15231
	15	22.7225	216.30981	0.16218
	16	24.2167	216.14764	0.17168
	17	25.7028	215.97596	0.18051
	18	27.1767	215.79545	0.18789
	19	28.6291	215.60757	0.19023
	20	30.0270	215.41733	—

[a] For $n_1 = 1.5$, $n_1/n_2 = 1.01$.

[b] *From* D. Marcuse, "Pulse propagation in multimode dielectric waveguides," *Bell. Syst. Tech. J.* **51**, 1199–1232 (1972). Copyright 1972, The American Telephone and Telegraph Co., reprinted with permission.

without having to pay an excessive loss penalty, since we need only to choose the spatial cutoff frequency in accordance with Eq. (5.5-89) for $M = 2$ to avoid radiation losses caused by coupling of guided modes to radiation modes due to second-order effects.

We close this section with a discussion of the loss penalty that has to be paid for intentional mode coupling, using the model of the symmetric slab waveguide as an example. The results obtained from the slab waveguide model have significance for round optical fibers. We have seen that the coupling coefficients are of equal order of magnitude for both types of waveguides. Comparison of Eqs. (1.3-89) and (5.5-84) (using $d = 2a$) shows that the compound-mode number $v + 2N$ of the round optical fiber replaces the single-mode number v of the slab waveguide. If this replacement is made, the analogy between the two types of waveguides is very close.

The discussion of the loss penalty incurred by coupled guided mode operation of the symmetric slab waveguide is based on first-order perturbation theory. If the power spectrum of $f(z)$ could be shaped as shown in Fig. 5.5.1, no loss penalty at all would result. However, it is impossible to devise an actual spectral distribution with an infinitely sharp cutoff. The loss penalty is thus caused by the tail of the power spectrum. We use symmetric slab waveguides that can support 5, 10, 20, and 40 guided modes. The eigenvalues of Eqs. (1.3-39) and (1.3-40) for even and odd TE modes are listed in Table 5.5.1 for the 5-, 10-, and 20-mode case [Me8].

Two types of power spectra are being considered. First we assume that the ideal spectrum of Fig. 5.5.1 is modified as shown in Fig. 5.5.2. The mathematical form of this power spectrum is given by the following equation:

$$\langle |F(\theta)|^2 \rangle = \begin{cases} \pi\bar{\sigma}^2/(\beta_{N-2} - \beta_{N-1}), & \text{for} \quad 0 < \theta \leqslant \beta_{N-2} - \beta_{N-1} \\ K\pi\bar{\sigma}^2/(\beta_{N-2} - \beta_{N-1}), & \text{for} \quad \theta = \beta_{N-1} - \beta_N \end{cases} \tag{5.5-90}$$

where $\bar{\sigma}$ is the rms value of the core boundary distortion. The symmetric slab waveguide supports N guided modes. The modes $v = 1$ through $v = N - 1$ are coupled strongly by means of the flat portion of the power spectrum shown in Fig. 5.5.2. Because of the increasing separation between the β_v values of the guided modes, all modes are coupled if the spectrum extends from zero

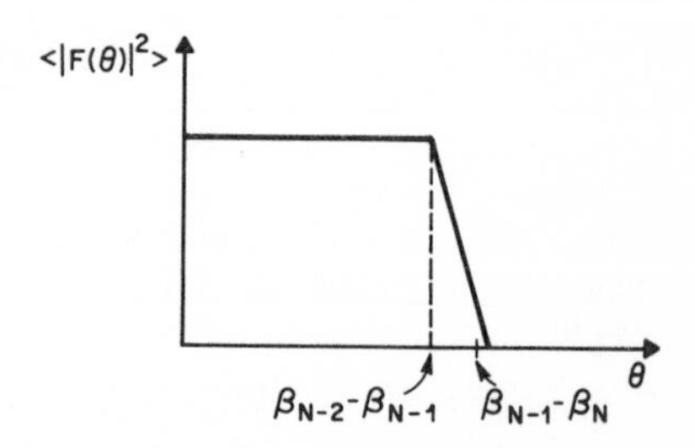

Fig. 5.5.2 *A more realistic model of the desirable core–cladding interface "power spectrum."* (From *D. Marcuse, "Pulse propagation in multimode dielectric slab waveguides,"* Bell Syst. Tech. J. ***51***, *1199–1232 (1970). Copyright 1972, The American Telephone and Telegraph Co., reprinted with permission.*)

spatial frequencies to the maximum frequency

$$\theta_m = \beta_{N-2} - \beta_{N-1} \tag{5.5-91}$$

The last mode is coupled to mode $N-1$ by the spatial frequency $\theta = \beta_{N-1} - \beta_N$, whose amplitude corresponds to the tail of the power distribution of Fig. 5.5.2. The amplitude that couples the highest-order mode, $v = N$, to its neighboring mode, $v = N-1$, is reduced by a factor K compared to the amplitude on the flat portion of the spectral distribution.

The power coupling coefficient for the guided slab waveguide modes is obtained from Eqs. (3.5-27) and (5.2-20) with the assumption that both faces of the core boundary are distorted with no mutual correlation but with identical power spectra:

$$h_{\mu\nu} = \{2\kappa_\mu{}^2\kappa_\nu{}^2/\beta_\mu\beta_\nu[d+(2/\gamma_\mu)][d+(2/\gamma_\nu)]\}\langle|F(\beta_\mu-\beta_\nu)|^2\rangle \tag{5.5-92}$$

Owing to the symmetry of the slab, we have $\gamma_\mu = \delta_\mu$. For comparison we also list the power coupling coefficients for the round fiber in the far from cutoff approximation using Eq. (3.6-16):

$$h_{\mu\nu} = (e_{\mu\nu m}^2/e_\nu e_\mu)(\kappa_\mu{}^2\kappa_\nu{}^2/4n^2k^2a^2)\langle|F(\beta_\mu-\beta_\nu)|^2\rangle \tag{5.5-93}$$

Far from cutoff we use $\beta = nk$ and neglect γ_μ and γ_ν in Eq. (5.5-92). Since we must use the correspondence $d = 2a$, we see that the power coupling coefficients for slabs and round fibers are very nearly the same except for a numerical factor. A slab waveguide with $N_s = (4/\pi)(V/2)$ TE and TM modes [see Eq. (1.3-70)] corresponds to a fiber with $N_f = 4V^2/\pi^2$. Because of their definitions, $V/2$ of the slab corresponds to V of the fiber. For comparable V numbers, the fiber supports roughly $N_f = N_s{}^2/4$ modes.

The model calculation for the slab waveguide includes radiation losses by assigning an arbitrary but large loss value to the highest-order mode N. The entire ensemble of waveguide modes loses power by being coupled to the lossy mode N. The actual loss coefficient of mode N is not important provided that this mode loses power at a higher rate than can be supplied by the neighbors. All results listed in the remainder of this section were obtained with the help of numerical computer evaluation of the theory.

Figure 5.5.3 [Me8] shows the product of the square of the improvement factor

$$R = T_1/\Delta\tau_{uc} = 4(\rho_2^{(1)})^{1/2}/[(1/v_N)-(1/v_1)](L_R)^{1/2} \tag{5.5-94}$$

times the loss penalty $\sigma^{(1)}L_R$ caused by mode coupling in the waveguide of length L_R. The subscript R serves as a reminder that the length L_R is associated with a pulse delay improvement R. The independent variable in Fig. 5.5.3 is

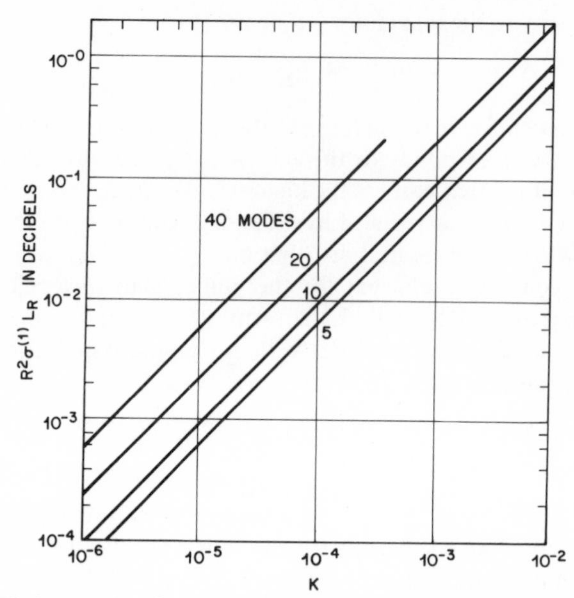

Fig. 5.5.3 *The loss penalty $R^2\sigma^{(1)}L_R$ (in dB) that must be paid for a pulse distortion reduction R ($R < 1$). The variable K is the fraction to which the "power spectrum" has decayed from its peak value [see Eq. (5.5-90)].* (From *D. Marcuse, "Pulse propagation in multimode dielectric waveguides,"* Bell Syst. Tech. J. ***51****, 1199–1232 (1972). Copyright 1972, The American Telephone and Telegraph Co., reprinted with permission.)*

the factor K appearing in Eq. (5.5-90). For a given value of the improvement factor R (R should be as small as possible, only $R < 1$ corresponds to an improved pulse delay distortion) the loss penalty increases with increasing mode number and increasing values of K. It may seem surprising at first glance that the rms deviation $\bar{\sigma}$ appearing in Eq. (5.5-90) does not occur in the variables involved in Fig. 5.5.3. The situation is similar to the two-mode formulas (5.5-80) and (5.5-83) that do not depend on the magnitude of the coupling coefficient h. If we divide Eq. (5.3-2) by $\bar{\sigma}^2$ we convert the eigenvalue $\sigma^{(j)}$ to the normalized eigenvalue $\sigma^{(j)}/\bar{\sigma}^2$. From Eqs. (5.5-55) and (5.5-94) we see that the product $R^2\sigma^{(1)}$ contains the ratio $\sigma^{(1)}/(\sigma^{(1)} - \sigma^{(j)})$. The normalizing factor $\bar{\sigma}^2$ thus cancels out and does not appear in $R^2\sigma^{(1)}L_R$. The loss penalty curves of Fig. 5.5.3 do not depend on the magnitude but only on the shape of the power spectrum curve.

The curves of Fig. 5.5.3 were computed for $n_1 = 1.5$ and $n_1/n_2 = 1.01$. They allow us to read off directly the loss penalty for any given improvement R. For example, if we want to obtain $R = 0.1$, which means that the equilibrium pulse for the coupled-mode case should have one tenth of the width of the pulse for uncoupled modes, we would have to pay a loss penalty of 1 dB in the 10-

mode case, 2.5 dB in the 20-mode case and 6 dB in the 40-mode case if we operate with a power spectrum with $K = 10^{-4}$. These numbers do not tell us how long a guide we need to get $R = 0.1$ or the loss penalty figures just stated. From Table 5.5.1 we find that in the 20-mode case the power spectrum must drop from its constant value at $\theta d/2 = (\beta_{18} - \beta_{19})d/2 = 0.188$ to $K = 10^{-4}$ of this peak value at $(\beta_{19} - \beta_{20})d/2 = 0.190$. This sharp dropoff must happen in a spectral interval that corresponds to 1% of the width of the spectrum. If the slope of the power spectrum is less abrupt, the loss penalty increases sharply, as can be seen from Fig. 5.5.3. An increase in K of a factor of 10 increases the loss penalty by the same amount.

Because of its normalization, Fig. 5.5.3 contains only a limited amount of information. If we knew the actual values of R we could calculate the actual amount of loss for any given length of waveguide. Table 5.5.2 supplies the

TABLE 5.5.2

Numerical Values for the Pulse Distortion in the Absence of Coupling[a] and the Normalized Second-Order Perturbation of the Eigenvalue[b] in the Presence of Coupling[c] for the "Power Spectrum" of Eq. (5.5-90)[d]

Number of modes	$kd/2$	$\frac{c}{n_1}\left(\frac{1}{v_N} - \frac{1}{v_1}\right) = \frac{c\,\Delta\tau_{uc}}{n_1 L}$	$\frac{2v_a{}^2\bar{\sigma}^2k^2\rho_2^{(1)}}{d}$
5	35	3.94×10^{-3}	0.02
10	70	5.02×10^{-3}	0.07
20	145	6.69×10^{-3}	0.4
40	290	8.13×10^{-3}	2.0

[a] Third column.
[b] Fourth column.
[c] The normalized second-order perturbation is nearly independent of K.
[d] *From* D. Marcuse, "Pulse propagation in multimode dielectric waveguides," *Bell. Syst. Tech. J.* **51**, 1199–1232 (1972). Copyright 1972, The American Telephone and Telegraph Co., reprinted with permission.

necessary information [Me8]. The first column gives the mode number; the second column gives the kd values that were used to calculate the eigenvalues of the modes; the third column gives the normalized pulse width for uncoupled multimode operation; and the last column supplies the second-order perturbation of the first eigenvalue of the coupled power eigenvalue problem. It is fortunate that $\rho_2^{(1)}$ is independent of K for the small K values appearing in Fig. 5.5.3. The numerical values for the uncoupled pulse width show that the approximate Eq. (5.5-64) is not very accurate for small mode numbers. With

$n_1 = 1.5$ and $n_1/n_2 = 1.01$, we obtain

$$(c/n_1 L)\Delta\tau_{uc} = [1-(n_2/n_1)] = 9.9 \times 10^{-3} \qquad (5.5\text{-}95)$$

The corresponding value 8.13×10^{-3} for the 40-mode case in Table 5.5.2 deviates from this approximation by 18%. However, it is apparent that the value of Eq. (5.5-95) is approached more closely for large mode numbers.

Some numerical values for a representative example are listed in Table 5.5.3. It is assumed that the improvement factor is $R = 0.1$, so that a tenfold improvement of the pulse width over the width of the uncoupled pulses results. We assumed a vacuum wavelength of $\lambda = 1$ μm, a waveguide length of $L = 1$ km and a factor $K = 10^{-4}$. The table lists the actual half width $d/2$ of the slab, the pulse width of the uncoupled pulses in seconds, the rms core–cladding interface deviation $\bar{\sigma}$ in micrometers, and the loss penalty $\bar{\sigma}^{(1)}L_R$ in decibels. We see that more core boundary roughness is required to achieve $R = 0.1$ for increasing mode number. However, all $\bar{\sigma}$ values are surprisingly small and show that the first-order perturbation theory, used for the derivation of the coupling coefficients, is indeed applicable.

TABLE 5.5.3

Numerical Values for the Example Listed in Table 5.5.2 Corresponding to the "Power Spectrum" of Eq. (5.5-90) or Fig. 5.5.2[a,b]

Number of modes	$d/2$ (μm)	$\Delta\tau_{uc}$ (sec)	$\bar{\sigma}$ (μm)	$\sigma^{(1)}L_R$ (dB)
5	5.57	1.97×10^{-8}	1.71×10^{-2}	0.63
10	11.1	2.51×10^{-8}	3.56×10^{-2}	1.0
20	23.1	3.35×10^{-8}	9.16×10^{-2}	2.3
40	46.2	4.07×10^{-8}	2.39×10^{-1}	6.0

[a] $R = 0.1$, $\lambda = 1$ μm, $K = 10^{-4}$, $L_R = 1$ km.

[b] *From* D. Marcuse, "Pulse propagation in multimode dielectric waveguides," *Bell. Syst. Tech. J.* **51**, 1199–1232 (1972). Copyright 1972, The American Telephone and Telegraph Co., reprinted with permission.

Finally we show the effect of the power spectrum

$$\langle|F(\theta)|^2\rangle = m\bar{\sigma}^2 \sin(\pi/m)/\Delta\beta(1 + |\theta/\Delta\beta|^m) \qquad (5.5\text{-}96)$$

on the loss penalty of the slab waveguide model. The function of Eq. (5.5-96) is plotted in Fig. 5.5.4 for different values of the exponent m. The power spectra of Fig. 5.5.4 are more realistic than the idealized power spectrum of Fig.

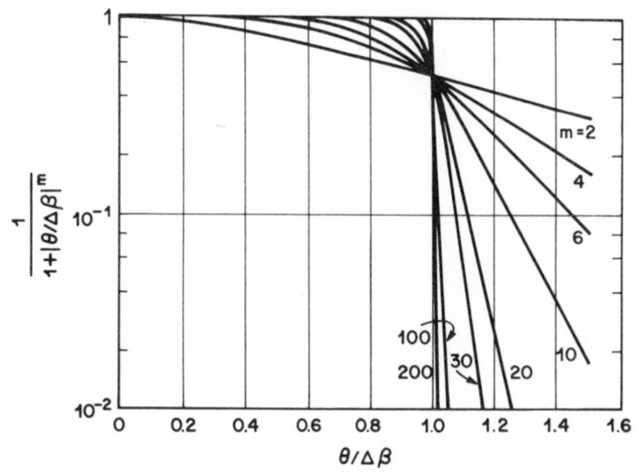

Fig. 5.5.4 *A simple power law model for the "power spectrum."* (From *D. Marcuse*, "*Pulse propagation in multimode dielectric waveguides*," Bell Syst. Tech. J. ***51****, 1199–1232 (1972). Copyright 1972, The American Telephone and Telegraph Co., reprinted with permission.)*

5.5.2, at least for small values of the exponent m. It may be possible to produce the mechanical power spectrum by modulating the pulling rate of a fiber by a suitable mechanism. When the pulling speed is increased, the fiber diameter decreases and *vice versa*. The modulation of the pulling speed may be derived from an electrical signal that may consist either of suitably filtered and amplified noise or may be synthesized from a superposition of sine waves. The loss penalty curves, which belong to the power spectra of Eq. (5.5-96) or Fig. 5.5.4, are shown for the 10-mode case in Fig. 5.5.5. The slab waveguide is still the same with $n_1 = 1.5$ and $n_1/n_2 = 1.01$. The most striking feature of the curves of Fig. 5.5.5 is the fact that a maximum exists as a function of the width parameter $\Delta\beta$ of the power spectra. For very narrow width, small values of $\Delta\beta$, the modes are coupled only *via* the tails of the power spectrum, so that the coupling is weak. The loss is correspondingly reduced since it depends on the strength of the coupling of the guided modes to the very lossy last mode, $\nu = N$. However, the delay distortion improvement is also slight for weak coupling. The net result is that the curves become nearly independent of the coupling strength for small values of $\Delta\beta$. With increasing width of the power spectrum, the coupling between nearest neighbors increases and spreads to other modes that are not nearest neighbors of each other. The losses increase more rapidly than the improvement in the delay distortion, causing a rise of the curves. For some values of $\Delta\beta$ a maximum is reached. The maximum of the curves separates two regions of different steady state behavior. To the left of the maxima all modes carry nearly equal power in the steady state. This situation is desirable from the point of view of efficient power transmission,

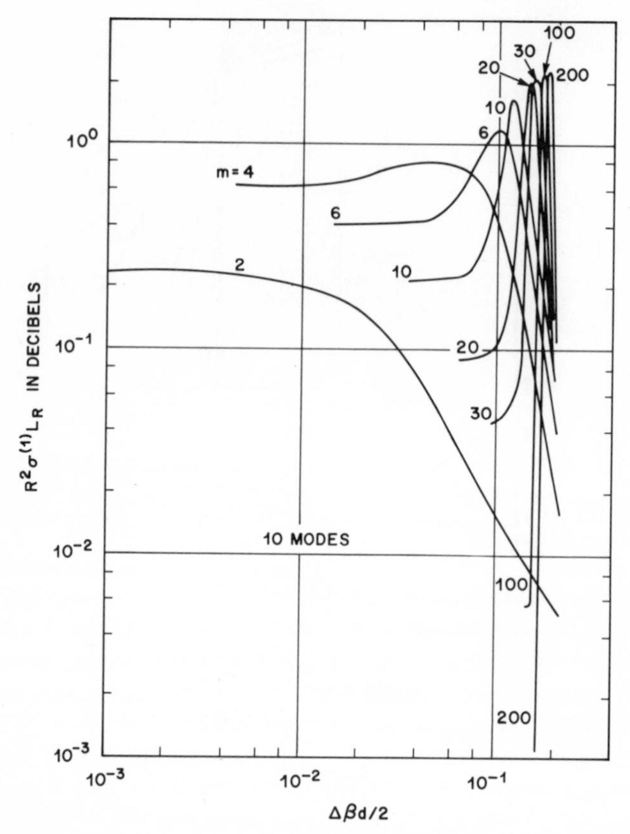

Fig. 5.5.5 *The loss penalty for the power spectrum of Fig. 5.5.4 for $n_1/n_2 = 1.01$.* (From *D. Marcuse, "Pulse propagation in multimode dielectric waveguides,"* Bell Syst. Tech. J. ***51***, *1199–1232 (1972). Copyright 1972, The American Telephone and Telegraph Co., reprinted with permission.)*

since there is no loss of power as the steady state distribution establishes itself. To the right of the maxima, the steady state distribution is altered; most of the power is now carried by the lowest-order mode. It is clear that the loss penalty is low in this case, since the pulse delay improvement is very substantial if only one or very few modes with nearly equal group velocities carry power. The curves in Fig. 5.5.3 all belong to the regime in which all modes share in the power transmission.

The choice on which side of the maxima the multimode waveguide is to be operated has to be a compromise between the desire for efficient power transmission, favoring operation to the left of the maxima, and the need for narrow equilibrium pulses that can be obtained to the right of the maxima. To the left of the maxima, in the region where all modes carry nearly equal amounts of power, a low-loss penalty is achievable only for a power spectrum with very steep slopes corresponding to a large exponent m.

The curves in Fig. 5.5.5 end to the left of the maxima. Before the endpoint is reached, the curves tend to level off. This leveling off is related to the fact that the loss penalty tends to be independent of the coupling strength, as we have observed earlier. The curves could not be extended beyond the points shown, because the matrix diagonalization routine used for the numerical computations failed to work. This failure of the numerical process has an important physical significance. The coupling between neighboring modes becomes weaker as the power spectrum narrows. If the power spectrum becomes too narrow, the coupling strength becomes insignificant and the modes remain effectively uncoupled. No steady state power distribution and no equilibrium pulse exists in this case. Any arbitrary power distribution that is launched at the beginning of the guide at $z = 0$ persists indefinitely. To the left of the endpoints of the curves in the figure, we are in the region where pulse propagation is described by the uncoupled pulse formula (5.5-63).

The second-order perturbation of the eigenvalue can not now be expressed by a single number as in Table 5.5.2, since it is strongly dependent on $\Delta\beta$. For estimates of the performance of pulsed multimode operation with coupled modes, it is best to consider the power spectra of Fig. 5.5.4 as approximations to the idealized power spectrum of Fig. 5.5.2 and calculate an effective factor K. This effective K can then be used to obtain the loss penalty from Fig. 5.5.3, and the entries of Table 5.5.2 can be used provided that K is sufficiently small.

5.6 Diffusion Theory of Coupled Modes

The coupled power theory described in the preceding sections accounts for the distribution of the average power among the guided modes of a multimode optical waveguide. If only one mode is excited initially, the theory predicts that the power will spread to the other guided modes. The distribution of power among the modes is reminiscent of a diffusion process. True diffusion requires that power distributes itself by being transferred only among immediate neighbors. Since, in general, power can be coupled directly from any guided mode to any other guided mode, we cannot hope to derive diffusion equations for the most general power coupling process. However, in some special cases, nearest neighbor coupling either predominates over other power transfer mechanisms or is actually the only possible way by which power can be distributed among the guided modes. The modes of an infinitely extended medium with the refractive index distribution

$$n = n_0(1-ax^2) \tag{5.6-1}$$

have the property that bends of the waveguide axis couple only nearest neighbors to each other. This type of optical waveguide is ideally suited for a description of the power coupling process in terms of diffusion equations.

The coupled power theory (5.2-19) or (5.5-5) can be evaluated only if the number of modes is not too large. The main part of the theory consists in the diagonalization of an $N \times N$ matrix for N guided modes. If N becomes too large, the numerical diagonalization of the matrix becomes impractical. For this reason we use the theory for discrete modes to derive a partial differential equation that is applicable for waveguides that support so many guided modes that we can regard the discrete set of modes as a quasi continuum. In addition to large mode numbers, the theory is restricted by the requirement that the coupling mechanism involves predominantly nearest neighbors. As an example, we use the modes of the round optical fiber. The extension of our coupled power theory to a diffusion theory was accomplished by Gloge [Ge3].

Instead of the integer numbers used to label the guided modes, we introduce a more directly observable quantity. The modes of an optical fiber can be observed by allowing the power to escape from the end of the fiber into free space. Each mode produces a characteristic far-field radiation pattern that can be calculated with the help of the Kirchhoff–Huygens diffraction theory [Me1]. Since our theory is applicable only to waveguides that support a very large number of modes, most of the modes must be far from cutoff, so that almost all of their power is contained in the waveguide core. The field of the mode can thus be described by Eq. (2.2-21):

$$\psi = E_y = AJ_\nu(\kappa\rho)\cos\nu\phi \tag{5.6-2}$$

The radiation far field from the end of a fiber is obtained from the Fraunhofer approximation of the diffraction integral (see, for example, Eq. (2.2-30) in [Me1, p. 38]):

$$\psi' = (i/\lambda z')\int_0^a\int_0^{2\pi}\psi(r,\phi)e^{-ikr}\rho\,d\rho\,d\phi \tag{5.6-3}$$

with

$$r = [(x'-x)^2+(y'-y)^2+z'^2]^{1/2} \tag{5.6-4}$$

The Fraunhofer approximation consists in neglecting the squares of x and y and expanding r as follows:

$$r = r' - (xx'+yy')/z' \tag{5.6-5}$$

with

$$r' = (z'^2+x'^2+y'^2)^{1/2} \tag{5.6-6}$$

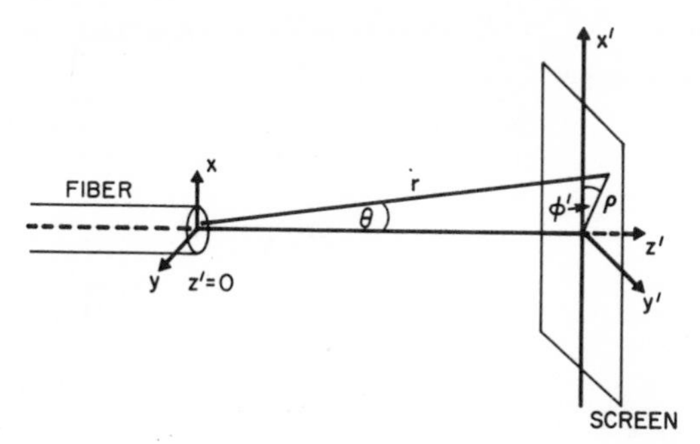

Fig. 5.6.1 *Far field radiation pattern from the end of a fiber displayed on a screen.*

The geometry of the radiation experiment is sketched in Fig. 5.6.1. Since the integral is most easily evaluated in polar coordinates, we introduce

$$\left.\begin{aligned} x &= \rho\cos\phi, \qquad & x' &= \rho'\cos\phi' \\ y &= \rho\sin\phi, \qquad & y' &= \rho'\sin\phi' \end{aligned}\right\} \tag{5.6-7}$$

and express the distance r between integration point and field point on the screen as

$$r = r' - (\rho\rho'/z')\cos(\phi - \phi') \tag{5.6-8}$$

Using Eqs. (5.6-2) and (5.6-8), we obtain the radiation far-field pattern from Eq. (5.6-3):

$$\begin{aligned} \psi' &= (i/\lambda z')A\exp(-ikr')\int_0^a\int_0^{2\pi} J_\nu(\kappa\rho)\exp[(ik\rho\rho'/z')\cos(\phi-\phi')] \\ &\quad \cdot \cos\nu\phi\rho\, d\rho\, d\phi \\ &= (k/z')i^{\nu+1}A\exp(-ikr')\cos\nu\phi'\int_0^a \rho J_\nu(\kappa\rho)J_\nu(k\rho\rho'/z')\, d\rho \\ &= (k/z')i^{\nu+1}A\exp(-ikr')\kappa a\{J_{\nu-1}(\kappa a)J_\nu(k\rho' a/z')/[(k\rho'/z')^2-\kappa^2]\}\cos\nu\phi' \end{aligned} \tag{5.6-9}$$

In the spirit of the far from cutoff approximation, we have used Eq. (2.2-66) to obtain this result. Since $J_\nu(\kappa a) = 0$, function (5.6-9) remains finite at the point where the denominator vanishes. However, it assumes its maximum value at this point. In fact, for large values of ka, the maximum is sharply peaked so that the far-field radiation pattern creates a narrow, bright ring on the screen at an angle that is given by the equation

$$\theta \approx \tan\theta = \rho'/z' = \kappa/k \tag{5.6-10}$$

With the help of Eq. (2.2-71) we can relate the far-field radiation angle θ to the compound mode number

$$M = \nu + 2m \tag{5.6-11}$$

as follows:

$$\theta = \tfrac{1}{2}\pi M/ka \tag{5.6-12}$$

The term $\frac{1}{2}$ was neglected in Eq. (2.2-71), since for most modes M is a very large number. The compound mode number M can be obtained by observing the radiation far-field angle θ. The number ν is also directly observable, since the term $\cos \nu\phi'$ in Eq. (5.6-9) causes the bright ring on the screen to break up into

$$q = 2\nu \tag{5.6-13}$$

bright or dark spots. Observation of the far-field pattern of a single guided mode provides a simple means of mode identification. If the mode field depends on the $\sin \nu\phi$ instead of the $\cos \nu\phi$ function, the position of the bright and dark spots inside the bright ring shifts, as the $\cos \nu\phi'$ function in Eq. (5.6-9) is accordingly replaced with $\sin \nu\phi'$. Complete mode identification requires measuring the polarization of the radiation field.

The maximum angle that can occur is given by Eqs. (2.2-70), (2.2-71), and (5.6-12):

$$\theta_{\max} = V/ka = (n_1{}^2 - n_2{}^2)^{1/2} \approx [2n(n_1 - n_2)]^{1/2} \tag{5.6-14}$$

For weakly guiding fibers with $n_1 - n_2 \ll 1$, Eq. (5.6-14) is always small, and approximation (5.6-10) is justified. The angle θ has another simple physical interpretation. In Sect. 1.2, Eq. (1.2-6), we associated each slab waveguide mode with a mode angle θ',

$$\kappa = n_1 k \sin\theta' \approx n_1 k\theta' \tag{5.6-15}$$

The angle θ' is subtended by the direction of propagation of each plane wave component of the guided mode and the waveguide axis. The same relation can be used for the modes of the optical fiber. When the field leaves the waveguide through the end face, the rays that can be associated with the phase fronts are broken according to Snell's law. With $n_2 = 1$, $\alpha_1 = \theta'$, and $\alpha_2 = \theta$, we obtain from the small-angle approximation of Eq. (1.2-1)

$$\kappa = n_1 k\theta' = k\theta \tag{5.6-16}$$

The comparison with Eq. (5.6-10) shows that the far-field radiation angle θ can be interpreted as the mode angle outside the waveguide. Instead of the diffraction theory we could have used ray optics to find relation (5.6-10). However, our derivation justified this method that, otherwise, would have been speculative.

From Eqs. (5.5-85) and (5.6-16) we obtain the propagation constant of the mode in terms of the far-field angle,

$$\beta = k(n_1{}^2 - \theta^2)^{1/2} \approx n_1 k[1 - (\theta^2/2n_1{}^2)] \qquad (5.6\text{-}17)$$

The power coupling coefficient $h_{\mu\nu}$ between two guided modes of the fiber can now be expressed in terms of the angle θ. From Eqs. (5.5-93), (5.6-16), and (5.6-17), we obtain

$$h_{\mu\nu} = (k^2\theta_\mu{}^2\theta_\nu{}^2/4n_1{}^2a^2)\langle|F[(k/2n_1)(\theta_\nu{}^2 - \theta_\mu{}^2)]|^2\rangle \qquad (5.6\text{-}18)$$

Table 3.6.1 shows that for the vast majority of modes we have $e_{\mu\nu m}^2/(e_\mu e_\nu) = 1$, which is the reason why this factor is absent from Eq. (5.6-18). The subscripts μ and ν are added to the far-field angle θ to distinguish the two modes.

The far-field angles, like the mode numbers, are discrete quantities. However, for fibers supporting a large number of modes, these angles appear to the experimenter as a continuum. It is thus natural to try to convert the discrete set of coupled power equations into a partial differential equation using the far-field radiation angle θ as an additional variable. With the assumption that only nearest neighbors are coupled to each other, we obtain from Eq. (5.2-19), neglecting the loss term,

$$\begin{aligned}\partial P_{\nu m}/\partial z = {} & h_{\nu+1,m,\nu m}(P_{\nu+1,m} - P_{\nu m}) + h_{\nu-1,m+1,\nu,m}(P_{\nu-1,m+1} - P_{\nu m}) \\ & - h_{\nu m,\nu-1,m}(P_{\nu m} - P_{\nu-1,m}) - h_{\nu m,\nu+1,m-1}(P_{\nu m} - P_{\nu+1,m-1})\end{aligned} \qquad (5.6\text{-}19)$$

The numbers ν and m combine to form the compound mode number (5.6-11). We have tacitly assumed that mode coupling is caused by random curvature changes of the waveguide, so that a mode with the label ν can couple only to modes with $\nu+1$ and $\nu-1$.

The power coupling coefficient is only a function of the compound mode number M. The set of labels ν, m corresponds to M, while $\nu+1, m$ as well as $\nu-1, m+1$ correspond to $M+1$; and $\nu-1, m$ as well as $\nu+1, m-1$ correspond to $M-1$. We introduce the notation

$$h_{\nu+1,m,\nu,m} = h_{\nu-1,m+1,\nu m} = h_{M+1,M} \qquad (5.6\text{-}20)$$

$$h_{\nu,m,\nu-1,m} = h_{\nu,m,\nu+1,m-1} = h_{M,M-1} \qquad (5.6\text{-}21)$$

$$P_{\nu,m} = P_M \qquad (5.6\text{-}22)$$

$$P_{\nu+1,m} = P_{\nu-1,m+1} = P_{M+1} \qquad (5.6\text{-}23)$$

$$P_{\nu-1,m} = P_{\nu+1,m-1} = P_{M-1} \qquad (5.6\text{-}24)$$

The symbol P_M indicates the power of one mode with compound mode number M. The total power of all the modes with fixed M is

$$\hat{P}_M = MP_M \tag{5.6-25}$$

because for each value of M there are M different combinations of v and m for two orthogonally polarized modes. If we add all the equations of form (5.6-19) with fixed label M, we obtain [Ge3]

$$M\,\partial P_M/\partial z = 2(M+1)\,h_{M+1,M}(P_{M+1}-P_M) - 2Mh_{M,M-1}(P_M - P_{M-1}) \tag{5.6-26}$$

It was assumed that all modes with the same value of M carry the same amount of power. On the left-hand side we now have the total power contained in all the modes with fixed M. These modes interact with $M+1$ modes with mode number $M+1$, and with M modes with modes number $M-1$.

We use the approximations

$$P_{M+1} - P_M = (\partial P_{M+1}/\partial M)\,\Delta M \tag{5.6-27}$$

and

$$P_M - P_{M-1} = (\partial P_M/\partial M)\,\Delta M \tag{5.6-28}$$

By replacing the right-hand side of Eq. (5.6-26) with the help of Eqs. (5.6-27) and (5.6-28) and using

$$\begin{aligned}(M+1)\,h_{M+1,M}\,\partial P_{M+1}/\partial M - Mh_{M,M-1}\,\partial P_M/\partial M \\ = \Delta M\,\partial(Mh_{M,M-1}\,\partial P_M/\partial M)/\partial M\end{aligned} \tag{5.6-29}$$

we obtain

$$\partial P_M/\partial z = [2(\Delta M)^2/M]\partial(Mh_{M,M-1}\,\partial P_M/\partial M)/\partial M \tag{5.6-30}$$

We have $\Delta M = 1$ and obtain from Eq. (5.6-12)

$$\partial/\partial M = (\pi/2ka)\,\partial/\partial\theta \tag{5.6-31}$$

This relation and Eq. (5.6-12) allow us to write Eq. (5.6-30) as a partial differential equation in terms of the cartesian coordinate z and the far-field mode angle θ:

$$\frac{\partial P(z,\theta)}{\partial z} \doteq -2\alpha P(z,\theta) + \frac{\pi^2}{2k^2a^2}\frac{1}{\theta}\frac{\partial}{\partial\theta}\left[\theta h(\theta)\frac{\partial P(z,\theta)}{\partial\theta}\right] \tag{5.6-32}$$

The loss term appearing in Eq. (5.2-19) has been reinstated.

Equation (5.6-32) has the form of a diffusion equation. For particle diffusion, the diffusion equation reads

$$\nabla\cdot(\kappa\nabla\rho) = \partial\rho/\partial t \tag{5.6-33}$$

where the variable t is the time coordinate, κ the diffusion coefficient, and ρ the particle density. If the particle density is only a function of the radial distance r in a cylindrical geometry, Eq. (5.6-33) assumes the form

$$(1/r)\,\partial(r\kappa\,\partial\rho/\partial r)/\partial r = \partial\rho/\partial t \tag{5.6-34}$$

Comparison with Eq. (5.6-32) shows clearly that the power density $P(z,\theta)$ is taking the place of the particle density, and time is replaced by the length coordinate z. The diffusion current is proportional to

$$I = h\,\partial P/\partial\theta \tag{5.6-35}$$

In analogy to Eq. (5.5-5), we obtain for the time-dependent coupled-mode problem:

$$\frac{\partial P}{\partial z} + \frac{1}{v(\theta)}\frac{\partial P}{\partial t} = -2\alpha P + \frac{\pi^2}{2k^2a^2}\frac{1}{\theta}\frac{\partial}{\partial\theta}\left(\theta h\frac{\partial P}{\partial\theta}\right) \tag{5.6-36}$$

with

$$P = P(z,t,\theta) \tag{5.6-37}$$

These partial differential equations hold only for very large mode numbers. The range of the θ variable is limited by Eq. (5.6-14) to

$$0 \leqslant \theta \leqslant [2n(n_1-n_2)]^{1/2} = \theta_{\max} \tag{5.6-38}$$

Since we had to assume nearest neighbor coupling for the derivation of the partial differential equations, we may assume that only the highest-order mode is coupled to the radiation field. We can then treat radiation losses in the same way as in Sect. 5.5. We assume that $\alpha(\theta) = \text{const.}$, for θ inside interval (5.6-38), and

$$\alpha(\theta) = \infty, \qquad \text{at} \quad \theta = \theta_{\max} \tag{5.6-39}$$

Since a mode with infinitely high loss cannot carry power, we require

$$P(z,t,\theta) = 0, \qquad \text{at} \quad \theta = \theta_{\max} \tag{5.6-40}$$

The loss is thus replaced by a boundary condition. The partial differential equations are of second order in the variable θ. For a complete determination of the permissible functions of θ, we need two initial or boundary conditions. The second condition is obtained from the realization that power is not being generated at the point $\theta = 0$. Boundary condition (5.6-40) implies power outflow from guided to radiation modes at the upper boundary of interval (5.6-38). The rate of power diffusion is determined by the current (5.6-35). We must thus require

$$h\,\partial P/\partial\theta = 0, \qquad \text{at} \quad \theta = 0 \tag{5.6-41}$$

Multimode Waveguide with Random Bends

As an application of the diffusion theory of coupled mode power, we consider a fiber with randomly bent axis. The coupling coefficient for random bends was derived in Sect. 4.6, Eq. (4.6-14). From Eqs. (4.6-14) and (5.2-20), we obtain the power coupling coefficient

$$h_{\mu\nu} = [\kappa_\mu{}^2\kappa_\nu{}^2/4n^2k^2a^2(\beta_\mu-\beta_\nu)^4]\langle|H|^2\rangle \tag{5.6-42}$$

where H is the Fourier component of the curvature d^2f/dz^2. We express κ_μ and β_μ in terms of the (quasi-) continuous variable θ with the help of Eqs. (5.6-10), (5.6-12), (5.6-16), and (5.6-17) and obtain with

$$\beta_\mu - \beta_\nu = (k/n)\theta\Delta\theta = \pi\theta/2na \tag{5.6-43}$$

[where $\Delta\theta = \pi/(2ka)$] the following expression for the power coupling coefficient:

$$h(\theta) = 4n^2a^2k^2C(\theta)/\pi^4 \tag{5.6-44}$$

We use the abbreviation

$$\langle|H|^2\rangle = C(\theta) \tag{5.6-45}$$

for the power spectrum of the curvature function. Since ν and μ belong to neighboring modes, $\theta_\nu \approx \theta_\mu \approx \theta$ has been used to obtain Eq. (5.6-44).

If we knew the power spectrum of the curvature function we could compute the losses of the multimode waveguide and the width of the impulse response function. Since the power spectrum is determined by the process that is used to make the optical fiber, no *a priori* knowledge of its shape is available. For want of more information, we assume that the power spectrum Eq. (5.6-45) is constant over the range of spatial frequencies $\Delta\beta = k\theta_{\max}\Delta\theta/n$ [see Eq. (5.6-43)]. In analogy with Eq. (5.5-90), we obtain, with Eq. (5.6-14) and $\Delta\theta = \pi/(2ka)$,

$$C(\theta) = \pi\bar{\kappa}^2/\Delta\beta = \sqrt{2}\,a\bar{\kappa}^2/[(n_1/n_2)-1]^{1/2} \tag{5.6-46}$$

The rms value of the curvature is designated as $\bar{\kappa}$.

The power coupling coefficient (5.6-44) with the power spectrum (5.6-46) is independent of θ. This means that the coupling strength is independent of the compound mode number. The coupling mechanism that we use in this model does not assure that only nearest neighbors are coupled. However, we see from Eqs. (5.6-42) and (5.6-43) that the coupling strength decreases with the inverse fourth power of the separation between the modes. This means that nearest neighbors are coupled most strongly and that the coupling strength decreases rapidly with increasing mode separation. The assumption of nearest neighbor coupling is thus not exactly correct, but it is a reasonable approximation.

We substitute the trial solution

$$P = e^{-(\sigma+2\alpha)z} G(\theta) \tag{5.6-47}$$

into Eq. (5.6-32), use (5.6-44) and (5.6-46), and consider α and the group velocity as constants. We thus obtain

$$(d^2G/d\theta^2) + (1/\theta)(dG/d\theta) + \Omega^2 G = 0 \tag{5.6-48}$$

with

$$\Omega^2 = \pi^2\sigma[(n_1/n_2)-1]^{1/2}/2\sqrt{2}\,n^2 a\bar{\kappa}^2 \tag{5.6-49}$$

The solution of Eq. (5.6-48) satisfying (5.6-41) is

$$G_j(\theta) = \sqrt{2}\,J_0(\Omega_j\theta)/\theta_{\max} J_1(\Omega_j\theta_{\max}) \tag{5.6-50}$$

where J_0 and J_1 are the Bessel functions of zero and first order. Condition (5.6-40) leads to

$$\Omega_j\theta_{\max} = w_j \tag{5.6-51}$$

with w_j defined as the jth root of the equation

$$J_0(w_j) = 0 \tag{5.6-52}$$

The eigenfunctions (5.6-50) form a complete orthogonal set satisfying the orthogonality condition

$$\int_0^{\theta_{\max}} \theta G_i(\theta)\, G_j(\theta)\, d\theta = \delta_{ij} \tag{5.6-53}$$

The eigenvalues are obtained from Eqs. (5.6-14), (5.6-49), and (5.6-51):

$$\sigma^{(j)} = \sqrt{2}\,a w_j^{\,2}\bar{\kappa}^2/\pi^2\,[(n_1/n_2)-1]^{3/2} \tag{5.6-54}$$

With $w_1 = 2.405$ we obtain for the lowest eigenvalue, which represents the steady state power loss of the multimode waveguide,

$$\sigma^{(1)} = 8.18 a\bar{\kappa}^2/\pi^2\,[(n_1/n_2)-1]^{3/2} \tag{5.6-55}$$

In order to get a feeling for the sensitivity of the multimode waveguide to random curvature of its axis, we assume that we can tolerate an excess loss of 10 dB/km corresponding to $\sigma_1 = 2.3\times10^{-5}\ \mathrm{cm}^{-1}$. With $n_1/n_2 = 1.01$ and a core radius $a = 50\ \mu\mathrm{m} = 5\times10^{-3}$ cm, we find from Eq. (5.6-55)

$$1/\bar{\kappa} = 4.24 \quad \mathrm{m} \tag{5.6-56}$$

The inverse average curvature has the dimension of length and is related to the average radius of curvature of the waveguide axis. The number in Eq. (5.6-56) provides an order of magnitude estimate of the permissible average radius of curvature that leads to 10 dB/km radiation loss. However, this estimate was based on arbitrary assumptions about the width of the power spectrum of the curvature function. By assuming that the power spectrum is constant over the narrow spatial frequency range that contributes to mode coupling but zero outside this range, we obtain an estimate of $\bar{\kappa}$ that is as small as possible (with the constraint that the power spectrum is flat over a limited range). Had we allowed the width of the power spectrum to be larger, we would have found a smaller value for Eq. (5.6-56). Thus it appears that the estimate (5.6-56) is pessimistic and that the actual permissible average radii of curvature may be smaller.

The width of the impulse response is obtained from Eqs. (5.5-25) and (5.5-55). For the continuum of modes we must write, instead of Eqs. (5.5-45) and (5.5-66),

$$V_{ij} = \int_0^{\theta_{max}} \theta G_i(\theta) \{[1/v(\theta)] - (1/v_0)\} G_j(\theta)\, d\theta \tag{5.6-57}$$

This expression is properly normalized. Equation (5.5-45) results in $V_{jj} = 1$ if we choose $V_{\mu\nu} = \delta_{\mu\nu}$. Correspondingly, we must obtain $V_{jj} = 1$ from Eq. (5.6-57) if the difference of the inverse group velocities is set equal to unity. However, this normalization is assured according to Eq. (5.6-53).

The inverse group velocity is obtained from Eq. (5.6-17):

$$\frac{1}{v(\theta)} = \frac{1}{c}\frac{d\beta}{dk} = \frac{n_1}{c} - \frac{\theta^2}{2cn_1} - \frac{\theta k}{cn_1}\frac{d\theta}{dk} \tag{5.6-58}$$

The derivative of θ follows from Eq. (5.6-12):

$$d\theta/dk = -\pi M/2k^2 a = -\theta/k \tag{5.6-59}$$

If we use $v_0 = c/n_1$, we obtain, from Eqs. (5.6-58) and (5.6-59),

$$[1/v(\theta)] - (1/v_0) = \theta^2/2cn_1 \tag{5.6-60}$$

From Eqs. (5.6-14), (5.6-50), (5.6-51), (5.6-57), and (5.6-60), we find

$$V_{1j} = 4\Omega_1\Omega_j/cn(\Omega_1{}^2 - \Omega_j{}^2)^2 = 8n[(n_1/n_2) - 1]\, w_1 w_j/c(w_1{}^2 - w_j{}^2)^2 \tag{5.6-61}$$

We are interested only in the second-order perturbation of the first eigenvalue. We thus have, from Eqs. (5.5-10) and (5.5-55),

$$\rho_2^{(1)} = \sum_{j=2}^{N} V_{1j}^2/(\sigma^{(j)} - \sigma^{(1)}) \tag{5.6-62}$$

With the help of Eqs. (5.6-54) and (5.6-61), Eq. (5.6-62) can be expressed in the form

$$\begin{aligned}\rho_2^{(1)} &= \{64\pi^2[(n_1/n_2) - 1]^{7/2} n^2/\sqrt{2}\, ac^2\bar{\kappa}^2\} \sum_{j=2}^{\infty} w_1{}^2 w_j{}^2/(w_j{}^2 - w_1{}^2)^5 \\ &= 8.71 \times 10^{-3} n^2 [(n_1/n_2) - 1]^{7/2}/ac^2\bar{\kappa}^2\end{aligned} \tag{5.6-63}$$

The width of the impulse response follows from Eqs. (5.5-25) and (5.6-63):

$$T_1 = 0.373\{n[(n_1/n_2)-1]^{7/4}/c\bar{\kappa}\}(L/a)^{1/2} \qquad (5.6\text{-}64)$$

Using Eq. (5.5-64) we obtain the improvement factor R of Eq. (5.5-70) in the following form:

$$R = 0.373[(n_1/n_2)-1]^{3/4}/(La)^{1/2}\bar{\kappa} \qquad (5.6\text{-}65)$$

The quantity of most interest is the loss penalty that must be paid for a given improvement of the impulse response. The loss of the equilibrium pulse is $\sigma_1 L$. We thus find, from Eqs. (5.6-55) and (5.6-65),

$$R^2\sigma_1 L = 0.115 = 0.5 \quad \text{dB} \qquad (5.6\text{-}66)$$

This is a very interesting result. It shows that the loss penalty depends only on the amount of relative reduction of the width of the impulse response. It is independent of the waveguide parameters and the rms value of the waveguide curvature. If we want a reduction of the width of the uncoupled impulse response by a factor ten, $R = 0.1$, we find from Eq. (5.6-66) that we must pay for this improvement with an additional loss of 50 dB. An improvement of $R = 0.316$ costs only a loss penalty of 5 dB. These figures do not tell us the amount of waveguide curvature or the length that is required to achieve this improvement. Using $n_1/n_2 - 1 = 0.01$, $a = 5\times10^{-3}$ cm, $L = 1$ km, we find from Eq. (5.6-65) that an average curvature of $\bar{\kappa} = 5.3\times10^{-3}$ cm^{-1} is required for $R = 0.1$. The average radius of curvature is $R_c = 1/\bar{\kappa} = 1.9$ m.

The argument can, of course, be turned around. If we know that a certain amount of excess loss is caused by random curvature of the waveguide axis, we can use this information to predict the amount of width reduction of the impulse response that is achieved by this mode coupling mechanism.

The diffusion theory provides solutions in closed form for multimode operation with a very large number of modes. By comparison, the theory of coupled power equations yielded a solution in closed form only for the two-mode case. For general shapes of the power spectrum of the distortion function $f(z)$, we are usually not able to solve the partial differential equation. However, for those cases where solutions of Eq. (5.6-32) can be obtained, this method is certainly very useful.

5.7 Power Coupling between Waves Traveling in Opposite Directions

In Sect. 5.2 we derived coupled power equations whose applicability is restricted to waves traveling in the same direction. In some applications we are interested in the problem of power coupling between two waves that travel in opposite directions. Even in a waveguide that can support only one guided mode, we are able to transfer power from the wave that travels in positive

z direction to the wave traveling in negative z direction. According to Eq. (4.2-4), coupling between waves traveling in opposite direction is caused by spatial frequencies θ of the distortion function that are twice as large as the propagation constant β of the waves,

$$\theta = 2\beta \tag{5.7-1}$$

The problem of two randomly coupled modes has been treated by several authors (see, for example, [Pu1, MPK, Re1, Yg1, RY1]). Rowe [Re1] considered waves traveling in opposite directions, but he assumed that the amplitude of mode 1 is known at the far end of the waveguide and imposed the boundary condition that the amplitude of mode 2, traveling in the opposite direction, is zero at the far end. These peculiar boundary conditions enabled him to treat the problem by means of a matrix approach. However, the physically interesting case must use boundary conditions at opposite ends of the waveguide. Experimentally, the amplitude of wave 1 is usually known at the near end of the waveguide, while it is also known that no wave is launched at the far end. Rowe's boundary conditions solve only the problem of the average amplitude and power of the subensemble of those input waves that are known to arrive at the far end with a given preselected amplitude.

Both of these statistical problems are solved approximately in this section. We are using the same perturbation technique for deriving coupled power equations that we employed in Sect. 5.2. But since the coupled waves are now traveling in opposite directions, it is not as obvious that we can assume lack of correlation between wave amplitudes and coupling function at different points in space. The derivation of coupled power equations for backward waves thus does not rest on quite as sound a basis as our derivation of coupled power equations for waves traveling in the same direction. However, computer simulated experiments are in good agreement with the predictions of the author's theory. For this reason the author feels that this simple theory is worth discussing. A particularly interesting feature of the coupled power equations of waves traveling in opposite directions is the fact that we obtain two different differential equations for the two statistical problems mentioned above. The physically meaningful case of a wave that is launched in the positive direction with a given amplitude, so that boundary conditions must be imposed at opposite ends of the waveguide, results in a different differential equation than Rowe's problem with both boundary conditions at the far end of the guide. We are thus faced with the unusual situation that a different differential equation is required for each set of boundary conditions.

Our starting point is the following set of coupled wave equations [Me9]:

$$dc_1/dz = K_{11} c_1 + K_{12} c_2 \exp[i(\beta_1 + \beta_2) z] \tag{5.7-2}$$

$$dc_2/dz = K_{21} c_1 \exp[-i(\beta_1 + \beta_2) z] + K_{22} c_2 \tag{5.7-3}$$

Equations (5.7-2) and (5.7-3) are identical with (3.2-49) and (3.2-50) if we restrict the number of modes to one forward and one backward traveling wave. We introduce the coupling function $f(z)$ *via*

$$K_{\nu\mu} = \hat{K}_{\nu\mu} f(z) \tag{5.7-4}$$

Coupling between the two waves is accomplished through the spectral component of $f(z)$ at the spatial frequency θ that is determined from

$$\theta = \beta_1 + \beta_2 \tag{5.7-5}$$

It is apparent that the diagonal and off-diagonal terms in Eqs. (5.7-2) and (5.7-3) have very different z dependence. The off-diagonal terms contain slowly varying parts that provide the main contribution to the coupled equations system. The diagonal terms vary rapidly with the spatial frequency θ. Their contributions cancel, so that we are allowed to neglect these terms. We thus consider the simpler equations

$$dc_1/dz = K_{12} c_2 \exp[i(\beta_1 + \beta_2)z] \tag{5.7-6}$$

$$dc_2/dz = K_{21} c_1 \exp[-i(\beta_1 + \beta_2)z] \tag{5.7-7}$$

Conservation of power resulted in condition (5.2-17) for the coupling coefficients of forward traveling waves. The corresponding condition for waves traveling in opposite directions is different. In order to derive it we form the expressions for the power carried by the waves,

$$Q_\nu = c_\nu c_\nu^* \tag{5.7-8}$$

If wave 1 travels in positive z direction, an increase in power requires that

$$dQ_1/dz > 0 \tag{5.7-9}$$

Mode 2 travels in negative z direction. Power increase for this mode is expressed by the condition

$$dQ_2/dz < 0 \tag{5.7-10}$$

Conservation of power for the combined system of the two waves is thus assured by the relation

$$(dQ_1/dz) - (dQ_2/dz) = 0 \tag{5.7-11}$$

Using definition (5.7-8) and the differential Eqs. (5.7-6) and (5.7-7), we obtain from (5.7-11)

$$(K_{12} - K_{21}^*) c_1^* c_2 \exp[i(\beta_1 + \beta_2)z] + \text{cc} = 0 \tag{5.7-12}$$

where the notation cc indicates that the complex conjugate of the expression must be added. Owing to our freedom of choice of initial conditions for the amplitudes of the two waves, Eq. (5.7-12) can only be satisfied if

$$K_{12} = K_{21}^* \tag{5.7-13}$$

Since $f(z)$ is a real function, we can substitute Eq. (5.7-4) into Eq. (5.7-13) and find

$$\hat{K}_{12} = \hat{K}_{21}^* \tag{5.7-14}$$

After this preliminary discussion we are now ready to commence with the derivation of the coupled power equations for two waves traveling in opposite directions. Since we consider lossless systems, we need to derive only the power equation for the first wave, since the derivative of the second wave follows from Eq. (5.7-11), which holds also for the ensemble average of the power

$$P_\nu = \langle Q_\nu \rangle = \langle c_\nu c_\nu{}^* \rangle \tag{5.7-15}$$

Taking the z derivative of Eq. (5.7-15) and using Eqs. (5.7-4) and (5.7-6) yields

$$dP_1/dz = \hat{K}_{12} \langle c_1{}^* c_2 f(z) \rangle \exp[i(\beta_1 + \beta_2)z] + \mathrm{cc} \tag{5.7-16}$$

Because of its success in Sect. 5.2 we again use the technique of substituting a perturbation solution of the coupled wave equations into Eq. (5.7-16). However, the situation is different in our present case. Wave 1 approaches point z from the left, while wave 2 approaches from the right. It appears natural to express the amplitude of wave 1 in terms of its values at a point $z' < z$ but express the amplitude of wave 2 in terms of its amplitude at $z'' > z$. Thus we use the following perturbations solutions of Eqs. (5.7-6) and (5.7-7):

$$c_1(z) = c_1(z') + \hat{K}_{12} c_2(z') \int_{z'}^{z} f(x) \exp[i(\beta_1 + \beta_2)x] \, dx \tag{5.7-17}$$

$$c_2(z) = c_2(z'') + \hat{K}_{21} c_1(z'') \int_{z''}^{z} f(x) \exp[-i(\beta_1 + \beta_2)x] \, dx \tag{5.7-18}$$

We use the assumption that the autocorrelation function

$$R(u) = \langle f(z) f(z \pm u) \rangle \tag{5.7-19}$$

has a finite correlation length D. When both waves traveled in the same direction (Sect. 5.2), we used the fact that the wave amplitudes at the point $z' < z - D$ are uncorrelated with $f(z)$. In case of two waves traveling in opposite directions, this assumption is no longer precisely true. We are now using the assumption that the wave amplitude $c_1(z')$ is uncorrelated with $f(z)$, and also that $c_2(z'')$ is uncorrelated with $f(z)$. However, the wave amplitude c_1 at $z = z'$ is influenced by the amplitude c_2, whose value, in turn, is affected by $f(z)$. Likewise, c_2 at $z = z''$ is influenced by c_1, whose value is dependent on $f(z)$. There is thus a certain correlation of $c_1(z')$ and $f(z)$, and also of $c_2(z'')$

and $f(z)$. However, we are urging that this correlation is only slight. We have already used the assumption that $c_1(z)$ is nearly constant over the distance $z-z'$ and that c_2 is nearly constant over the distance $z''-z$ when we used approximations (5.7-17) and (5.7-18). Since $f(z)$ is correlated with itself only over the smaller distance D, it is reasonable to assume that $c_1(z')$ and $c_2(z'')$ are almost uncorrelated with $f(z)$. The property of uncorrelated wave amplitudes is not as certain as it was in Sect. 5.2 but we can, nevertheless, assume that the correlation remains so small as to be negligible to the approximation involved in using Eqs. (5.7-17) and (5.7-18). The field amplitudes are mostly influenced by the cumulative effect of wave coupling over large distances outside the interval of length $z''-z'$. For this reason we use

$$\langle c_1{}^*(z')\, c_2(z'')\, f(z)\rangle = 0 \tag{5.7-20}$$

and

$$\langle c_1{}^*(z')\, c_1(z'')\, f(z)\, f(x)\rangle \approx \langle |c_1(z)|^2\rangle\, R(z-x) \tag{5.7-21}$$

Equation (5.7-20) is based on the assumption that the average value of $f(z)$ vanishes. Equation (5.7-21) is based on the discussion of the lack of correlation between the wave amplitudes and the function $f(z)$ and on the assumption that the field amplitudes are nearly constant over the range between z' and z. The autocorrelation function $R(u)$ was introduced through its definition (5.7-19). The assumptions expressed in Eqs. (5.7-20) and (5.7-21) allow us to obtain, by substitution of (5.7-17) and (5.7-18) into (5.7-16):

$$\begin{aligned} dP_1/dz = &-\hat{K}_{12}\hat{K}_{21}P_1\int_{-\infty}^{0} R(u)\exp[i(\beta_1+\beta_2)u]\, du \\ &+\hat{K}_{12}\hat{K}_{12}^{*}P_2\int_{0}^{\infty} R(u)\exp[i(\beta_1+\beta_2)u]\, du + \mathrm{cc} \end{aligned} \tag{5.7-22}$$

Terms of third order in the coupling coefficient have been neglected. The integration limits have been extended to infinity since the autocorrelation function vanishes outside the interval from z' to z''. Finally, we use Eq. (5.7-14), and after adding the complex conjugate term indicated in (5.7-22) we make use of (5.2-18) to obtain

$$dP_1/dz = h(P_2-P_1) \tag{5.7-23}$$

with the power coupling coefficient

$$h = |\hat{K}_{12}|^2\langle |F(\beta_1+\beta_2)|^2\rangle \tag{5.7-24}$$

From Eq. (5.7-11) we also have

$$dP_2/dz = h(P_2-P_1) \tag{5.7-25}$$

The coupled power Eqs. (5.7-23) and (5.7-25) for waves traveling in opposite directions have a simple intuitive meaning. Equation (5.7-23) shows that wave 1 gains power from wave 2 if $P_2 > P_1$, and loses power to wave 2 in case that $P_2 < P_1$. Power gain of wave 2 is associated with a negative derivative in Eq. (5.7-25). Wave 2 is thus seen to gain power if $P_1 > P_2$ and to lose power if $P_2 > P_1$.

The power coupling coefficient h is constant. The two Eqs. (5.7-23) and (5.7-25) have the simple solution

$$P_1(z) = P_1(0)[1 + (L-z)h]/(1+hL) \tag{5.7-26}$$

and

$$P_2(z) = P_1(0)(L-z)h/(1+hL) \tag{5.7-27}$$

These solutions satisfy the boundary conditions $P_1(z) = P_1(0)$ at $z = 0$, and $P_2(L) = 0$. This means that no power is injected at the far end into mode 2. The total amount of power is

$$P_1(L) + P_2(0) = P_1(0) \tag{5.7-28}$$

Our solution states that for $L \to \infty$ all the power is reflected; no power emerges in mode 1 at the end of the waveguide. In this respect, the random coupling process behaves just like a waveguide with a periodic perturbation causing a stopband for waves satisfying condition (5.7-5), with θ now indicating the spatial frequency of the sinusoidal perturbation. However, there is an important difference. Solutions (4.2-18) and (4.2-19) of the purely sinusoidally distorted waveguide tell us that the power decreases exponentially with increasing z, whereas solutions (5.7-26) and (5.7-27) indicate that this decrease is linear in case of random coupling. The random deformation of the waveguide is not quite as effective as a purely sinusoidal deformation.

In order to check the theory the author performed a computer simulated experiment [Me9]. It is easy to solve the coupled wave Eqs. (5.7-6) and (5.7-7) for constant values of the coupling function $f(z)$. If the sign of the piecewise constant coupling function is changed randomly, a waveguide with random coupling results. The correlation function for a coupling function $f(z)$ with constant value $\pm\varepsilon$ over an interval of length D but with randomly varying sign is given by the expression

$$R(u) = \begin{cases} \varepsilon^2(D-|u|)/D, & \text{for} \quad |u| \leqslant D \\ 0, & \text{for} \quad |u| \geqslant D \end{cases} \tag{5.7-29}$$

The power spectrum follows from Eq. (5.2-18):

$$\langle |F(\beta_1+\beta_2)|^2 \rangle = [2\varepsilon^2/D(\beta_1+\beta_2)^2][1-\cos(\beta_1+\beta_2)D] \tag{5.7-30}$$

The power coupling coefficient is obtained from Eq. (5.7-24):

$$h = [2\kappa^2/D(\beta_1+\beta_2)^2][1-\cos(\beta_1+\beta_2)D] \tag{5.7-31}$$

with

$$\kappa = \varepsilon|\hat{K}_{12}| \tag{5.7-32}$$

The amplitudes of the two waves at the beginning and end of the waveguide are related by the matrix equation

$$\begin{pmatrix} a_1(L) \\ a_2(L) \end{pmatrix} = M_N M_{N-1} \cdots M_2 M_1 \begin{pmatrix} a_1(0) \\ a_2(0) \end{pmatrix} \tag{5.7-33}$$

(where $a_1 = c_1 e^{-i\beta z}$, and $a_2 = c_2 e^{i\beta z}$) with each of the matrices given by

$$M_j = \begin{pmatrix} \cos\beta' D - i(\beta/\beta')\sin\beta' D & \pm(\kappa/\beta')\sin\beta' D \\ \pm(\kappa/\beta')\sin\beta' D & \cos\beta' D + i(\beta/\beta')\sin\beta' D \end{pmatrix} \tag{5.7-34}$$

with randomly varying signs indicated by plus or minus in Eq. (5.7-34). For simplicity we have used

$$\beta_1 = \beta_2 = \beta \tag{5.7-35}$$

The parameter β' is defined by

$$\beta' = (\beta^2-\kappa^2)^{1/2} \tag{5.7-36}$$

In order to verify the statistical coupled power theory, random waveguides were simulated by generating random plus and minus sign sequences and computing the product matrix in Eq. (5.7-33) numerically. Once the product matrix is known for a given simulated waveguide, it is easy to compute $a_1(L)$ and $a_2(0)$ by assuming a given value for $a_1(0)$ and $a_2(L) = 0$. This experiment was repeated 10 times for a given value of κ. The resulting wave amplitudes were used to compute the average powers $P_1(L)$ and $P_2(0)$. For simplicity it was assumed that $2\beta D = \pi/2$. However, it was also confirmed that the computer simulated results agreed with formula (5.7-31) for more general choices of the parameter $2\beta D$. Figure 5.7.1 shows a comparison between theory (solid lines) and experiment (crosses). The waveguide that was used for this simulated experiment had a length of $L = 500D$, with D the length of the sections with constant coupling. With Eq. (5.7-35) and $2\beta D = \pi/2$, we obtain from Eq. (5.7-31)

$$hL = 500(2\pi\kappa/\beta)^2/8\pi^2 \tag{5.7-37}$$

Figure 5.7.1 shows remarkably good agreement between the coupled power theory and the results of the computer simulated experiment for values of the coupling constant that remain below 0.4. The derivation of the coupled power

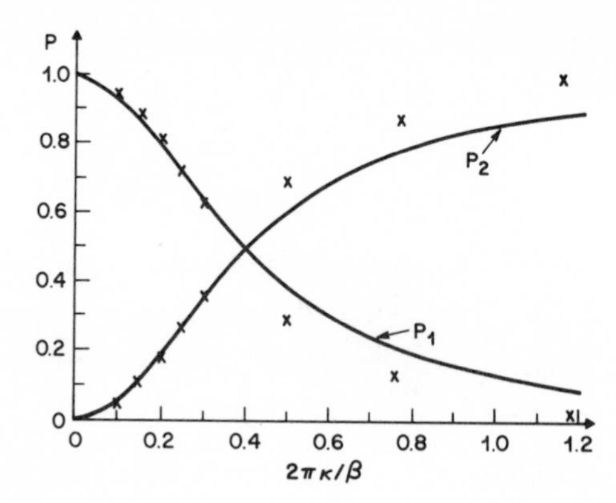

Fig. 5.7.1 *The power P_1 at the far end and the reflected power P_2 at the input end of the waveguide. The crosses are the results of a computer simulated experiment.*

Eqs. (5.7-23) and (5.7-25) is valid only for weak coupling. It seems remarkable that the theory works so well even for values as large as $2\pi\kappa/\beta = 0.4$. For very strong coupling we see clearly the limitations of the approximate theory. For $2\pi\kappa/\beta = 1$, the value for the reflected power $P_2(0)$ is in error by 10%.

It is tempting to apply the same theory to Rowe's case [Re1]. He considered only those members of the ensemble that resulted in a preselected value of the mode amplitude $a_1(L)$ and computed the ensemble average of the input power $P_1(0)$. The computer simulation can handle this case easily by assuming $a_1(L) = 1$ and $a_2(L) = 0$. The averages $P_1(0)$ and $P_2(0)$ can then be computed from the averages of the squares of the wave amplitudes by means of the product matrix in Eq. (5.7-33). The corresponding solutions of the coupled power equations are $P_1 = 1 + h(L-z)$ and $P_2 = h(L-z)$. It was found that these solutions are in disagreement with the results of the computer simulation. The coupled power theory, Eqs. (5.7-23) and (5.7-25), does not work for Rowe's case.

In order to explore this situation it seemed tempting to try to modify the derivation of the theory. Since Rowe's boundary conditions assume knowledge of the wave amplitudes at the end of the waveguide, it appeared plausible to try to derive coupled power equations by computing the amplitudes of both waves in terms of their values at the point $z'' > z$. By replacing z' in Eq. (5.7-17) with z'' and proceeding in the same manner as before, the following set of coupled power equations is obtained:

$$dP_1/dz = -h(P_1 + P_2) \tag{5.7-38}$$

and

$$dP_2/dz = -h(P_1 + P_2) \tag{5.7-39}$$

For Rowe's boundary conditions $P_1(L) = 1$ and $P_2(L) = 0$, we obtain the following solutions:

$$P_1(z) = \tfrac{1}{2}(e^{2h(L-z)} + 1) \tag{5.7-40}$$

and

$$P_2(z) = \tfrac{1}{2}(e^{2h(L-z)} - 1) \tag{5.7-41}$$

These solutions of the statistical coupled power equations are in reasonably good agreement with the corresponding computer simulated experiment. The agreement is poorer in this case, because the results of the individual samples scatter far more than they do in the case shown in Fig. 5.7.1. The agreement between theory and an ensemble average over only 10 samples is not very good. However, the two results are definitely not contradicting each other. The solutions of Eqs. (5.7-38) and (5.7-39) that satisfy the conventional boundary conditions $P_1(0) = 1$ and $P_2(L) = 0$ are

$$P_1(z) = (e^{-2hz} + e^{-2hL})/(1 + e^{-2hL})$$

and

$$P_2(z) = (e^{-2hz} - e^{-2hL})/(1 + e^{-2hL})$$

These solutions do not agree well with the experimental results shown by the crosses in Fig. 5.7.1. Solutions (5.7-26) and (5.7-27) are in much better agreement with the simulated experiment.

The coupled power Eqs. (5.7-38) and (5.7-39) follow from Rowe's theory. They seem to be the correct description of the statistical problem posed by Rowe.

We are thus faced with the fact that two types of differential equations describe the two experimental conditions. Each differential equation works only with the boundary conditions that are appropriate to the statistical assumptions. This result, that the differential equations are intimately related to the boundary conditions of the case of interest, is unusual in the context of ordinary, linear, deterministic theories. However, the statistical theory is different in nature. When we ask for the solutions of the case where $a_1(0)$ and $a_2(L)$ are prescribed, we are considering very different experimental conditions than when we ask for solutions of the case where $a_1(L)$ and $a_2(L)$ are prescribed. In the first case, we are simply considering the ensemble average of the outcomes of similar experiments that are conducted by launching the same amount of power into the input end of a waveguide. In the other case, we must conduct a much larger number of experiments, launching different amounts of power into the input ends of similar waveguides and considering only those results that yield the same amount of output power for wave 1 at the end of the waveguide. After this subensemble is selected, we take the average of the input powers that were necessary to yield the fixed amount of output power. The two cases are statistically very different from each other. It is thus not surprising that different differential equations are needed to describe them. The fact that the boundary conditions are also different for the two cases is also related to the different nature of the two experiments. The boundary conditions as well as the differential equations are an integral part of the statistical theory. Neither can be selected independently.

References

AS1 Abromovitz, M., and Stegun, I. A., "Handbook of Mathematical Functions with Formulas, Graphs, and Mathematical Tables," National Bureau of Standards Applied Mathematics Series, Vol. 55. National Bureau of Standards, Washington, D.C., 1965.

Bn1 Beckmann, P., "Probability in Communication Engineering." Harcourt, New York, 1967.

DS1 Dyott, R. B., and Stern, J. R., "Group Delay in glass fibre waveguide," Conference on Trunk Telecommunications by Guided Waves, IEE Conference Publication. No. 17, pp. 176–181, Institute of Electrical Engineers, London, 1970.

Ge1 Gloge, D., "Weakly guiding fibers," *Appl. Opt.* **10**, 2252–2258 (1971).

Ge2 Gloge, D., "Dispersion in weakly guiding fibers," *Appl. Opt.* **10**, 2442–2445 (1971).

Ge3 Gloge, D., "Optical power flow in multimode fibers," *Bell Syst. Tech. J.* **51**, 1767–1783 (1972).

Gl1 Goell, J. E., "A circular-harmonic computer analysis of rectangular dielectric waveguides," *Bell Syst. Tech. J.* **48**, 2133–2160 (1969).

GR1 Grashteyn, I. S., and Ryzhik, I. M., "Tables of Integrals, Series, and Products," 4th Ed. Academic Press, New York, 1965.

JE1 Jahnke, E., and Emde, P., "Tables of Functions with Formulas and Curves," 4th Ed. Dover, New York, 1945.

Js1 Jones, A. L., "Coupling of optical fibers and scattering in fibers," *J. Opt. Soc. Am.*, **55**, 261–271 (1965).

KH1 Kao, K. C., and Hockham, G. A., "Dielectric fibre surface waveguide for optical frequencies," *Proc. IEE* **113**, 1151–1158 (1966).

Ky1 Kapany, N. S., "Fiber Optics." Academic Press, New York, 1967.

KB1 Kapany, N. S., and Burke, J. J., "Optical Waveguides." Academic Press, New York, 1972.

KKM Kapron, F. P., Keck, D. B., and Maurer, R. D., "Radiation losses in glass optical waveguides," *Appl. Phys. Lett.* **17**, 423–425 (1970).

KSZ Keck, D. B., Schultz, P. C., and Zimar, F., "Attenuation of multimode glass optical waveguides," *Appl. Phys. Lett.* **21**, 215–217 (1972).

Kr1 Kerker, M., "The Scattering of Light and Other Electromagnetic Radiation." Academic Press, New York, 1969.

KS1 Kogelnik, H., and Shank, C. V., "Stimulated emission in a periodic structure," *Appl. Phys. Lett.* **18**, 152–154 (1971).

KS2 Kogelnik, H., and Shank, C. V., "Coupled wave theory of stimulated emission in a periodic structure," *J. Appl. Phys.* **43**, 2327–2335 (1972).

Mi1 Marcatili, E. A. J., "Dielectric rectangular waveguide and directional coupler for integrated optics," *Bell Syst. Tech. J.* **48**, 2071–2102 (1969).

Mi2 Marcatili, E. A. J., "Bends in optical dielectric guides," *Bell Syst. Tech. J.* **48**, 2103–2132 (1969).

MS1 Marcatili, E. A. J., and Schmeltzer, R. A., "Hollow metallic and dielectric waveguides for long distance optical transmission and lasers," *Bell Syst. Tech. J.* **43**, 1783–1809 (1964).

Me1 Marcuse, D., "Light Transmission Optics." Van Nostrand Reinhold, Princeton, New Jersey, 1972.

Me2 Marcuse, D., "Radiation losses of the dominant mode in round dielectric waveguides," *Bell Syst. Tech. J.* **49**, 1665–1693 (1970).

Me3 Marcuse, D., "Hollow dielectric waveguides for distributed feedback lasers," *IEEE J. Quantum Electron.* **QE-8**, 661–669 (1972).

Me4 Marcuse, D., "Derivation of coupled power equations," *Bell Syst. Tech. J.* **51**, 229–237 (1972).

Me5 Marcuse, D., "Power distribution and radiation losses in multimode dielectric slab waveguides," *Bell Syst. Tech. J.* **51**, 429–454 (1972).

Me6 Marcuse, D., "Higher-order scattering losses in dielectric waveguides," *Bell Syst. Tech. J.* **51**, 1801–1817 (1972).

Me7 Marcuse, D., "Higher-order loss processes and the loss penalty of multimode operation," *Bell Syst. Tech. J.* **51**, 1819–1836 (1972).

Me8 Marcuse, D., "Pulse propagation in multimode dielectric waveguides," *Bell Syst. Tech. J.* **51**, 1199–1232 (1972).

Me9 Marcuse, D., "Coupled power equations for backward waves," *IEEE Trans. Microwave Theory Tech.* **MTT-20**, 541–546 (1972).

Me10 Marcuse, D., "Engineering Quantum Electrodynamics." Harcourt, New York, 1970.

Me11 Marcuse, D., "Pulse propagation in two-mode waveguide," *Bell Syst. Tech. J.* **51**, 1785–1791 (1972).

Me12 Marcuse, D., "Fluctuations of the power of coupled modes," *Bell Syst. Tech. J.* **51**, 1793–1800 (1972).

Me13 Marcuse, D., "Coupling coefficients for imperfect asymmetric slab waveguides," *Bell Syst. Tech. J.* **52**, 63–82 (1973).

Me14 Marcuse, D. (ed.), "Integrated Optics." IEEE Press, New York, 1973.

MW1 Mathews, J., and Walker, R. L., "Mathematical Methods of Physics." Benjamin, New York, 1965.

MS2 Maurer, R. D., and Schultz, P. C., "Fused Silica Optical Waveguide." U.S. Patent 3,659,915.

MK1 McKenna, J., "The excitation of planar dielectric waveguides at *p–n* junctions I," *Bell Syst. Tech. J.* **46**, 1491–1566 (1967).

Mr1 Miller, S. E., "Integrated optics: an introduction," *Bell Syst. Tech. J.* **48**, 2059–2069 (1969).

Mr2 Miller, S. E., "Coupled wave theory and waveguide applications," *Bell Syst. Tech. J.* **33**, 661–720 (1954).

Mr3 Miller, S. E., "A survey of integrated optics," *IEEE J. Quantum Electron.* **QE-8**, 199–205 (1972).

MPK Morrison, J. A., Papanicolaou, G. C., and Keller, J. B., "Mean power transmission through a slab of random medium," *Comm. Pure Appl. Math.* **24**, 473–489 (1971).

NM1 Nelson, D. F., and McKenna, J., "Electromagnetic modes of anisotropic dielectric waveguides at p–n junctions," *J. Appl. Phys.* **38**, 4057–4074 (1967).

OCS Ogawa, K., Chang, W. S. C., Sopori, B. L., and Rosenbaum, F. J., "A theoretical analysis of etched grating couplers for integrated optics," *IEEE J. Quantum Electron.* **QE-9**, 29–42 (1973).

Pu1 Papanicolaou, G. C., "Wave propagation in a one-dimensional random medium," *SIAM J. Appl. Math.* **21**, 13–18 (1971).

Pk1 Personick, S. D., "Time dispersion in dielectric waveguides," *Bell Syst. Tech. J.* **50**, 843–859 (1971).

Rn1 Rawson, E. G., "Measurement of angular distribution of light scattered from a glass fiber optical waveguide," *Appl. Opt.* **11**, 2477–2481 (1972).

Re1 Rowe, H. E., "Propagation in a one-dimensional random medium," *IEEE Trans. Microwave Theory Tech.* **MTT-19**, 73–80 (1971).

RW1 Rowe, H. E., and Warters, W. D., "Transmission in multimode waveguide with random imperfections," *Bell. Syst. Tech. J.* **41**, 1031–1170 (1962).

RY1 Rowe, H. E., and Young, D. T., "Transmission distortion in multimode random waveguides," *IEEE Trans. Microwave Theory Tech.* **MTT-20**, 229–237 (1972).

SSS Schinke, D. P., Smith, R. G., Spencer, E. G., and Galvin, M. F., "Thin-film distributed-feedback laser fabrication by ion milling," *Appl. Phys. Lett.* **21**, 494–496 (1972).

SBK Shank, C. V., Bjorkholm, J. E., and Kogelnik, H., "Tunable distributed-feedback dye laser," *Appl. Phys. Lett.* **18**, 395–396 (1971).

So1 Shevchenko, V. V., "Continuous Transitions in Open Waveguides." Golem Press, Boulder, Colarado, 1971.

Sh1 Smith, P. W., "A waveguide laser," *Appl. Phys. Lett.* **19**, 132–134 (1971).

Sn1 Snitzer, E., "Cylindrical dielectric waveguide modes," *J. Opt. Soc. Am.* **51**, 491–498 (1961).

SO1 Snitzer, E., and Osterberg, H., "Observed dielectric waveguide modes in the visible spectrum," *J. Opt. Soc. Am.* **5**, 499–505 (1961).

Sr1 Snyder, A. W., "Asymptotic expressions for eigenfunctions and eigenvalues of a dielectric or optical waveguide," *IEEE Trans. Microwave Theory Tech.* **MTT-17**, 1130–1138 (1969).

Sr2 Snyder, A. W., "Coupled mode theory for optical fibers," *J. Opt. Soc. Am.* **62**, 1267–1277 (1972).

Sr3 Snyder, A. W., "Coupling of modes on a tapered dielectric cylinder," *IEEE Trans. Microwave Theory Tech.* **MTT-18**, 383–392 (1970).

Sr4 Snyder, A. W., "Mode propagation in a nonuniform cylindrical medium," *IEEE Trans. Microwave Theory Tech.* **MTT-19**, 402–403 (1971).

Sr5 Snyder, A. W., "Continuous mode spectrum of a circular dielectric rod," *IEEE Trans. Microwave Theory Tech.* **MTT-19**, 720–727 (1971).

Sr6 Snyder, A. W., "Radiation losses due to variation of radius on dielectric or optical fibers," *IEEE Trans. Microwave Theory Tech.* **MTT-18**, 608–615 (1970).

Sr7 Snyder, A. W., and De La Rue, R., "Asymptotic solutions of eigenvalue equations for surface waveguide structures," *IEEE Trans. Microwave Theory Tech.* **MTT-18**, 650–651 (1970).

Sr8 Snyder, A. W., "Excitation and scattering of modes on a dielectric or optical fiber," *IEEE Trans. Microwave Theory Tech.* **MTT-17**, 1138–1144 (1969).

Se1 Stone, J., "Optical transmission in liquid-core quartz fibers," *Appl. Phys. Lett.* **20**, 239–240 (1972).

Ti1 Tien, P. K., "Light waves in thin films and integrated optics," *Appl. Opt.* **10**, 2395–2413 (1971).

Tn1 Tolman, R. C., "The principles of Statistical Mechanics." Oxford University Press, London and New York, 1938.

Yg1 Young, D. T., "Model for relating coupled power equations to coupled amplitude equations," *Bell Syst. Tech. J.* **42**, 2761–2764 (1963).

Index

I

K

L

M

N

O

S

T

U

V

W

B 7
C 8
D 9
E 0
F 1
G 2
H 3
I 4
J 5